TRANSGENIC PLANTS

BOOKS IN SOILS, PLANTS, AND THE ENVIRONMENT

Soil Biochemistry, Volume 1, edited by A. D. McLaren and G. H. Peterson
Soil Biochemistry, Volume 2, edited by A. D. McLaren and J. Skujiņš
Soil Biochemistry, Volume 3, edited by E. A. Paul and A. D. McLaren
Soil Biochemistry, Volume 4, edited by E. A. Paul and A. D. Mclaren
Soil Biochemistry, Volume 5, edited by E. A. Paul and J. N. Ladd
Soil Biochemistry, Volume 6, edited by Jean-Marc Bollag and G. Stotzky
Soil Biochemistry, Volume 7, edited by G. Stotzky and Jean-Marc Bollag
Soil Biochemistry, Volume 8, edited by Jean-Marc Bollag and G. Stotzky

Organic Chemicals in the Soil Environment, Volumes 1 and 2, edited by C. A. I. Goring and J. W. Hamaker
Humic Substances in the Environment, M. Schnitzer and S. U. Khan
Microbial Life in the Soil: An Introduction, T. Hattori
Principles of Soil Chemistry, Kim H. Tan
Soil Analysis: Instrumental Techniques and Related Procedures, edited by Keith A. Smith
Soil Reclamation Processes: Microbiological Analyses and Applications, edited by Robert L. Tate III and Donald A. Klein
Symbiotic Nitrogen Fixation Technology, edited by Gerald H. Elkan
Soil-Water Interactions: Mechanisms and Applications, edited by Shingo Iwata, Toshio Tabuchi, and Benno P. Warkentin
Soil Analysis: Modern Instrumental Techniques, Second Edition, edited by Keith A. Smith
Soil Analysis: Physical Methods, edited by Keith A. Smith and Chris E. Mullins
Growth and Mineral Nutrition of Field Crops, N. K. Fageria, V. C. Baligar, and Charles Allan Jones
Semiarid Lands and Deserts: Soil Resource and Reclamation, edited by J. Skujiņš
Plant Roots: The Hidden Half, edited by Yoav Waisel, Amram Eshel, and Uzi Kafkafi
Plant Biochemical Regulators, edited by Harold W. Gausman
Maximizing Crop Yields, N. K. Fageria
Transgenic Plants: Fundamentals and Applications, edited by Andrew Hiatt
Soil Microbial Ecology: Applications in Agricultural and Environmental Management, edited by F. Blaine Metting, Jr.

Additional Volumes in Preparation

Principles of Soil Chemistry: Second Edition, Kim H. Tan

Fundamentals
and
Applications

edited by
Andrew Hiatt
The Scripps Research Institute
La Jolla, California

Marcel Dekker, Inc. New York • Basel • Hong Kong

Library of Congress Cataloging-in-Publication Data

Transgenic plants : fundamentals and applications / edited by Andrew Hiatt.
p. cm. — (Book in soils, plants, and the environment)
Includes bibliographical references (p.) and index.
ISBN 0-8247-8766-8
1. Transgenic plants. I. Hiatt, Andrew. II. Series.
SB123.57.T732 1992
631.5′23—dc20 92-26046
CIP

This book is printed on acid-free paper.

Marcel Dekker, Inc.
270 Madison Avenue, New York, New York 10016

Current printing (last digit):
10 9 8 7 6 5 4 3 2 1

PRINTED IN THE UNITED STATES OF AMERICA

Preface

Transgenic plants expressing bacterial genes, mammalian genes, or other plant genes are becoming commonplace. Most transgenic plants are generated for fairly specific basic research purposes such as the evaluation of tissue-specific expression of genes, processing and secretion signals, and enhancer elements. A growing effort is being made to use transgenic plants for broader objectives such as insect and herbicide resistance and expression of pharmacologically useful proteins. To reflect these developments, this book emphasizes the types of transgenic plants that have the potential for influencing agricultural practices as well as producing valuable proteins.

In planning this book, it became apparent that a number of new results in plant genetic engineering have the potential to influence specific areas of agriculture as well as initiate new practices utilizing plants as bioreactors. Specifically, major advances have been made in transgenic pathogen protection and in understanding the mechanisms involved in this effect. Production of pharmaceuticals in transgenic plants and transgenic manipulation of plant physiology are two additional areas that hold great promise. In addition, plant transformation is an area of research that has great potential for developing agriculturally important practices. These include transformation of new types of photosynthetic organisms (*Chlamydomonas reinhardtii*), transformation and genetic engineering of extremely important crop plants such as corn and rice, and new vector strategies for expression of foreign genes in plants. Finally, emerging techniques from other areas of biology—in particular, antisense and catalytic RNAs—are successfully being used or will be used in the near future in agricultural research.

Since planning this book, it has become apparent that new developments in making novel transgenic plants are rapidly occurring. Inhibition of fruit ripening, creation of male sterility, and manipulation of carbon metabolism in transgenic plants are three examples of exciting results published while this book was being assembled. I am certain this rapid rate of progress will continue for many years.

Although the level of scientific information is quite high, an effort was made to write chapters that are generally accessible to scientists not involved in plant biology or molecular biology. *Transgenic Plants* will be of value to plant physiologists and molecular biologists as well as to scientists involved in virtually all other areas of agricultural biology.

Andrew Hiatt

Contents

II. Viral Pathogen Resistance

III. Techniques for Transformation of Photosynthetic Organisms

IV. Molecular Farming

V. Ribozymes and Antisense RNA

Contributors

Charles L. Armstrong Monsanto, St. Louis, Missouri

Roger N. Beachy Department of Cell Biology, The Scripps Research Institute, La Jolla, California

Hassan Bolkan Campbell Institute for Research and Technology, Davis, California

William O. Dawson Citrus Research and Education Center, University of Florida, Lake Alfred, Florida

Wenlian Deng Department of Agronomy, University of Kentucky, Lexington, Kentucky

Richard A. Dixon Plant Biology Division, The Samuel Roberts Noble Foundation, Ardmore, Oklahoma

Brent V. Edington* Plant Biology Division, The Samuel Roberts Noble Foundation, Ardmore, Oklahoma

**Present affiliation*: Department of Cell Biology/Virology, Ribozyme Pharmaceuticals, Inc., Cleveland, Ohio

Michael Fromm Department of Plant Sciences, Monsanto, St. Louis, Missouri

W. Scott Grayburn Department of Agronomy, University of Kentucky, Lexington, Kentucky

Don Grierson Department of Physiology and Environmental Science, Nottingham University, Sutton Bonington, England

Andrew Hiatt Department of Cell Biology, The Scripps Research Institute, La Jolla, California

William R. Hiatt Research and Development, Calgene Fresh, Inc., Davis, California

David F. Hildebrand Department of Agronomy, University of Kentucky, Lexington, Kentucky

André Hoekema Crop Improvement, Mogen International NV, Leiden, The Netherlands

Harry Klee Monsanto, St. Louis, Missouri

Matthew Kramer Product Development, Calgene Fresh, Inc., Davis, California

Enno Krebbers Plant Genetic Systems, Ghent, Belgium

Junko Kyozuka Plantech Research Institute, Yokohama, Japan

Moshe Lapidot Department of Cell Biology, The Scripps Research Institute, La Jolla, California

George P. Lomonossoff Department of Virus Research, John Innes Institute, John Innes Centre, Norwich, England

Stephen P. Mayfield Department of Cell Biology, The Scripps Research Institute, La Jolla, California

Fionnuala Morrish New Products Division, Monsanto, St. Louis, Missouri

Keith Mostov Departments of Anatomy, and Biochemistry and Biophysics, University of California, San Francisco, San Francisco, California

Richard S. Nelson Plant Biology Division, The Samuel Roberts Noble Foundation, Ardmore, Oklahoma

Jan Pen Mogen International NV, Leiden, The Netherlands

Debra L. Robertson Departments of Chemistry and Molecular Biology, The Scripps Research Institute, La Jolla, California

Charles P. Romano Monsanto, St. Louis, Missouri

Rick A. Sanders Product Development, Calgene Fresh, Inc., Davis, California

Raymond E. Sheehy Calgene Fresh, Inc., Davis, California

Ko Shimamoto Plantech Research Institute, Yokohama, Japan

Peter C. Sijmons Mogen International NV, Leiden, The Netherlands

David D. Songstad* Monsanto, St. Louis, Missouri

Ann P. Sturtevant Department of Cell Biology, The Scripps Research Institute, La Jolla, California

Thomas H. Turpen Biosource Genetics Corporation, Vacaville, California

Joël Vandekerckhove Laboratory of Physiological Chemistry, University of Ghent, Ghent, Belgium

Albert J. J. van Ooijen Heterologous Gene Expression, Gist-brocades NV, Delft, The Netherlands

Jan Van Rompaey Technology Planning and Protection, Plant Genetic Systems, Ghent, Belgium

Curtis M. Waters Department of Plant Pathology, Campbell Institute for Research and Technology, Davis, California

C. F. Watson Department of Physiology and Environmental Science, Nottingham University, Sutton Bonington, England

---

**Present affiliation*: Department of Biotechnology, Pioneer Hi-Bred International, Inc., Johnston, Iowa

I

TRANSGENIC MANIPULATION OF METABOLISM

1

Use of Genetic Transformation to Alter Fatty Acid Metabolism in Plants

David F. Hildebrand, Wenlian Deng, and W. Scott Grayburn

University of Kentucky, Lexington, Kentucky

There is interest in altering the metabolism of fatty acids in plant tissues because of the influence of fatty foods on food quality and their significance in biological processes. The properties of fats and oils are determined by the fatty acid composition, which affects nutritional quality and oxidative stability. There has recently been considerable interest in reducing the saturated fatty acid content (in addition to the total lipid level) of foods. However, high levels of polyunsaturated fatty acids can result in low oxidative stability. Efforts have therefore been initiated to develop vegetable oils with higher monounsaturated fatty acid contents. There is also interest in increasing the dietary intake of ω-3 fatty acids relative to ω-6 fatty acids.

The properties of biological membranes are also influenced by the fatty acid composition of the membranes. Changes in membrane lipid composition are associated with development of tolerance to stresses, such as chilling, freezing, and drought. Polyunsaturated fatty acids are subject to enzymatic peroxidation, such as that catalyzed by LOXs in plant tissues or protein-containing preparations derived from plants. Unsaturated fatty acids can undergo nonenzymatic or auto-oxidation in extracted lipids, such as vegetable oils and lecithin. Fatty acid oxidation products include important flavor and aroma compounds. Fatty acid peroxidation products can also react with other molecules, such as membrane proteins, and further influence membrane function and affect the physical properties of foods.

I. FATTY ACID BIOSYNTHESIS

The use of plant breeding for altering the fatty acid composition of vegetable oils to improve their food value has been highly successful in some crop plants, such as sunflowers and canola. In other crop plants, such as soybeans, breeding for altered oil composition has been limited by a lack of genetic diversity. Recent developments in biotechnology provide alternative approaches for altering the fatty acid composition of vegetable oils. The biosynthesis of the major fatty acids found in plant membranes and vegetable oils is fairly well understood. Various enzymatic steps in the biosynthesis of fatty acids can be targets for molecular genetic manipulation. Several of the genes controlling key biosynthetic steps in fatty acid biosynthesis have been cloned. These include genes encoding proteins involved in saturated fatty acid biosynthesis, such as acetyl CoA carboxylase, fatty acid synthase, and acyl carrier protein (ACP) (1–3). Genes encoding enzymes involved in fatty acid chain length determination, such as thioesterases, have also been cloned (4,5). There has been considerable interest in a type of thioesterase, thioesterase II, which releases the fatty acids from the fatty acid synthase complex before elongation to palmitate (C16:0) (e.g., C8 and C10). Over-expression of thioesterase II in transgenic plant seeds is expected to result in accumulation of these shorter fatty acids, which are important in detergent manufacturing.

There are three major unsaturated fatty acids, found in varying levels, in most plant tissues: oleic (18:1), linoleic (18:2), and α-linolenic (C18:3) acids. These are synthesized by consecutive desaturations from stearic (C18:0) acid → C18:1 → C18:2 → C18:3. These desaturation reactions involve enzymes (desaturases) that catalyze the hydrogen-removal and electron-transport components. The first desaturation step in plants is catalyzed by a soluble chloroplast-localized desaturase. In animals and fungi this reaction is catalyzed by an endoplasmic reticulum-associated enzyme. The subsequent desaturation reactions can occur in both plastids and the endoplasmic reticulum, with different genes apparently encoding the enzymes found at the different subcellular locations. Somerville and Browse (6) estimate that there are at least eight genes that control the activity of specific desaturases in plant (*Arabidopsis*) leaves. Genes encoding the desaturase catalyzing the C18:0 to C18:1 step have been cloned from animals, plants, and yeast (7–10). The double bonds in oleic acids are at the Δ-9 (measured from the carboxyl end) or ω-9 (measured from the methyl end) position. Additional double bonds are at the ω-6 position in linoleic acid and at the ω-6 and ω-3 positions in α-linolenic acid. Vertebrate animals require, but do not have the capacity to synthesize, ω-6 and ω-3 fatty acids and therefore must rely on dietary sources ultimately derived from plants. The ω-6 desaturase gene has been cloned from a cyanobacterium (11). No gene encoding the ω-3 desaturase has yet been reported, but a polypeptide associated with control of linolenate levels in plant tissues has been identified (12).

Transformation of plants with a cloned lipid biosynthetic gene has been reported by Post-Beittenmiller et al. (3). They expressed a spinach ACP at two to three times the level of the endogenous protein in transgenic tobacco plant plastids. This overexpression of ACP did not result in any significant alterations of lipid biosynthesis, suggesting that native levels of this protein are not limiting for fatty acid biosynthesis. Preliminary results indicate that expression of an animal Δ-9 desaturase in plants can significantly increase conversion of saturated to monounsaturated fatty acids in certain lipid classes.

II. FATTY ACID PEROXIDATION

There are at least four pathways in which fatty acids are oxidatively metabolized. The LOX pathway is probably of greatest significance to food quality. Many plant and animal tissues contain LOX, which catalyzes the hydroperoxidation of polyunsaturated fatty acids. The fatty acid hydroperoxide products from the LOX reaction can be converted by other enzymes into a number of additional products, with known or hypothesized biological activity. Wound healing and pest defense appear to involve LOX. Consistent with this hypothesized role is the fact that formation of LOX products can be greatly increased by damage to plant tissues (Fig. 1). Some steps in food processing, such as freezing and thawing or homogenization, can cause massive increases in the formation of C_6-aldehydes from LOX/lyase action in plant tissues [and fish (13)]. Such increases occur during the production of potato granules (14) and the freezing of certain vegetables (15).

The flavor of a number of food products is affected either positively or negatively by LOX (16–18). They are positive contributors to the flavor of certain fruits, such as melons and tomatoes. In soybeans, on the other hand, some isozymes result in the formation of hexanal, which has a highly undesirable flavor and aroma and limits the use of soybeans in some food applications. The LOX-catalyzed peroxidation of fatty acids can also affect the

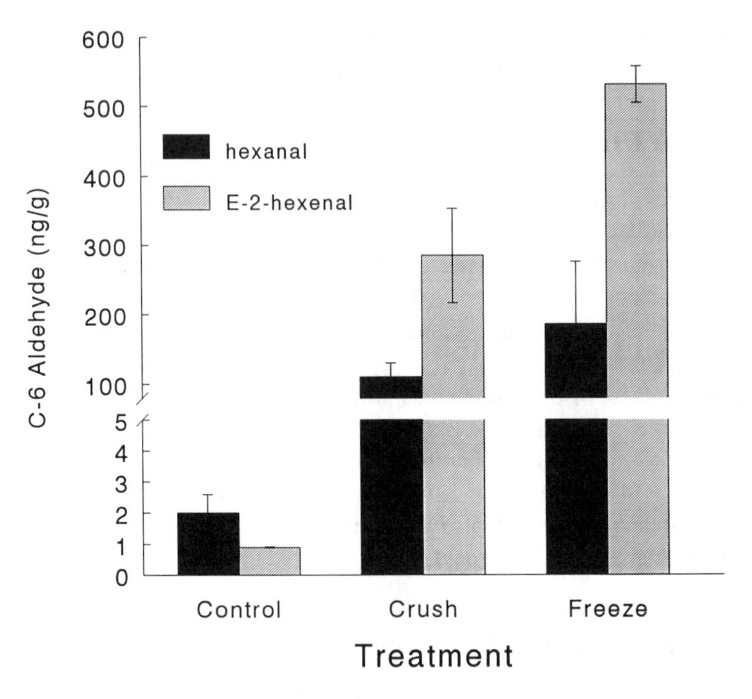

Figure 1 Effects of crushing and freezing on the formation of the C_6-aldehydes hexanal and E-2-hexenal in wheat leaves (which are typical of many plant tissues).

rheological properties of foods, such as breads (19), and co-oxidation reactions can cause bleaching of chlorophylls and carotenoids (20). Fatty acid peroxidation catalyzed by LOX appears to be a major factor in the deterioration of many fruits and vegetables after harvest.

The physiological roles of LOXs in plants have not been well established, although evidence is accumulating that they are involved in pest defense, growth and development, and storage product accumulation. In plant pest defense responses, LOX activity is often dramatically increased (21,22). The LOX/lyase products hexanal and *trans*-2-hexenal have been shown to be detrimental to the growth and reproduction of certain bacteria, fungi, and arthropods (23,24). In addition to their possible role in direct pest defense, certain volatile LOX products, such as *cis*-3- and *trans*-2-hexenal, can play a secondary role in plant defense by attracting natural enemies of herbivore pests

(25). Recently it was shown by Farmer and Ryan (26) that methyl jasmonate, when applied to tomato plants, induced the synthesis of a defensive proteinase inhibitor protein in the treated plants and in nearby plants as well, strongly suggesting that jasmonic acid (JA) (or me-JA) may act as a signal or elicit signal production and activate plant defensive responses.

Studies of mammalian physiology have shown that leukotrienes and lipoxins, derived from oxygenated arachidonic acid (catalyzed by LOX) function in the regulation of cell growth and differentiation (27). In recent years some studies on plant LOXs have shown an association of LOX or LOX products with plant growth and development. The highest LOX activity is found in rapidly growing tissue, such as embryonic axes or young leaves (28,29). Further, JA, a LOX product, is structurally similar to abscisic acid and appears to have some similar effects (30). It can specifically induce the expression of a number of proteins in a wide range of plants (14,31–33). It is not yet clear whether these proteins are important for plant growth and development. It has also been suggested that intracellular free radicals and fatty acid hydroperoxides produced by LOX are involved in plant senescence (34,35). Although some components of the LOX pathway have been elucidated (36,37), other important questions remain. For example, the subcellular location of LOX and the identity of in vivo substrates for the LOX pathway remain ambiguous (38–40).

The best studies of LOXs have used soybean seeds, in which there are usually three distinct isozymes, designated LOX 1, LOX 2, and LOX 3. The LOX 1 (also known as type I LOX) isozyme has a strong preference for charged fatty acid substrates and therefore has a high pH optimum. In contrast, LOX 2 and LOX 3 (type II LOX) can effectively peroxidize neutral substrates and have pH optima near 7. Null mutants that are missing LOX 1, LOX 2, or LOX 3 have been found. It was found that these LOX isozymes differed in their effects on C_6-aldehyde formation, with LOX 2 being most active in the formation of these compounds (41–43). Surprisingly, LOX 3 was found to reduce formation of C_6-aldehydes by LOX 1 or LOX 2, apparently by conversion of the hydroperoxides into products that are not substrates for hydroperoxide lyase (42,43). The LOX isozymes expressed in vegetative tissues of the soybean plant have been found to be different from those in embryos (28,44,45).

The genes for LOX 1–3 have been cloned and sequenced (46,47). To better understand the effects of altered LOX on formation of products in the LOX pathway and its biological role, we produced transgenic plants with altered LOX activity. Tobacco was used because of the utility of this system in plant gene transfer studies. The LOX 2 plus-sense coding sequence was cloned into a plant transformation vector 3′ to the duplicated 35S CaMV promoter and 5′ to a pea RUBISCO polyadenylation signal (Fig. 2). This plasmid contains neomycin phosphotransferase (conferring kanamycin resistance) as the selectable marker. The LOX 2/vector construct was transferred into *Agrobacterium tumefaciens* cells, which were then used in tobacco plant transformation with standard procedures. Transformed tobacco cells were selected and regenerated into plants or grown as selected transformed calli. The regenerated transformed plants were screened for expression of the introduced LOX 2 gene product by using monoclonal antibodies that showed no cross-reaction to tobacco LOXs or other LOXs of soybeans.

As shown in Fig. 3, many of the transgenic plants showed clear expression of the introduced LOX 2, with several of them showing expression at about 0.1% of the extracted protein. Transgenic calli showed higher expression (about 0.2%) of the introduced LOX 2 protein (Fig. 4). The expression of the introduced LOX 2 was confirmed by S1 nuclease protection analysis (Fig. 5). The highest-expressing transgenic calli had a highly significant increase in LOX activity relative to controls transformed with the vector alone (Table 1). On the other hand, leaves from regenerated transformed plants did not show a significant increase in LOX activity with linoleic acid as substrate (Table 2). Arachidonic acid is reported to be a better substrate than linoleic acid for LOX 2. However, our data also showed no increase in LOX activity of LOX-expressing leaves over controls with arachidonic acid as the substrate (Table 2). This may be due to the high background of LOX activity in the leaves of the controls. Without addition of linoleic acid, transformed calli did not show an increase in the production of the LOX/lyase product hexanal. When exogenous free fatty acid (FFA) substrate was added, however, up to a five- to six-fold increase over the controls was seen (Fig. 6A). About a five-fold increase in C_6-aldehyde product formation was seen with the highest-expressing transgenic leaves without exogenous FFA substrate addition (Fig. 6B). These data suggest that FFAs are limiting for C_6-aldehyde formation with calli but not leaves. These results are consistent with other studies indicating relatively high endogenous FFA levels in leaves (unpublished). The significant increase in C_6-aldehyde production but not LOX activity of transgenic plant leaves demonstrates the greater propensity for LOX 2 in formation of C_6-aldehyde products as compared with fatty acid hydroperoxide production. These results also confirm the importance of LOX in the formation of fatty acid peroxidation products, but they also implicate other enzymatic controls.

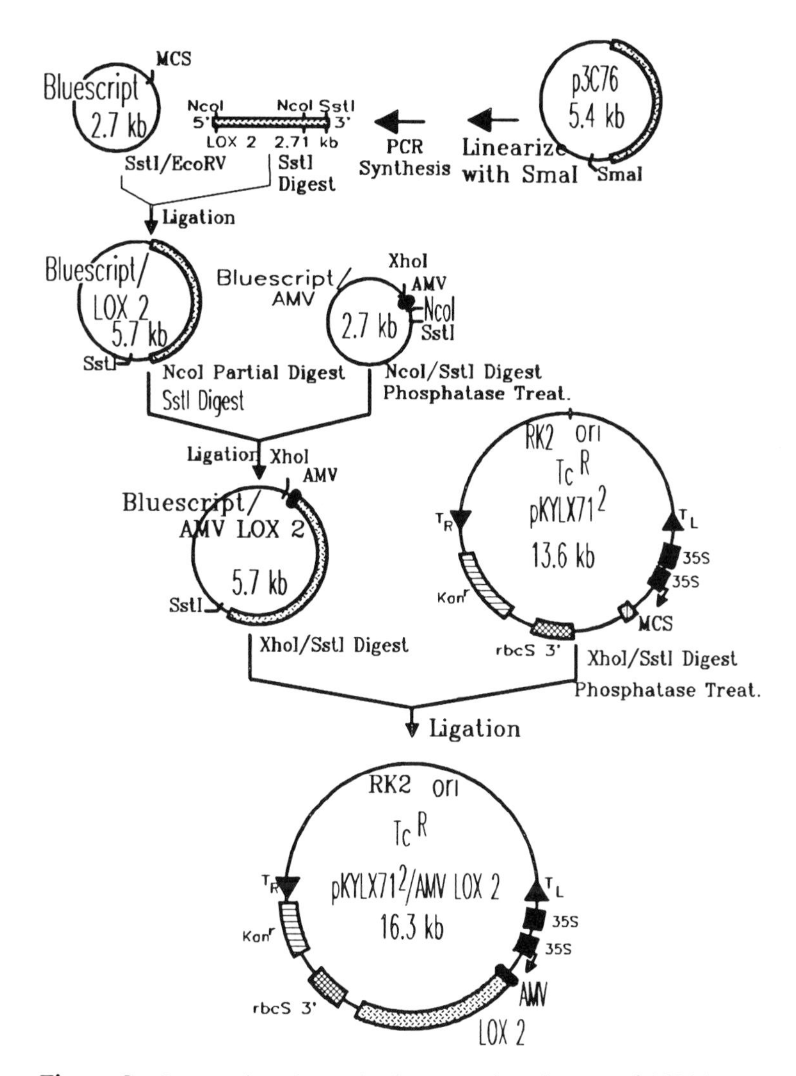

Figure 2 Construction of plasmids for expression of soybean LOX 2 in tobacco. The PCR-amplified soybean embryo LOX 2 cDNA (2592 bp coding sequence plus 87 bp 3′ untranslated region) was cut with *Sst*I, then cloned into BS (cut with *Sst*I and *Eco*RV), from which LOX 2 DNA with *Sst*I and *Nco*I cohesive ends at its 3′ and 5′ ends was obtained. The LOX 2/*Nco*I/*Sst*I fragment was ligated to BS/AMV cut with *Nco*I and *Sst*I. The BS/AMV-LOX 2 plasmids were restricted with *Xho*I, *Sst*I, and *Pvu*I again to release the AMV-LOX 2 DNA, which was cloned into the gene vector pKYLX71^2, which is a derivative of pKYLX7 (48) that contains a duplicated rather than a single 35S promoter. The pKYLX71^2 gene vector also contains a kanamycin-resistance gene for selection of transformants such as pKYLX7 (48). The pKYLX71^2/AMV-LOX 2 plasmid was transferred into *Agrobacterium* A281 by the triparental mating procedure (49). A culture of A281 containing the full-size AMV-LOX 2 was used to transform tobacco cotyledons and leaves.

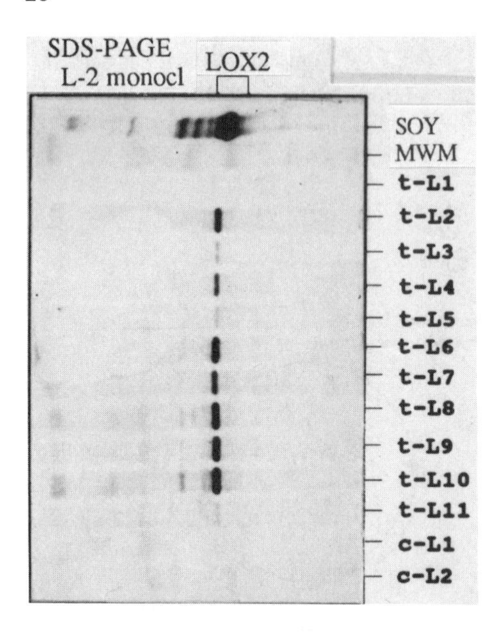

Figure 3 Western immunoblot of SDS-PAGE of proteins from tobacco plant leaves immunodecorated with soybean embryo LOX 2 monoclonal antibody. Sixty micrograms of soluble proteins from transgenic tobacco plant leaves, molecular weight markers (MWM), and a soybean seed extract (soy-B) were subjected to SDS-PAGE. t-L: leaf samples from the plants transformed by pKYLX71[2]/AMV-LOX 2. c-L: leaf samples from the plants transformed by pKYLX71[2]. SDS-PAGE electrophoresis was performed according to Laemmli (50) by using 8.3% running and 4% stacking gels. Native IEF was performed in a 6% polyacrylamide gel by using an LKB flat-bed system as described by Funk et al. (51). The SDS gels were analyzed by Western blotting as described by Wang and Hildebrand (52) with the following modifications: After blotting, the Immobilon-P (Millipore Co.) transfer membranes were washed with TBS for 10–15 min, then incubated with 10% dry milk in TBS for 1 hr. The LOX 2 monoclonal antibody was added and incubated with the blot for 5–6 hr at room temperature. After being washed four times (10 min each time) in TBS containing 0.25% gelatin and 0.5% Tween 20 and one time in TBS for 10 min, the blots were incubated with alkaline-phosphatase-conjugated goat anti-mouse IgG in 10% milk–TBS overnight, then rinsed as above. They were then stained with 5-bromo-4-chloro-3-indolyl phosphate *p*-toluidine salt and nitro blue tetrazolium chloride staining solution. The MWM bands were 200, 97.4, 68, 43, and 29 kDa.

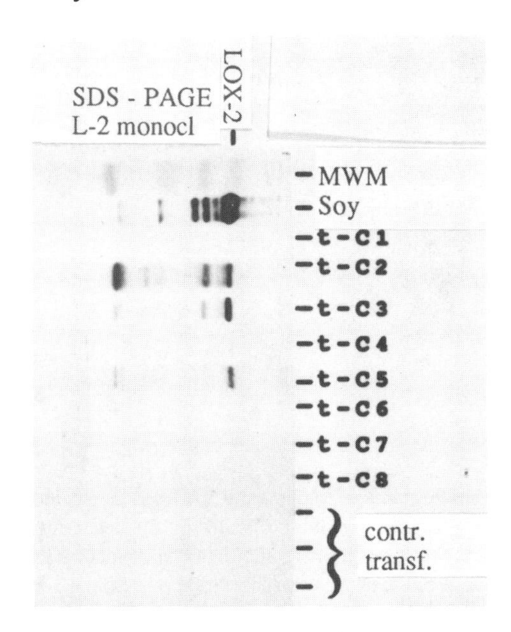

Figure 4 Western immunoblot of SDS-PAGE immunodecorated with soybean embryo LOX 2 monoclonal antibody of proteins from tobacco calli. Sixty micrograms of soluble proteins from tobacco calli, plus MWM and soy-B were subjected to SDS-PAGE (see legend to Fig. 3 for additional information). T-C: tobacco calli transformed by pKYLX71[2]/AMV-LOX 2. Contr. transf.: tobacco calli transformed by pKYLX71[2].

Recently, LOXs have been found to be under stringent metabolic control in plant leaves. Jasmonate, a product of the LOX/dehydrase pathway, another branch of the LOX pathway (37), has been found to increase LOX activity in soybean and tobacco leaves. We have found that treatment of tobacco leaves with low levels of methyl jasmonate can result in a 10-fold increase in LOX activity (Fig. 7A) and in production of the C_6-aldehydes hexanal and *trans*-2-hexenal and the C_6-aldehyde derivative *cis*-3-hexenyl acetate (Figs. 7B and 8). This suggests that a product of one major LOX pathway can increase the activity of LOX itself and the activity of a determining enzyme of another major branch pathway (Fig. 8).

We have been able to obtain a much greater increase in LOX activity and C_6-aldehyde formation by treating tobacco plants with methyl jasmonate than

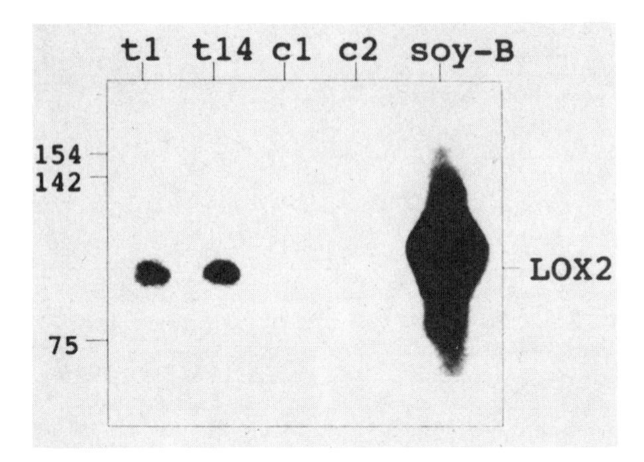

Figure 5 S1 nuclease analysis. Ten micrograms of total plant RNA from transgenic tobacco plants and soybean seed (soy-B) or a negative control of 10 μg of wheat germ tRNA were combined with a volume of probe corresponding of 120,000 cpm, incubated at 75°C for 15 min, 37°C overnight; 1.5 μl of S1 nuclease (600 U, Boehinger Mannheim) was added to degrade the unhybridized RNA at 37°C for 1 hr; the undigested RNA–DNA hybrid was analyzed on polyacrylamide gels (19:1, acrylamide:bis) containing 4% acrylamide, 7 *M* urea, and TBE buffer. c-1 and c-2 are two control plants transformed with pKYLX71[2]. t-1 and t-14 are two plants transformed with AMV-LOX 2. The DNA size markers are indicated in base pairs.

by introducing the soybean LOX 2 gene under control of a strong constitutive promoter. This indicates that in genetically engineering plants for altered formation of a metabolic product it may be important to manipulate factors involved in control of that metabolic pathway in addition to enzymes directly involved in that pathway.

Table 1 LOX Activity in Transgenic Tobacco Calli[a]

Source[b]	Mean LOX activity (nkat/g)[c]
Control	2.83
t-A	5.64
t-E	5.71

[a]Activity based on fresh tissue weight. LOX activity was determined by spectrophotometric measurement of the formation of conjugated dienes at 235 nm (53). The activity of LOX 2 (together with LOXs with similar pH optima) was determined at pH 6.8 by using potassium linoleate as a substrate or at pH 6.1 by using potassium arachidonate as a substrate. The substrates were prepared immediately before the assay and were adjusted to an absorbance of 0.45 A235 against an appropriate buffer blank, standardizing hydroperoxy fatty acid levels that would otherwise vary because of auto-oxidation.
[b]The control average is the mean of six independent control calli, three analyses each, for 18 determinations; t-A and t-E are two independent transformed calli, and the means each represent three analyses.
[c]FLSD = 2.25 at $\alpha = 0.05$.

Table 2 LOX Activity in Transgenic Tobacco Leaves[a]

	Mean LOX activity (nkat/g)	
Source[b]	Linoleic acid substrate[c]	Arachidonic acid substrate[d]
C1	23.0	25.5
C2	33.9	32.8
t14	20.1	26.5
t18	19.9	25.2

[a]Measurement of activity is described in Table 1. The measurements were performed three times for each sample.
[b]C1 and C2 represent leaf samples from two control tobacco plants transformed with pKYLX71[2], and t14 and t18 represent leaf samples from two tobacco plants transformed with pKYLX71[2]/AMV-LOX 2.
[c]FLSD = 14.68 at $\alpha = 0.05$.
[d]FLSD = 22.68 at $\alpha = 0.05$.

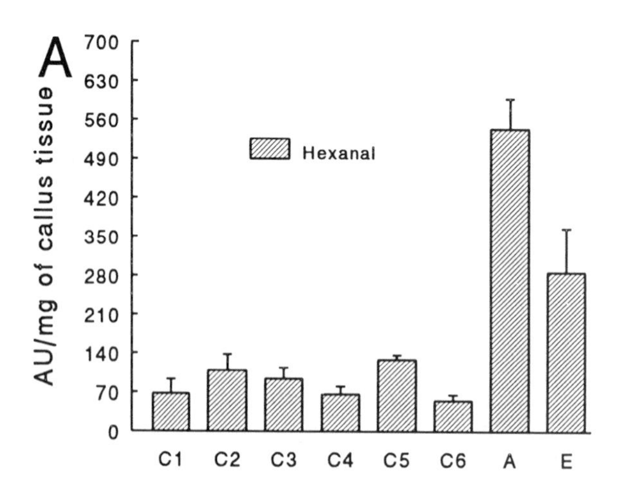

Figure 6 Comparison of C_6-aldehyde formation in transgenic tobacco tissues. (A) Formation of the LOX product hexanal in calli. Controls were six calli independently transformed with pKYLX71[2]. A and E are two calli transformed by AMV-LOX 2. (B) Formation of LOX products hexanal and *trans*-2-hexenal (E-2-hexenal) in leaves. Controls c1 and c2 are leaf samples from two plants independently transformed with pKYLX71[2]. T14 and T18 are two plants transformed by AMV/LOX 2. The data are means of at least three measurements ± SE. C_6-aldehyde (E-2-hexenal and hexanal) levels were determined by gas chromatographic analysis of the headspace vapor of 1.8-ml screwtop vials. For assay of C_6-aldehydes of calli, 150 μl of substrate (1 m*M*

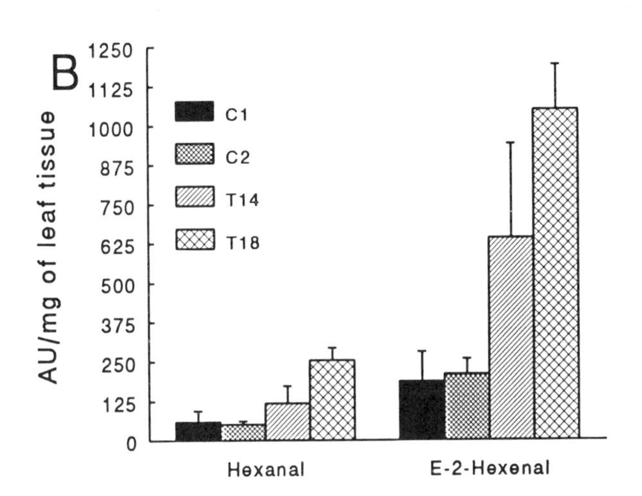

sodium linoleate) and 50 μl of 0.2 *M* sodium phosphate buffer (pH 6.8) were added to the calli. Both calli and leaves (100–150 mg) were frozen in the vials with liquid N_2 for 1–2 min, then incubated at 30°C for 20 min. Calli were stirred 10–15 sec with magnetic stir bars, heated in a microwave oven for 30 sec (for leaves) or 1 min (for calli), and kept at 80°C for 5 min; then 250 μl of the headspace vapor was injected onto a 30 m × 0.53 mm DB-wax (polyethylene glycol) fused silica capillary column operated under the following conditions: column oven 50°C for 5 min, then programmed at 3°C per min to 150°C; injector 220°C; flame ionization detector 240°C. The flow rate of the carrier, helium, was 6 m/min.

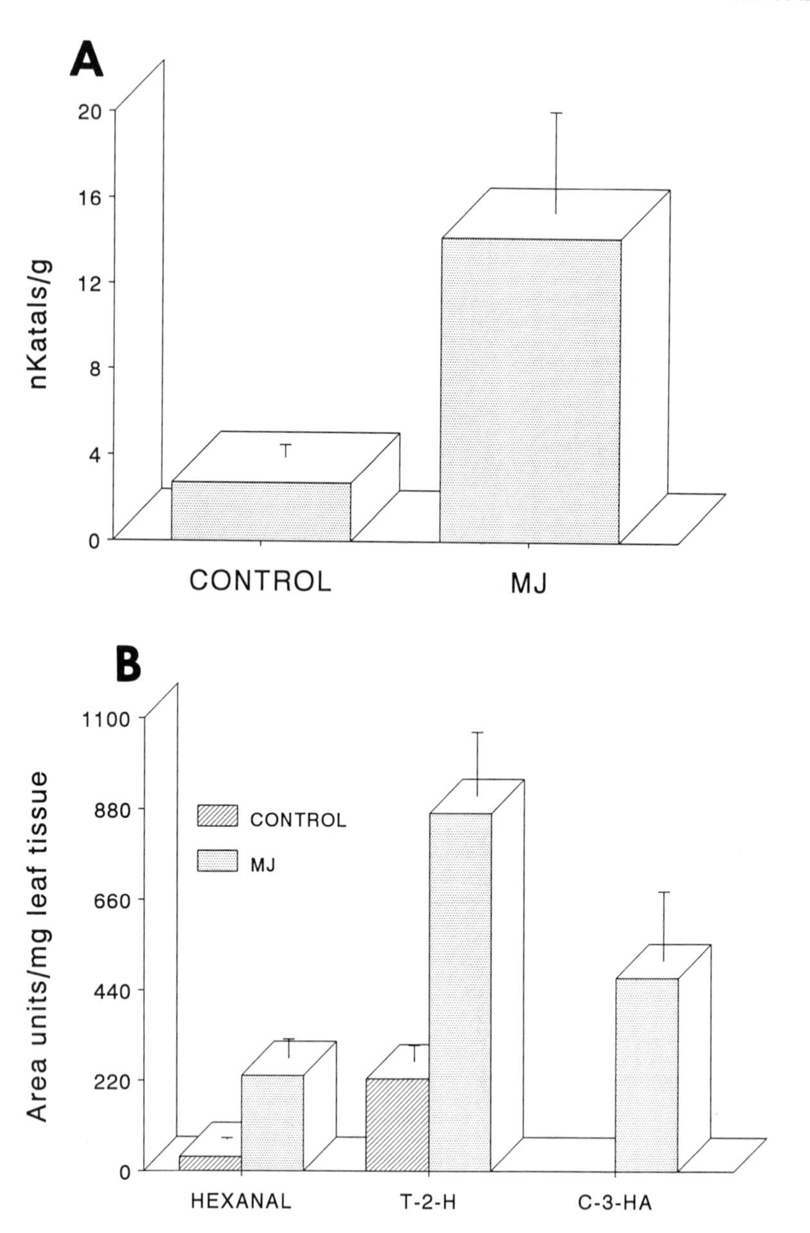

Figure 7 Effects of methyl jasmonate on LOX activity (A) and LOX/lyase product production (B). MJ = methyl jasmonate; T-2-H = *trans*-2-hexenal; C-3-HA = *cis*-3-hexenyl acetate.

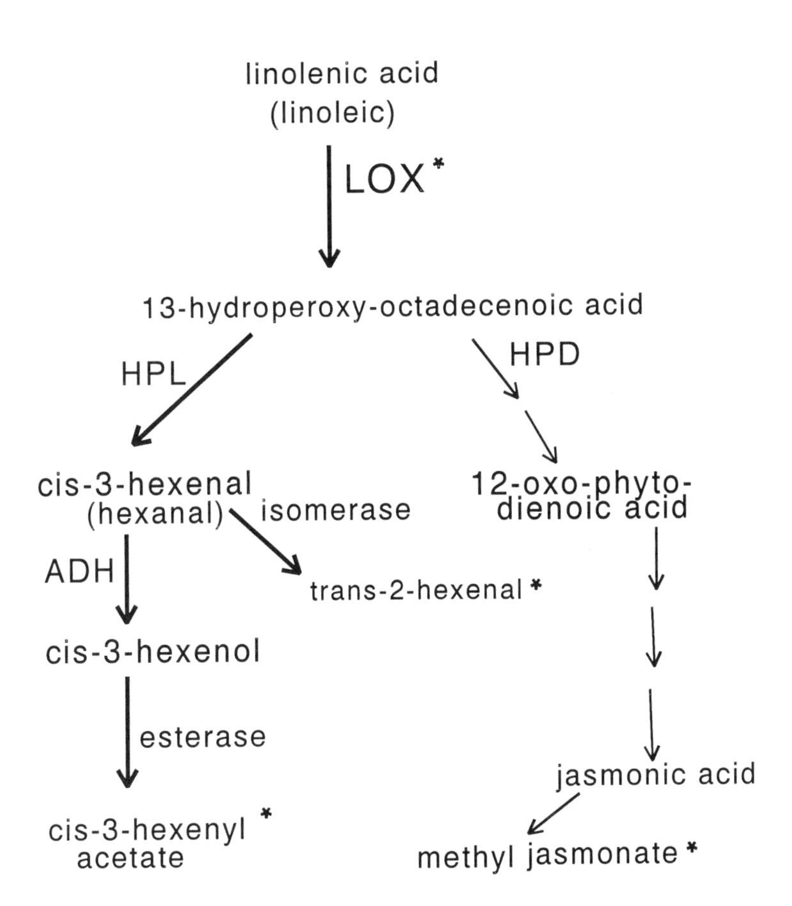

Figure 8 Model of the effects of methyl jasmonate on the regulation of the LOX pathway. HPL = hydroperoxide lyase; HPD = hydroperoxide dehydrase (or allene oxide synthase); ADH = alcohol dehydrogenase. Methyl jasmonate was found to increase levels of LOX and the LOX/lyase metabolites marked with an asterisk (*) as well as hexanal, which is derived from linoleic acid.

ACKNOWLEDGMENTS

We thank Bernard Axelrod and Daisuke Shibata for supplying the LOX 2 cDNA and Art Hunt for supplying the pKYLX71[2] and pBS/AMV. The advice of Art Hunt, Chris Schardl, Sadik Tuzun, Dwight Tomes, and Tom Hamilton-Kemp is greatly appreciated, as is the technical assistance of Udaya Chand, Shaohui Yin, and Robert Versluys.

REFERENCES

1. Braddock, M., and Hardie, D. G. Cloning of cDNA to rat mammary-gland fatty acid synthase mRNA, *Biochem. J., 249*: 603, 1988.
2. Luo, X., Park, K., Lopez-Casillas, F., and Kim, K.-H. Structural features of the acetyl-CoA carboxylase gene: mechanisms for the generation of mRNA's with 5′ end heterogeneity, *Proc. Natl. Acad. Sci. USA, 86*: 4042, 1989.
3. Post-Beittenmiller, M. A., Schmid, K. M., and Ohlrogge, J. B. Expression of holo and apo forms of spinach acyl carrier protein-I in leaves of transgenic tobacco, *Plant Cell, 1*: 889, 1989.
4. Naggert, J., Witkowski, A., Wessa, B., and Smith, S. Expression in *Escherichia coli*, purification and characterization of two mammalian thioesterases involved in fatty acid synthesis, *Biochem. J., 273*: 787, 1991.
5. Sasaki, G. C., Cheesbrough, V., and Kolattukudy, P. E. Nucleotide sequence of the S-acyl fatty acid synthase thioesterase gene and its tissue-specific expression, *DNA, 7*: 449, 1988.
6. Somerville, C., and Browse, J. Plant lipids: metabolism, mutants, and membranes, *Science, 252*: 80, 1991.
7. Theide, M. A., Ozols, J., and Strittmatter, P. Construction and sequence of cDNA for rat liver stearyl coenzyme A desaturase, *J. Biol. Chem., 261*: 13230, 1986.
8. Shanklin, J., and Somerville, C. Stearoyl-acyl-carrier-protein desaturase from higher plants is structurally unrelated to the animal and fungal homologs, *Proc. Natl. Acad. Sci. USA, 88*: 2510, 1991.
9. Thompson, G. A., Scherer, D. E., Foxall-Van Aken, S., Kenny, J. W., Young, H. L., Shintani, D. K., Kridl, J. C., and Knauf, V. C. Primary structures of the precursor and mature forms of stearoyl-acyl carrier protein desaturase from safflower embryos and requirement of ferredoxin for enzyme activity, *Proc. Natl. Acad. Sci. USA, 88*: 2578, 1991.
10. Stukey, J. E., McDonough, V. M., and Martin, C. E. The *OLE1* gene of *Saccharomyces cerevisiae* encodes the Δ9 fatty acid desaturase and can be functionally replaced by the rat stearoyl-CoA desaturase gene, *J. Biol. Chem., 265*: 20144, 1990.
11. Wada, H., Gombos, Z., and Murata, N. Enhancement of chilling tolerance of a cyanobacterium by genetic manipulation of fatty acid desaturation, *Nature, 347*: 200, 1990.

12. Brockman, J. A., and Hildebrand, D. F. A polypeptide alteration associated with a low linolenate mutant of *Arabidopsis thaliana, Plant Physiol. Biochem., 28*: 11, 1990.
13. Hsieh, R. J., German, J. B., and Kinsella, J. E. Lipoxygenase in fish tissue: some properties of the 12-lipoxygenase from trout gill, *J. Agric. Food Chem., 36*: 680, 1988.
14. Hallberg, M. L., and Lingnert, H. Lipid oxidation in potato slices under conditions stimulating the production of potato granules, *J. Am. Oil Chem. Soc., 68*: 167, 1991.
15. Velasco, P. J., Lim, M. H., Pangborn, R. M., and Whitaker, J. R. Enzymes responsible for off-flavor and off-aroma in blanched and frozen-stored vegetables, *Biotechnol. Appl. Biochem., 11*: 118, 1989.
16. Arai, S., Noguchi, M., Kaji, M., Kato, H., and Fujimaki, M. n-Hexanal and some volatile alcohols—their distribution in raw soybean tissues and formation in crude soy protein concentrate by lipoxygenase, *Agric. Biol. Chem., 34*: 1420, 1970.
17. Buttery, R. G., Teranishi, R., and Ling, L. C. Fresh tomato aroma volatiles: a quantitative study, *J. Agric. Food Chem., 35*: 540, 1987.
18. Galliard, T., Phillips, D. R., and Reynolds, J. The formation of *cis*-3-nonenal, *trans*-2-nonenal and hexanal from linoleic acid hydroperoxide isomers by a hydroperoxide cleavage enzyme system in cucumber (*Cucumis sativus*) fruits, *Biochim. Biophys. Acta, 441*: 181, 1976.
19. Shiiba, K., Negishi, Y., Okada, K., and Nagao, S. Purification and characterization of lipoxygenase isozymes from wheat germ, *Cereal Chem., 68*: 115, 1991.
20. Eskin, N. A. M., Grossman, and Pinsky, A. Biochemistry of lipoxygenase EC-1,13,11,12 in relation to food quality, *Crit. Rev. Food Sci. Nutr., 9*: 1, 1977.
21. Ruzicska, P., Gombos, Z., and Farkas, G. L. Modification of the fatty acid composition of phospholipids during the hypersensitive reaction in tobacco, *Virology, 128*: 60, 1983.
22. Sekizawa, Y., Haruyama, T., Kano, H., Urushizaki, S., Saka, H., Matsumoto, K., and Haga, M. Dependence on ethylene of the induction of peroxidase and lipoxygenase activity in rice leaf infected with blast fungus, *Agric. Biol. Chem., 54*: 471, 1990.
23. Schildknecht, H., and Rauch, G. Report on the defensive substances of plants II: the chemical nature of the volatile phytocides of leafy plant, particularly of *Robinia pseudoacacia, Z. Naturforsch., 16B*: 422, 1961.
24. Lyr, H., and Banasiak, L. A. Volatile defense substances in plants, their properties and activities, *Acta Phytopathol. Hungaricae, 18*: 3, 1983.
25. Turlings, T. C., Tumlinson, J. H., and Levine, W. J. Exploitation of herbivore-induced plant odors by host-seeking parasitic wasp, *Science, 250*: 1251, 1990.
26. Farmer, E. E., and Ryan, C. A. Interplant communication: airbore methyl jasmonate induces synthesis of proteinase inhibitors in plant leaves, *Proc. Natl. Acad. Sci. USA, 87*: 7713, 1990.
27. Holtzman, M. J., Pentland, A., and Hansbrough, J. R. Heterogeneity of cellular expression of arachidonate 15 lipoxygenase: implication for biological activity, *Biochim. Biophys. Acta, 1003*: 204, 1989.

28. Grayburn, W. S., Schneider, G. R., Hamilton-Kemp, T. R., Bookjans, G., Ali, K., and Hildebrand, D. F. Soybean leaves contain multiple lipoxygenases, *Plant Physiol., 95*: 1214, 1991.
29. Hildebrand, D. F., Versluys, R. T., and Collins, G. B. Changes in lipoxygenase isozyme levels during soybean embryo development, *Plant Sci., 75*: 1, 1991.
30. Wilen, R. W., Van Rooijen, G. J. H., Pearce, D. W., Pharis, R. P., Holbrook, L. A., and Moloney, M. M. Effects of jasmonic acid on embryo-specific processes in Brassica and Linum oil seeds, *Plant Physiol., 95*: 399, 1991.
31. Anderson, J. M., Spilstro, S. R., Klauer, S. F., and Franceschi, V. R. Jasmonic acid-dependent increase in the level of vegetative storage proteins in soybean, *Plant Sci., 62*: 45, 1989.
32. Herrmann, G., Lehmann, J., Peterson, A., Sembdner, G., and Weidhase, R. A. Species and tissue specificity of jasmonate induced abundant proteins, *J. Plant Physiol., 134*: 703, 1989.
33. Mason, H. S., and Mullett, J. E. Expression of two soybean vegetative storage protein genes during development and in response to water deficit, wounding, and jasmonic acid, *Plant Cell, 2*: 2569, 1990.
34. Leshem, Y. Y., Wurzburger, J., Grossman, S., and Frimer, A. A. Cytokinin interaction with free radical metabolism and senescence: effects on endogenous lipoxygenase and purine oxidation, *Physiol. Plant., 53*: 9, 1981.
35. Lynch, D. V., and Thompson, J. E. Lipoxygenase mediated production of superoxide anion in senescing plant tissue, *FEBS Lett., 173*: 251, 1984.
36. Vick, B. A., and Zimmerman, D. C. Oxidative systems for modification of fatty acid: the lipoxygenase pathway, *The Biochemistry of Plants: A Comprehensive Treatise, Vol. 9* (P. K. Stumpf, ed.), Academic Press, Orlando, Fla., 1987, p. 53.
37. Hildebrand, D. F. Lipoxygenase, *Physiol. Plant., 76*: 249, 1989.
38. Wardale, D. A., and Galliard, T. Further study of the subcellular location of lipid-degrading enzymes, *Phytochem., 16*: 333, 1977.
39. Douillard, R., and Bergeron, E. Chloroplastic location of soluble lipoxygenase activity in young leaves, *Plant Sci. Lett., 22*: 263, 1981.
40. Song, S. L., Love, M. H., and Murphy, P. Subcellular location of lipoxygenase-1 and -2 in germinating soybean seeds and seedlings, *J. Am. Oil Chem. Soc., 67*: 961, 1990.
41. Matoba, T., Hidaka, H., Narita, H., Kitamura, K., Kaizuma, N., and Kito, M. Lipoxygenase-2 isozyme is responsible for generation of n-hexanal in soybean homogenate, *J. Agric. Food Chem., 33*: 852, 1985.
42. Hildebrand, D. F., Hamilton-Kemp, T. R., Loughrin, J. H., Ali, K., and Andersen, R. A. Lipoxygenase 3 reduces hexanal production from soybean seed homogenates, *J. Agric. Food Chem., 38*: 1934, 1990.
43. Zhuang, H., Hildebrand, D. F., Hamilton-Kemp, T. R., and Anderson, R. A. Effects of polyunsaturated free fatty acids and esterified linoleoyl derivatives on oxygen consumption and C_6-aldehide formation with soybean seed homogenates, *J. Agric. Food Chem.*, in press.

44. Park, T. K., and Polacco, J. C. Distinct lipoxygenase species appear in the hypocotyl/radicle of germinating soybean, *Plant Physiol., 90*: 285, 1989.
45. Shibata, D., Kato, T., and Tanaka, K. Nucleotide sequences of a soybean lipoxygenase gene and the short intergenic region between an upstream lipoxygenase gene, *Plant Mol. Biol., 16*: 353, 1991.
46. Shibata, D., Steczko, J., Dixon, J. E., Andrews, P. C., Hermodson, M., and Axelrod, B. Primary structure of soybean lipoxygenase L-2, *J. Biol. Chem., 263*: 6816, 1988.
47. Yenofsky, R. L., Fine, M., and Liu, C. Isolation and characterization of a soybean (*Glycine max*) lipoxygenase-3 gene, *Mol. Gen. Genet., 211*: 215, 1988.
48. Schardl, C. L., Byrd, A. D., Benzion, G., Altschuler, M. A., Hildebrand, D. F., and Hunt, A. G. Design and construction of a versatile system for the expression of foreign genes in plants, *Gene, 61*: 1, 1987.
49. Ditta, G., Stanfield, S., Corbin, D., and Helsinki, D. R. Broad host range DNA clone systems for gram-negative bacteria: construction of a gene bank of *Rhizobium meliloti, Proc. Natl. Acad. Sci. USA, 77*: 7347, 1980.
50. Laemmli, U. K. Cleavage of structural proteins during the assembly of the head of bacteriophage T4, *Nature, 227*: 680, 1970.
51. Funk, M. O., Whitney, M. A., Hausknecht, E. C., and O'Brian, E. M. Resolution of the isozymes of soybean lipoxygenase using isoelectric focusing and chromatofocusing, *Anal. Biochem., 146*: 246, 1985.
52. Wang, X. M., and Hildebrand, D. F. Effect of a substituted pyridazone on the decrease of lipoxygenase activity in soybean cotyledons, *Plant Sci., 51*: 29, 1987.
53. Axelrod, B., Cheesbrough, T. M., and Laakso, S. Lipoxygenase from soybeans, *Methods Enzymol., 71*: 441, 1981.

2

Hormone Manipulation in Transgenic Plants

Charles P. Romano and Harry Klee

Monsanto, St. Louis, Missouri

I. INTRODUCTION

Phytohormones are critical in coordinating plant growth and development. These small molecules have been implicated in the regulation of such diverse processes as seed dormancy, stem elongation, apical dominance, and fruit ripening (1). However, the mechanisms of plant hormone action and interaction are far from clear. For example, the isolation and unequivocal identification of a plant hormone receptor protein has yet to be reported. In the case of auxin, the enzymatic steps of biosynthesis await elucidation. Given our lack of knowledge in such fundamentally important areas, our ability to usefully direct plant growth and development through plant hormone perturbations is understandably limited.

Nonetheless, a variety of biochemical, genetic, and molecular biological approaches have rapidly advanced our understanding of how plant hormones work. Biochemistry has allowed several proteins that specifically bind auxins to be identified (2–4). At a genetic level, mutants defective in hormone synthesis and perception have been characterized (5). Finally, transgenic plants with defined perturbations in hormone levels have been examined (5). This chapter will review past and potential contributions of transgenic plant studies to our understanding of plant hormone action. As will become apparent, gene transfer techniques are a key component of a multidisciplinary approach to understanding plant hormone action.

The transgenic approach to hormone manipulations offers several distinct advantages over classical exogenous application techniques. First, since the hormones are produced endogenously, uncertainties associated with uptake and subsequent transport are eliminated. Second, both the timing and tissue specificity of hormone perturbation can be controlled by use of the appropriate transcriptional promoters (6). For example, the constitutive CaMV 35S promoter permits adjustment of hormone levels in virtually all cells throughout the life of the plant, whereas inducible heat shock promoters permit adjustment at specified stages of development (7,8). Moreover, tissue-specific promoters permit alteration of hormone levels in specific tissues. Such localized perturbations could never be achieved in application experiments, particularly in the case of internalized tissues such as the vascular cambium. Finally, the transgenic plants have reproducible and defined hormone perturbations that result from the action of characterized gene products. The degree of perturbation is established by the site of transgene insertion and is transmitted as a dominant Mendelian trait. Consequently, transgene-mediated hormone manipulations are far more reproducible and predictable than are application-mediated manipulations.

II. HORMONES AND GENES

A. Auxin Overproduction

The uncontrolled cell growth characteristic of *Agrobacterium tumefaciens*–mediated crown gall tumorigenesis is in part due to auxin overproduction by plant cells that are genetically transformed by the Ti plasmid (9,10). Auxin (indole-3-acetic acid) is formed by a two-step biosynthetic pathway that is not ordinarily found in plants. Tryptophan is first converted to indoleacetamide by the tryptophan mono-oxygenase encoded by the *Agrobacterium iaaM* gene (11). Indoleacetamide is then converted to indoleacetic acid (IAA) by the indoleacetamide hydrolase encoded by the *Agrobacterium iaaH* gene (12,13).

In transgenic petunia and tobacco, free IAA levels have been increased up to 10-fold through constitutive overexpression of the *iaaM* gene by the CaMV 19S promoter (8,14). Indoleacetamide accumulates to high levels and is evidently converted to IAA by spontaneous or nonspecific amidohydrolase-mediated hydrolysis in these plants. Auxin-overproducing plants display, among other effects, increased apical dominance, vascular differentiation, and adventitious rooting.

B. Auxin Inactivation

Another tumor-forming bacterial plant pathogen, *Pseudomonas syringae* pv. *savastanoi*, also has plasmid encoded genes that both synthesize and further modify IAA (15). The *iaaL* gene encodes an indoleacetic acid–lysine synthetase activity that conjugates lysine to the critical carboxyl moiety of IAA (16,17). Although the role of *iaaL* in *P. savastanoi* tumorigenesis is unclear, there is considerable evidence that plants reversibly inactivate and store auxin as amino acid conjugates (18,19).

Free IAA levels are reduced up to 20-fold in transgenic tobacco that constitutively express *iaaL* under CaMV 35S promoter control (14). Auxin-underproducing tobacco exhibits loss of apical dominance, inhibition of vascular differentiation, and severe leaf wrinkling (Fig. 1). The phenotypic effects of 35S-*iaaL*-mediated auxin inactivation can be reverted by the 19S-*iaaM* auxin-overproducing transgene. In contrast, tobacco expressing *iaaL* under heat shock promoter control only display a loss of apical dominance, indicating that this growth parameter is especially sensitive to transient decreases in auxin levels (Fig. 2).

C. Cytokinins

Cytokinin biosynthesis is also required in *Agrobacterium tumefaciens*–mediated crown gall tumorigenesis (9). The *ipt* gene encodes an isopentenyl transferase activity that condenses isopentenyl pyrophosphate and AMP to produce isopentenyl AMP (iPMP) (20,21). Once iPMP is formed, it is rapidly converted to a variety of zeatin-derived cytokinins of greater biological activity (8,22).

Recovery of whole transgenic plants expressing *ipt* has been difficult because constitutive expression yields transgenic shoots that do not root (23). Tobacco and potato shoots transformed with *ipt* have been successfully grafted to wild-type roots and have elevated cytokinin levels and aberrant growth patterns (24). However, fertile transgenic tobacco expressing *ipt* under heat shock promoter control have been obtained (8). Although heat shock results in cytokinin increases of up to 100-fold, these plants display significant increases in cytokinin levels of up to 7-fold even when grown at ambient temperatures. The loss of apical dominance observed in non-heat-shocked *hsp70-ipt* tobacco is similar to the effects of *iaaL*-mediated auxin inactivation.

Figure 1 Comparison of wild-type and auxin-inactivating plants. Wild-type plants (left) and CaMV 35S–IAA–lysine synthetase plants (right) are shown at maturity.

Figure 2 Heat-shock-inducible expression of IAA–lysine synthetase in tobacco. Transgenic plants were grown with (+) and without (–) a 2-wk period of daily 1-hr heat shock treatments. Wild-type control plants subjected to the same heat shock regimen did not display the loss of apical dominance observed in the heat-shocked HSP 70–IAA–lysine synthetase plants.

D. Ethylene

Ethylene gas is critical in the ripening of agronomically important fruits such as tomato and is an obvious target for hormone manipulations. In plants, *S*-adenosylmethionine is converted to 1-amino-cyclopropane-1-carboxylic acid (ACC) and then to ethylene by means of an ACC synthase and an ethylene-forming enzyme (EFE) that is an ACC oxidase (25–27). Ethylene production in transgenic tomato fruit is inhibited by up to 97% by CaMV 35S-promoter-driven expression of what has subsequently been identified as an EFE antisense RNA (26,28,29; see also Watson and Grierson chapter in this volume). Such fruit redden but do not overripen, as do wild-type controls.

Tomato fruit ripening is completely blocked by inhibiting ethylene production through expression of a fruit-specific ACC synthase antisense RNA from the CaMV 35S promoter (30). Ethylene exposure induced maturation of the transgenic fruit, demonstrating the causal role of ethylene in ripening (30). The success of this experiment is probably due to the fact that the two functional ACC synthase mRNAs expressed in tomato fruit share extensive regions of homology, resulting in reduced expression of both messages by a single antisense RNA (30–32).

Ethylene production has also been significantly reduced in both leaves and fruit of transgenic tomato plants by CaMV 35S-promoter-driven expression of a bacterial ACC deaminase (33). The ACC deaminase, which converts ACC to α-ketobutyric acid and ammonia, was first identified by selecting a *Pseudomonas* species capable of using ACC as a sole nitrogen source. The gene conferring the capacity to grow on ACC as a sole nitrogen source was then cloned from a genomic library derived from the *Pseudomonas* species by complementation in *E. coli.* Ethylene production rates are reduced by up to 97% in leaves and fruit of ACC-deaminase-expressing tomatoes. Both ripening and softening are significantly delayed in these transgenic tomatoes. However, all other aspects of growth and development are apparently unaffected by reduction of ethylene production throughout the plant. It is thus possible that ethylene is not required for processes other than fruit ripening. However, this hypothesis must be tested by creating plants that synthesize absolutely no ethylene.

Another approach to regulating ethylene responses would be to express the dominant ethylene resistance gene of *Arabidopsis, etr1* or *ein1*, in a transgenic plant (34,35). It will obviously first be necessary to clone these genes through one of the T-DNA tagging, AC transposon tagging, RFLP mapping/chromosome walking, or genomic subtraction techniques currently being developed in *Arabidopsis* (36–40). As these genes apparently act by inhibiting ethylene perception or signal transduction rather than synthesis, they may be especially useful in specifically eliminating ethylene responses in tissues located near other tissues that produce large quantities of this diffusible gas.

E. Abscisic Acid and Gibberellins

Transgenic plants with genetically engineered gibberellin (GA) or abscisic acid (ABA) imbalances await development. However, molecular tools for manipulation of GA or ABA levels should soon be available. In *Arabidopsis*, recessive mutations in the GA_1 gene block were an early step in GA biosynthesis, resulting in a nongerminating dwarf phenotype in the absence of exogenously supplied GAs (41,42). The GA_1 gene has recently been cloned by genomic subtraction (43). Antisense GA_1 T-DNA constructions should effectively inhibit GA biosynthesis. Cloning of the *Arabidopsis* genes defective in ABA production (ABA locus) or perception (ABI1, ABI2, ABI3) should also permit control of ABA production and response in transgenic plants (44,45).

III. MAJOR CONCLUSIONS OF TRANSGENIC PLANT STUDIES

A. Auxin/Cytokinin Ratios: Control of Apical Dominance and Vascular Proliferation in Intact Plants

Regulation of lateral bud growth or apical dominance by antagonistic auxin/cytokinin interactions is well established (46). However, several lines of evidence from transgenic plant studies suggest that apical dominance is controlled by the auxin-to-cytokinin ratio rather than by the absolute hormone levels. First, the extreme apical dominance of auxin overproducers can be overcome by outcrossing to a cytokinin overproducer or by applying cytokinin to dormant lateral buds (Klee, unpublished). Second, inactivation of endogenous auxin also results in reduced apical dominance (14). The auxin-to-cytokinin ratio also appears to be critical in vascular differentiation because the effects of cytokinin overproduction are similar to those induced by auxin inactivation. This auxin-to-cytokinin ratio hypothesis obviously must be critically analyzed by quantitative analysis of auxin and cytokinin levels in whole transgenic plants. At present, the effect of transgenic cytokinin overproduction on auxin levels has only been addressed in regenerating tissue in culture (47). In this study, increases in cytokinin levels of greater than 100-fold relative to controls resulted in less than 3-fold increases and decreases in auxin levels.

B. Apparent Insensitivity of Plants to Hormone Perturbations

Perhaps the most surprising finding of the hormone manipulations studies is that such plants are fertile and pass through a relatively normal life cycle in an essentially normal time period. For example, tobacco can withstand 10-fold surpluses and 20-fold deficits of auxin. Tobacco can also tolerate 100- to 200-fold increases in cytokinin levels. One potential explanation of this tolerance is that many tissue types are inherently insensitive to large differences in hormone levels. Another possibility is that certain tissue types can actively compensate for hormone imbalances by changing the levels of an antagonistic hormone, by sequestering or otherwise neutralizing excess hormone, or by modulating the activity of hormone reception and signal transduction pathways. Detailed biochemical analyses of the modified transgenic plants should determine which hypothesis is correct.

IV. FRONTIERS IN TRANSGENIC STUDIES OF PLANT HORMONES

A. Hormone Interactions

In whole plants, hormone effects are exerted through both antagonistic and synergistic interactions. For example, auxin is known to stimulate ethylene production by increasing ACC synthase levels (48,49). Thus, auxin overproduction in transgenic tobacco and petunia also results in 5-fold increases in ethylene production rates (Romano, Kretzmer, and Klee, unpublished). Consequently, the phenotypic abnormalities observed in auxin-overproducing plants could be due to increased auxin, increased ethylene, or both. These factors can now be distinguished from one another by crossing auxin-overproducing plants with transgenic plants deficient in ethylene production. Auxin and ethylene effects could also be uncoupled in *Arabidopsis* by placing an auxin-overproducing transgene in an ethylene-insensitive genetic background. Finally, the effect of overproducing ethylene without excess auxin can be tested by constitutive overexpression of ACC synthase.

B. Hormonal Homeostasis

Given the apparent importance of maintaining appropriate hormone levels, plants clearly must regulate both hormone synthesis and degradation. Transgenic plants with defined hormone imbalances provide a unique opportunity to examine how plants respond to hormone perturbations. Auxin metabolism has been examined in transgenic tobacco with modest increases in IAA production mediated by intact *Agrobacterium iaaM* and *iaaH* genes (50). In these plants, free IAA levels are increased by only 1.3-fold, while IAA conjugate levels are increased up to 4-fold, supporting the notion that plants can compensate for increases in free or active IAA by converting the excess to conjugated or inactive IAA. However, this study was restricted to phenotypically normal plants with limited perturbations in auxin levels. It will obviously be important to examine IAA metabolism in phenotypically abnormal plants with more dramatic increases and decreases in IAA levels. In these instances, it is clear that the homeostatic capacity of the plant to adjust the concentration of free IAA to normal levels has been exceeded.

C. Regeneration as an Obstacle

Because hormones are pivotal in root and shoot formation in culture, plants with massive hormone level alterations may never be recovered from standard *Agrobacterium*-mediated transformation experiments that rely on the regeneration of transformed callus. A classic example of this problem is seen in the failure of CaMV 35S-*ipt*–transformed shoots to form roots. Nonetheless, there are several strategies for overcoming this problem. One strategy is to introduce the genes directly into meristematic tissue by particle gun bombardment or *Agrobacterium*-mediated seed transformation (36,51). It is also possible to introduce a hormone gene that has been silenced by a transposon insertion, regenerate the plant in the absence of transgene activity, and then activate that gene in subsequent generations by crossing the regenerant to a line expressing active transposase (52). Finally, grossly affected plants expressing high levels of one hormone-modulating gene might be recovered by first transforming tissue from a characterized transgenic line with high levels of a counteracting hormone. The two compensatory transgenes could then be separated and analyzed independently in subsequent generations by segregation.

D. Tissue-Specific Expression: Potential and Problems

Given currently available promoters and hormone-modulating genes, hormone levels in specific tissues can theoretically be altered at will. Selected changes in a single tissue are likely to be the most informative and agronomically useful tools emerging from this technology. However, it must be remembered that most of the available hormone-modulating genes act by synthesizing excess hormone (*iaaM* and *ipt*), inhibiting endogenous synthesis (EFE and ACC synthase antisense and ACC deaminase), or by inactivating internally produced hormone (*iaaL*). All of these genes thus function in a non-cell-autonomous manner in that they are likely to either affect or be affected by adjacent nonexpressing cells. For example, tissue-specific expression of *iaaM* would create a localized auxin source that would raise auxin levels in surrounding tissues, whereas similar expression of *iaaL* might create an auxin sink that would decrease auxin levels in surrounding tissues. Consequently, such experiments must be interpreted with great caution, especially in view of our inability to accurately measure levels of any hormone on this scale. Monitoring the expression of endogenous-hormone-inducible genes (53) by in situ techniques may be an excellent way to assay the effects of tissue-specific hormone perturbations.

Tissue-specific expression of yet uncharacterized genes that change hormone responses in a cell-autonomous manner will ultimately yield the least equivocal results. Such genes may be obtained by cloning putative receptors or by isolating the dominant alleles of genes conferring hormone insensitivity.

V. SUMMARY: PROSPECTUS FOR APPLICATIONS AND FUTURE ADVANCEMENT

Transgenic plants that increase or inactivate auxin, decrease ethylene production, and increase cytokinin have been described. These studies clearly demonstrate the role of hormones in plant processes such as apical dominance, vascular differentiation, and fruit ripening. Moreover, our demonstrated ability to manipulate these processes will ultimately have practical consequences. Alteration of plant architecture (apical dominance) in desirable and predictable ways holds significant value in many agronomic and horticultural crops. The ability to control ethylene synthesis and sensitivity will directly affect the world's food supply by extending shelf life and reducing spoilage. This will be particularly true in countries lacking sophisticated and expensive storage and processing facilities.

Further analysis of these plants will also advance our understanding of hormone interactions and regulation in plant growth and development. Basic questions concerning interactions between different hormones, such as auxin and ethylene, will soon be addressed with already available molecular and genetic tools. Biochemical analyses of hormone metabolism in altered transgenic plants will also reveal how plants respond to hormonal imbalances. Finally, deliberate manipulation of hormone levels and sensitivity in specific tissues and at specific times will elucidate the role of these molecules in the ordered pattern of whole plant development.

REFERENCES

1. Davies, P. J. *Plant Hormones and Their Roles in Plant Growth and Development* (P. J. Davies, ed.), Martinus Nijhoff, Boston, 1987, pp. 1–23.
2. Hesse, T., Feldwisch, J., Balshuesemann, D., Bauw, G., Puype, M., Vanderckhove, J., Loebler, M., Klaembt, D., and Schell, J. Molecular cloning and structural analysis of a gene from *Zea mays* L. coding for a putative receptor for the plant hormone auxin, *EMBO J., 8*: 2453–2462, 1989.
3. Hicks, G. R., Rayle, D. L., and Lomax, T. L. The diageotropica mutant of tomato lacks high specific activity auxin binding sites, *Science, 245*: 52–54, 1989.
4. Prasad, P. V., and Jones, A. M. Putative receptor for the plant hormone auxin identified and characterized by anti-idiotype antibodies, *Proc. Natl. Acad. Sci. USA, 88*: 5479–5483, 1991.
5. Klee, H., and Estelle, M. Molecular genetic approaches to plant hormone biology, *Annu. Rev. Plant Physiol. Plant Mol. Biol., 42*: 529–551, 1991.
6. Benfey, P. N., and Chua, N.-H. Regulated genes in transgenic plants, *Science, 244*: 174–181, 1989.
7. Klee, H. J., Horsch, R. B., Hinchee, M. A., Hein, M. B., and Hoffmann, N. L. The effects of overproduction of two *Agrobacterium tumefaciens* T-DNA auxin biosynthetic gene products in transgenic petunia plants, *Genes Dev., 1*: 86–96, 1987.
8. Medford, J. I., Horgan, R., El-Sawi, Z., and Klee, H. J. Alterations of endogenous cytokinins in transgenic plants using a chimeric isopentenyl transferase gene, *Plant Cell, 4*: 403–413, 1984.
9. Morris, R. O. Genes specifying auxin and cytokinin biosynthesis in phytopathogens, *Annu. Rev. Plant Physiol., 37*: 509–538, 1987.
10. Zambryski, P., Tempe, J., and Schell, J. Transfer and function of T-DNA genes from *Agrobacterium* Ti and Ri plasmids in plants, *Cell, 56*: 193–201, 1989.
11. Thomashow, M. F., Hugly, S., Buchholz, W. G., and Thomashow, L. S. Molecular basis for the auxin-independent phenotype of crown gall tumor tissues, *Science, 2 231*: 616–618, 1986.

12. Schroder, G., Waffenschmidt, S., Weiler, E. W., and Schroder, J. The T-region of Ti plasmids codes for an enzyme synthesizing indole-3-acetic acid, *Eur. J. Biochem., 138*: 387–391, 1984.
13. Thomashow, L. S., Reeves, S., Thomashow, M. F. Crown gall oncogenesis: evidence that a T-DNA gene from the *Agrobacterium* Ti plasmid pTiA6 encodes an enzyme that catalyzes synthesis of indoleacetic acid, *Proc. Natl. Acad. Sci. USA, 81*: 5071–5075, 1984.
14. Romano, C. P., Hein, M. B., and Klee, H. J. Inactivation of auxin in tobacco transformed with the indoleacetic acid-lysine synthetase gene of *Pseudomonas savastanoi, Genes Dev., 5*: 438–446, 1991.
15. Nester, E. W., and Kosuge, T. Plasmids specifying plant hyperplasias, *Annu. Rev. Microbiol., 35*: 531–565, 1981.
16. Glass, N. J., and Kosuge, T. Cloning of the gene for indoleacetic acid-lysine synthetase from *Pseudomonas syringae subsp. savastanoi, J. Bacteriol., 166*: 598–602, 1986.
17. Roberto, F. F., Klee, H., White, F., Nordeen, R., and Kosuge, T. Expression and fine structure of the gene encoding N^{ϵ}-(indole-3-acetyl)-L-lysine synthetase from *Pseudomonas savastanoi, Proc. Natl. Acad. Sci. USA, 87*: 5797–5801, 1990.
18. Cohen, J. D., and Bialek, K. The biosynthesis of indole-3-acetic acid in higher plants, *The Biosynthesis and Metabolism of Plant Hormones* (A. Crozier and J. R. Hillman, eds.), Cambridge University Press, Cambridge, England, U.K., 1984 , pp. 165–181.
19. Bialek, K., and Cohen, J. D. Free and conjugated indole-3-acetic acid in developing bean seeds, *Plant Physiol., 91*: 775–779, 1989.
20. Akiyoshi, D., Klee, H., Amasino, R., Nester, E. W., and Gordon, M. P. T-DNA of *Agrobacterium tumefaciens* encodes an enzyme of cytokinin biosynthesis, *Proc. Natl. Acad. Sci. USA, 81*: 5994–5998, 1984.
21. Barry, G. F., Rogers, S. G., Fraley, R. T., and Brand, L. Identification of a cloned cytokinin biosynthetic gene, *Proc. Natl. Acad. Sci. USA, 81*: 4776–4780, 1984.
22. Scott, I. M., and Horgan, R. Mass spectrometric quantification of cytokinin nucleotides in tobacco crown gall tissue, *Planta, 161*: 345–354, 1984.
23. Klee, H. J., Horsch, R. B., and Rogers, S. G. *Agrobacterium*-mediated plant transformation and its further applications to plant biology, *Annu. Rev. Plant Physiol., 38*: 467–468, 1987.
24. Ooms, G., and Lenton, J. R. T-DNA genes to study plant development: precocious tuberisation and enhanced cytokinins in *A. tumefaciens* transformed potato, *Plant Mol. Biol., 5*: 205–212, 1985.
25. Yang, S. F., and Hoffman, N. E. Ethylene biosynthesis and its regulation in higher plants, *Annu. Rev. Plant Physiol., 35*: 155–189, 1984.
26. Hamilton, A. J., Lycett, G. W., and Grierson, D. Antisense gene that inhibits synthesis of the hormone ethylene in transgenic plants, *Nature, 346*: 284–287, 1990.
27. Ververidis, P., and John, P. Complete recovery *in vitro* of ethylene-forming enzyme activity, *Phytochem., 30*: 725–727, 1991.

28. Spanu, P., Reinhardt, D., and Boller, T. Analysis and cloning of the ethylene-forming enzyme from tomato by functional expression of its mRNA in *Xenopus laevis* oocytes, *EMBO J.*, *10*: 2007–2013, 1991.
29. Hamilton, A. J., Bouzayen, M., and Grierson, D. Identification of a tomato gene for the ethylene-forming enzyme by expression in yeast. *Proc. Natl. Acad. Sci. USA*, *88*: 7434–7437, 1991.
30. Oeller, P. W., Min-Wong, L., Pike, D. A., and Theologis, A. Reversible inhibition of tomato fruit senescence by antisense RNA, *Science*, *254*: 437–439, 1991.
31. Van Der Straeten, D., Van Wiemeersch, L., Goodman, H. M., and Van Montagu, M. Cloning and sequence of two different cDNAs encoding 1-aminocyclopropane-1-carboxylate synthase in tomato, *Proc. Natl. Acad. Sci. USA*, *87*: 4859–4863, 1990.
32. Olson, D. C., White, J. A., Edelman, L., Harkins, R. N., and Kende, H. Differential expression of two genes for 1-aminocyclopropane-1-carboxylate synthase in tomato fruits, *Proc. Natl. Acad. Sci. USA*, *88*: 5340–5344, 1991.
33. Klee, H. J., Hayford, M. B., Kretzmer, K. A., Barry, G. F., and Kishore, G. M. Control of ethylene synthesis by expression of a bacterial enzyme in transgenic tomato plants, *Plant Cell*, *3*: 1187–1193 (1991).
34. Bleecker, A. B., Estelle, M. A., Somerville, C., and Kende, H. Insensitivity to ethylene conferred by a dominant mutation in *Arabidopsis thaliana*, *Science*, *241*: 1086–1089, 1988.
35. Guzman, P., and Ecker, J. R. Exploiting the triple response of *Arabidopsis* to identify ethylene-related mutants, *Plant Cell*, *2*: 513–523, 1990.
36. Feldman, K. A., Marks, M. D., Christianson, M. L., and Quatrano, R. S. A dwarf mutant of *Arabidopsis* generated by T-DNA insertion mutagenesis, *Science*, *243*: 1351–1354, 1989.
37. Van Sluys, M. A., Tempe, J., and Federoff, N. Studies on the introduction and mobility of the maize *Activator* element in *Arabidopsis thaliana* and *Daucus carota*, *EMBO J.*, *6*: 3881–3889, 1987.
38. Chang, C., Bowman, J. L., DeJohn, A. W., Lander, E. S., and Meyerowitz, E. M. Restriction fragment length polymorphism linkage map for *Arabidopsis thaliana*, *Proc. Natl. Acad. Sci. USA*, *85*: 6856–6860, 1988.
39. Nam, H.-G., Giraudat, J., Boer, B., Moonan, F., Loos, W. D. B., Hauge, B. M., and Goodman, H. M. Restriction fragment length polymorphism linkage map of *Arabidopsis thaliana*, *Plant Cell*, *1*: 699–705, 1989.
40. Straus, D., and Ausubel, F. M. Genomic subtraction for cloning DNA corresponding to deletion mutations, *Proc. Natl. Acad. Sci. USA*, *87*: 1889–1993, 1990.
41. Barendse, G. W. M., Kepczynski, J., Karssen, C. M., and Koorneef, M. The role of endogenous gibberellins during fruit and seed development: studies on the gibberellin-deficient genotypes of *Arabidopsis thaliana* (L.) Heyhn, *Physiol. Plant*, *67*: 315–319, 1986.
42. Koorneef, M., and van Der Veen, J. H. Induction and analysis of gibberellin-sensitive mutants in *Arabidopsis thaliana* (L.) Heyhn, *Theor. Appl. Genet.*, *58*: 257–263, 1980.

43. Sun, T.-p., Goodman, H. M., and Ausubel, F. M. Cloning the *Arabidopsis* GA1 locus by genomic subtraction. *Plant Cell, 4*: 119–128 (1992).
44. Koornneef, M., Jorna, M. L., Brinkhorst-van der Swan, D. L. C., Karssen, C. M. The isolation of abscisic acid (ABA) deficient mutants by selection of induced revertants in non-germinating gibberellin sensitive lines of *Arabidopsis thaliana* (L.) Heyhn, *Theor. Appl. Genet., 61*: 385–393, 1982.
45. Finkelstein, R. R., Somerville, C. R. Three classes of abscisic acid (ABA)-insensitive mutations of *Arabidopsis* define genes which control overlapping subsets of ABA responses. *Plant Physiol., 94*: 1172–1179, 1990.
46. Tamas, I. A. Hormonal regulation of apical dominance, in *Plant Hormones and Their Roles in Plant Growth and Development* (P. J. Davies, ed.), Martinus Nijhoff, Boston, 1987, pp. 393–410.
47. Smigocki, A., and Owens, L. Cytokinin-to-auxin ratios and morphology of shoots and tissues transformed by a chimeric isopentenyl transferase gene. *Plant Physiol., 91*: 808–811, 1989.
48. Sato, T., and Theologis, A. Cloning the mRNA encoding 1-aminocyclopropane-1-carboxylate synthase, the key enzyme for ethylene synthesis in plants. *Proc. Natl. Acad. Sci. USA, 86*: 6621–6625, 1989.
49. Huang, P.-L., Parks, J. E., Rottman, W. H., and Theologis, A. Two genes encoding 1-aminocyclopropane-1-carboxylate synthase in zucchini (*Cucurbita pepo*) are clustered and similar but differentially regulated, *Proc. Natl. Acad. Sci. USA, 88*: 7021–7025, 1991.
50. Sitbon, F., Sundberg, B., Olsson, O., and Sandberg, G. Free and conjugated indoleacetic acid (IAA) contents in transgenic tobacco plants expressing the *iaa*M and *iaa*H IAA biosynthesis genes from *Agrobacterium tumefaciens, Plant Physiol., 95*: 480–485, 1991.
51. McCabe, D. E., Swain, W. F., Martinell, B. J., and Christou, P. Stable transformation of soybean (*Glycine max*) by particle acceleration, *Biotechnol., 6*: 923–926, 1991.
52. Spena, A., Aalen, R. B., and Schulze, S. C. Cell autonomous behavior of the *rol*C gene of *Agrobacterium rhizogenes* during leaf development: a visual assay for transposon excision in transgenic plants, *Plant Cell., 1*: 1157–1164, 1989.
53. Key, J. Modulation of gene expression by auxin. *Bioessays, 11*: 52–57, 1989.

3

Expression of Modified Seed Storage Proteins in Transgenic Plants

Enno Krebbers and Jan Van Rompaey

Plant Genetic Systems, Ghent, Belgium

Joël Vandekerckhove

University of Ghent, Ghent, Belgium

I. INTRODUCTION

The seeds of higher plants contain large quantities of storage proteins, which have been defined by Higgins (1) as proteins accumulated in significant quantities in the developing seed which upon germination are hydrolyzed to provide a source of nitrogen in the early stages of seedling growth. Seeds of different species contain a variety of such proteins, which are conventionally classified on the basis of their size and solubility in various solvents (2). Although there are prominent exceptions, albumins (soluble in water) and globulins (soluble in salt solutions) are found primarily in the dicots, whereas prolamins (alcohol soluble) and glutelins (soluble in acid or basic solutions) are found in monocots. Proteins of the same class are often highly homologous across different species. Individual classes may represent up to 60% of total seed protein, or 15–30% of total seed weight.

Seeds provide a significant portion of the human diet [reviewed in (3)]; in turn, their proteins are nutritionally significant. The practical importance of seed storage proteins, their high levels of expression (and resultant ease of gene cloning), and their ability to serve as models for the study of both developmentally regulated gene expression (4) and intracellular transport of proteins (5) have resulted in their being the objects of extensive study by plant molecular and cellular biologists. These studies, combined with advances in transformation described elsewhere in this volume, have now made it possible to express both native and modified forms of these proteins in seeds of transgenic plants. The purposes of this chapter are to consider what the objectives of such modifications are, to outline the problems involved and the approaches being taken to overcome them, and to review the progress made so far. It is beyond the scope of this chapter to review all the work on the expression and biosynthesis of seed storage proteins, so at appropriate points the reader will be referred to the other reviews covering different aspects of the field.

II. WHY MODIFY SEED STORAGE PROTEINS?

A. Changing Seed Amino Acid Composition

Apart from studies of fundamental interest, there are two major potential applications for the modification of seed storage proteins. The most prominent goal has been to alter the amino acid composition of the total seed protein. This is part of a larger overall effort to improve the nutritional properties of seeds, recently reviewed by de Lumen (3). Other objectives include reducing the levels of antinutritional factors, modifying starch content, and, for both nutritional and industrial purposes, modifying oil quality and content. In general, the seeds of leguminous plants tend to be poor in the sulfur-containing amino acids and tryptophan, whereas the seeds of the cereals have low levels of tryptophan, lysine, and threonine (1). The details for individual crops vary, and it is instructive to examine an individual case. The dicot *Brassica napus* (rapeseed) is grown in North America and Europe for its oil, which is second only to olive oil in content of monounsaturated fatty acids. Plant breeders have succeeded, using classical methods, in developing lines of rapeseed, dubbed canola, low in both erucic acid and glucosinolates, which are antinutritional factors. The presence of erucic acid had limited the use of rapeseed oil, and the high levels of glucosinolates had restricted the use of the seed cake (meal) for livestock feed after oil extraction. Research has shown that canola meal can be successfully used in increased quantities in livestock feed (6,7). This is of particular interest in Europe, where large quantities of canola are now grown but where soybean, another major component of livestock feed, is not. On a 100% protein basis, canola meal contains 2.48%, 2.10%, and 5.95% of cysteine, methionine, and lysine, respectively, whereas the corresponding values for soybean are 1.61%, 1.59%, and 6.24% (6). Canola is thus richer in the sulfur-containing amino acids but poorer in lysine and in total energy content than soybean. Relatively small increases in the lysine content of canola meal would thus allow the use of higher percentages of canola meal (which is less expensive in Europe and some parts of North America) relative to soybean meal in livestock feed. It should be emphasized that many other factors, which are beyond the scope of this review, affect the determination of the optimal composition of feed for different livestock species, but similar reasoning can be done for different combinations of crop, livestock species, and geographical area. In areas where human diets depend greatly on single crops, altering seed protein composition could directly affect food quality.

Two approaches are being taken in altering the amino acid composition of seed by molecular means. One is to find a naturally occurring seed storage protein with high levels of the desired amino acid, clone the corresponding gene, and express this gene at high levels in the species of interest. This method, which is limited to the availability of suitable genes, will not be considered further in this review, but approaches and recent results are described elsewhere (3,8). The other approach being taken is to modify seed storage protein genes by using recombinant DNA or in vitro mutagenesis techniques so that they encode proteins that are similar to wild-type proteins but contain higher levels of essential amino acids.

It should be noted that although most of the effort so far in this area has concentrated on amino acid composition, there are other reasons, related to nutritional or food processing characteristics, to modify seed storage proteins. For example, the structure of the glutelins in wheat is critical in the quality of the bread made with grain from different lines of wheat plants; this is related to the effect of the storage proteins on the viscoelasticity of wheat flour (9). In other cases storage protein solubility or gel-forming ability may be important.

B. Producing Biological Peptides

Modification of seed storage proteins can also be contemplated for an entirely separate objective, the production of biological peptides using the seed storage proteins as a vehicle. A large part of the work in our laboratories has focused on this objective. Other chapters in this volume (see Chapters 11, 12), as well as Krebbers et al. (10), have discussed the advantages of plants as a production system for foreign proteins or peptides. In short, they are inexpensive to grow, and the amount grown can easily be varied and is not tied to increments of fermentor capacity, as is the case for prokaryotic or eukaryotic cell culture. The technology for harvesting and processing plants or parts of plants on medium-to-very-large scales already exists. Of particular relevance here is that plants have natural storage organs (seeds, tubers, specialized roots) in which high levels of particular proteins are stockpiled. It is hoped that the protein composition of these storage organs can be modified without affecting the overall health and vigor of the plant. In the case of peptide production there are potential advantages, which appear to be borne out by the results obtained so far, to expressing the peptide as part of a modified seed storage protein. The storage protein is already equipped with information resulting in its being targeted to a specific intracellular location, relieving the biotechnologist of the decision as to where in the cell the peptide would best be directed. Seed storage proteins are stable over long periods, and if the peptide does not disrupt this stability the modified form should also be stably stored. The harvest and initial processing of seeds is well established and economical, and changes in the protein content of the seed seem unlikely to affect the health of the rest of the plant. Finally, if the high expression levels typical of seed storage proteins can also be attained with modified forms, the procedure has a good chance of being economically feasible.

The technical problems and approaches to seed storage protein modification are similar regardless of whether altered amino acid composition or peptide production is the final goal. The two will thus be considered together, after which additional issues specific to peptide production will be addressed.

III. PRACTICAL CONSIDERATIONS AND CONSTRAINTS

Among the criteria for useful expression of modified seed storage proteins are the following:

1. The modified protein should be stably expressed and, as the name suggests, stored in the cell, preferably in its natural storage organelle. This implies that modification should not interfere with correct folding, intracellular transport, targeting, posttranslational modifications, or protein storage vacuole formation and packing. It will be seen that these are not trivial constraints.
2. The expression level should be high enough to have the desired effect, either a significant change in amino acid composition or levels of peptide-containing storage proteins high enough to allow economical production.
3. Other qualities of the plants should not be negatively affected. The plants should have normal agronomic characteristics (rate of growth, days to flowering, seed set, and so on), and the yield of harvested material should be normal.

4. The seeds should be otherwise normal. Thus, there should not be a reduction in other amino acids, the oil quality and content should be unchanged, and the levels of antinutritional factors should remain low. In other words, the use of seed containing modified seed storage proteins should result in feed or food that is of equal or higher quality in all respects. These criteria are less important if seed storage proteins are to be used for peptide production.

A. Gene Expression

The large body of work on the organization and expression of seed storage protein genes cannot all be chronicled here; the reader is referred to Refs. 1, 4, 11, and 12. Suffice it to say that tissue-specific, developmentally regulated expression of both dicot and monocot seed storage protein genes has been extensively demonstrated in dicot transgenic plants. The list of genes so tested grows almost by the month, and any cataloging would soon be out of date. The examples tried so far include both complete genes transferred from one species to another and chimeric genes.* Recent advances in the transformation of monocots (13,14) should soon result in similar experiments being done in the cereal crops. A striking exception to the pattern are the zein genes of maize, which are not correctly regulated in petunia (15). Zein genes can be expressed in dicots if a dicot promoter is used (16).

Two caveats must be appended to the preceding discussion. Even in dicots, most of the experiments have been carried out in model plants, largely tobacco or petunia. However, this situation is changing; *Brassica napus* is now routinely transformable, and both chimeric and modified storage proteins have been expressed in this species (17–19). Work is also underway in soybean (20) and other dicots. A more serious problem is that, whereas expression at the mRNA level indeed appears to be correctly regulated in most cases, protein levels are not always what might be expected. Guerche et al. (17) expressed a chimeric gene in *Brassica*, Okamuro et al. (21) expressed an unmodified soybean gene in tobacco, and Williamson et al. (22) expressed a chimeric phaseolin-zein gene in tobacco, and these investigators all found lower than expected protein levels that did not necessarily correlate with the observed mRNA levels. These results may reflect problems with protein quantitation or

*The term chimeric gene will refer both to genes composed of the promoter of one seed storage protein and the protein coding region of another, perhaps with a fusion in the region of the signal peptide, which account for the majority of cases, and to modified storage protein genes. The latter will be considered to be chimeric genes with significant changes within the open reading frame.

extraction as much as a biological problem. Vandekerckhove et al. (18) observed higher levels of expression at the protein level (1–3%) when quantitation was done on the basis of sequencing data rather than Western blots.

How high an expression level is needed? Altenbach et al. (23) have reported raising the methionine content of tobacco seeds from 3.60% to 4.74% of total protein by expressing at least five copies of a chimeric methionine-rich (18%) gene. The five copies together gave approximately 8% of total seed protein, a significant achievement. A similar increase in the lysine content of canola meal would be four times that required to bring it up to that of soybean meal. Plant breeders will prefer to have such traits behave as single loci, but this can be arranged by putting multiple genes in a single T-DNA. Thus, expression levels on the order of 1–5% of total protein of proteins with 10–20% of the desired amino acid should make significant changes, assuming no shifts in the expression of other proteins. The latter assumption may be naive, but little work has been done in this area. However, many workers in the field have observed expression levels below 1%. Whether this is due to promoter choice or other factors is not yet clear. Obtaining significant expression levels of chimeric seed storage proteins, modified or not, on a consistent basis remains a major challenge.

B. Protein Transport and Compartmentalization

It was initially hoped that because seed storage proteins have no apparent enzymatic function they would be relatively amenable to modifications. Hoffman et al. (24) showed that this is not necessarily the case. They made an insertion of 45 nucleotides, including six methionine codons, into the β-phaseolin (a 7S globulin) gene. It was demonstrated that both mRNA and protein were made but that the modified protein was degraded during or after transport to the protein storage vacuoles. Subsequent to this work, the phaseolin protein was crystallized (25), and it became apparent that the modification was made in a structurally important region. This work was an important warning that seed storage proteins will not tolerate just any change. It is thus useful to briefly outline the biosynthesis of these proteins. The reader is referred to Refs. 1, 5, and 26 for more details. Seed storage protein mRNAs are translated on the rough endoplasmic reticulum (ER), and the products are cotranslationally translocated into the ER, the signal peptide being removed in the process, as it is for secretory proteins. Within the ER, the (pro)peptides fold, disulfide bridges (if any) are formed, and many but not all are glycosylated (5), assemble into complex-higher order structures, or both (27,28). At this point the pathways of the albumins and globulins diverge from that of the prolamins. The former are translated to the Golgi (20), where, as with other glycosylated proteins, the side chains may be modified. On the basis of specific (30) but as yet not completely characterized targeting information they are then diverted from the secretory pathway to the protein storage vacuoles (PSV, also referred to as protein bodies), which are bounded by tonoplast-derived membranes. Many classes of seed storage proteins undergo further posttranslational processing, and pulse chase experiments (27) suggest that this occurs in the PSV. This process may involve a simple cleavage into two subunits or be quite extreme; in the case of the diminutive 2S albumins, 20% of the 143 amino acid propeptide (in the case of at2S1, see Ref. 31) is removed in three separate pieces from the amino terminal end, the carboxyl terminal end, and an internal location. Within the PSV there appear to be specific structural arrangements (32), suggesting that they are not just a random "bag" of proteins.

In monocots PSVs are also nonrandom arrangements of storage proteins. Lending and Larkins (26) showed that different classes of zeins are segregated in different locations of maize endosperm PSVs, and they proposed a model in which specific zeins interact to assemble the protein body. As previously stated, the transport of storage proteins to the PSV is fundamentally different for the prolamins. In principle it is much simpler: the storage proteins assemble into PSVs within the ER, budding directly off the ER. The pathway through the Golgi is thus avoided altogether. In rice the two systems coexist; the prolamins, as expected, avoid the Golgi pathway, whereas rice glutelins are transported into separate PSVs via the Golgi (33).

The role of glycosylation in those seed storage proteins that are glycosylated is not clear. It was originally thought that the side chains might be involved in the diversion of the proteins from the secretory pathway, but inhibitor (34) and mutagenesis (35–37) experiments showed this not to be the case. Bustos et al. (37) did detect differences in accumulation levels of different glycoforms of phaseolin in tobacco, suggesting that at least in this case glycosylation may play some other role in the stability of the protein. Voelker et al. (35) reported less dramatic effects with a double glycosylation mutant (but not with two single glycosylation mutants) of phytohemagglutinin expressed in tobacco seeds. Wilkins et al. (36) showed that the kinetics of posttranslational processing and intracellular transport differed between glycoforms of barley lectin expressed in tobacco; they were faster for the molecule lacking the glycosylation site. No significant differences in final accumulation levels were observed in this case. It should be noted that the studies with barley lectin involved expression in tobacco leaves, not seeds. It is not yet possible to draw general conclusions about the importance of maintaining glycosylation sites when modifying seed storage proteins.

It is clear that in modification of seed storage proteins care must be taken not to disrupt any stage of the complex process of seed storage protein biosynthesis. Unfortunately, despite great effort, most of the steps are not yet completely understood. The regions of the polypeptide chain carrying the information responsible for targeting to the protein body have yet to be completely defined, and it is already clear that these will be different in different classes of storage proteins. In two cases a discrete peptide fragment necessary and sufficient for targeting has been identified. Bednarek and Raikhel (64) have demonstrated that the carboxyl terminal propeptide (CTPP) of barley lectin is a vacuolar sorting determinant. This 15 amino acid fragment is removed from the lectin in a posttranslational processing step. Similarly, Matsuoka and Nakamura (65) have demonstrated that the 16 amino acid propeptide of sporamin (a storage protein found in the tuberous root of the sweet potato), which is present at the amino terminal end of the protein after removal of the signal peptide but is later itself removed, is required for vacuolar targeting. There is little apparent sequence homology between the barley lectin CTPP and the sporamin propeptide. In these two cases, then, targeting depends on a short discrete sequence (albeit two different ones), a situation analogous to the ER retention signals found at the carboxyl terminus of ER-located proteins (38). However, it is already clear that in other cases the situation is more complex, and no such fragment could be identified. Chrispeels and Tague (5,66) have suggested that a "patch" composed of residues from different parts of the protein may contain the targeting information. Saalbach et al. (39) have recently published further data supporting this concept; by using a series of amino and carboxyl terminal legumin-invertase fusion proteins, they demonstrated that in yeast the vacuolar targeting information is functionally only when large segments of the seed storage proteins are included and that such information is found in different regions of the protein. The latter suggests that there may be redundant signals. Similar, albeit less extensive, results (not all of the fusions were stably expressed) were obtained in transgenic tobacco plants. It is difficult when designing modifications to avoid disrupting a targeting label whose location is not yet known, and Saalbach et al. suggested that the failure of a modified 11S globulin to accumulate in transgenic plants may be related to difficulties in intracellular transport (see Section IV). Similarly, the formation of tertiary structures is only partially understood for individual classes (28 and Section IV), and, as stated above, the importance of maintaining glycosylation sites is not clear. Finally, the secondary structure of the storage protein should not be disrupted, as suggested by the results of Hoffman et al. (24), and yet crystal structures are available for only a few storage proteins. In the next section we will discuss some of the experimental approaches that, despite these limitations, have been or might be taken to explore the effects of modifications.

IV. EXPERIMENTAL APPROACHES

A major experimental constraint in the modification of seed storage proteins is the time required to regenerate and obtain seeds from transformed plants. Even in species for which transformation and regeneration are routine, the time from explant to seed is usually a matter of months and in some cases may approach a year. This is in contrast to bacterial or cell culture systems, in which a series of modified proteins can be tested in a matter of days or weeks. Those interested in the modification of seed proteins must thus plan their experiments with care and, if possible, test modified proteins in some other system before moving to plants themselves.

The most rational approach to modifying seed storage proteins would be to modify on the basis of the known structure of the protein. This would allow computer modeling of changes before any biological experiments are performed. Unfortunately, the crystal structures of most seed storage proteins have not been elucidated. An exception is phaseolin, a vicilinlike 7S globulin from *Phaseolus vulgaris* (the French bean), which is representative of a class of storage proteins found in several legumes. The structure of this protein has been determined at 3-Å resolution (25). As already mentioned briefly, the elucidation of this structure suggested what had gone wrong in the modification carried out by Hoffman et al. (24); the modification was made in what turned out to be an important structural component of the protein, a helical portion of a helix-turn-helix motif (25). On the basis of the known structure, the same group (40) went on to do computer modeling that suggested a series of possible modifications of the phaseolin protein that would not be expected to disrupt its tertiary or quaternary structure or the complexes it makes. These include changes that increase the methionine content, add a glycosylation site, and enhance the stability of the protein to enzymes with trypsinlike specificities. At the time of writing no in vivo results with these modifications had been presented, but this approach is perhaps the most risk free possible.

Less precise structural information can be obtained by other means. Perhaps the most common method is a comparison of the sequences of related proteins from different species. It is then assumed that conserved regions are more likely to be structurally important than regions that vary. Argos et al. (41) used such an approach to define a variable region at the carboxyl end of the acidic domain of the glycinins that might tolerate changes; this was later supported by in vitro data (see the following). Wright (42) carried out similar comparisons and proposed a somewhat different model. This model was tested by Kim et al. (43) in a bacterial system (see the following); it was concluded that, at least in that system, not all the variable regions defined by Wright were equally insensitive to changes. Vandekerckhove et al. (18) compared the sequences of 2S albumins from several species and defined a variable region in the protein, which was subsequently used to make a series of modifications that were stably expressed in plants, as will be described. In that case a hypothetical structural model could be made on the basis of the protein's small size and knowledge of the arrangement of the cysteine bridges (Van Damme and Vandekerckhove, unpublished results). The conclusions of circular dichroism studies, combined with acetylation and succinylation studies (44), are consistent with such a model. Circular dichroism has also been used to make a model for the structure of the 19-kDa zeins from maize (45), a model that was used as the basis for the modifications carried out by Wallace et al. (46) and tested in a heterologous biological system.

To avoid the long procedures of plant transformation and regeneration, proposed modifications can be evaluated in either in vitro or substitute biological systems. Such experiments may not allow the entire biosynthetic process to be examined, but they can determine if one or more of the posttranslational maturation steps summarized in the previous sections are carried out correctly. Perhaps one of the most elegant examples of this is found in the experiments with soybean glycinins done by the group at Purdue University. Glycinins are 12S globulins composed of 60-kDa subunits that assemble first into 9S trimers and then into 12S hexamers, which in turn combine to form insoluble aggregates. Dickinson et al. (28,47,48) developed an in vitro system that allowed them to follow the formation of 9S and 12S oligomers. This was done by in vitro transcription and translation of normal or modified glycinin subunits in the presence of ^{3}H-labeled amino acids. Labeled subunits were then allowed to assemble with themselves or to reassemble with unlabeled, unmodified glycinin monomers or oligomers. Assembly or reassembly was monitored on sucrose gradients. Dickinson et al. were able to show that, although sequences in the basic portion of the monomer are critical for assembly, this is not so for the acidic domain. Furthermore, they were able to confirm a previous prediction that a hypervariable region defined by sequence comparisons might tolerate changes; mutations made in this region, including some in which methionine residues were inserted, assembled properly (48).

Work in our laboratories (K. D'Hondt, D. Bosch, J. Van Damme, J. Vandekerckhove, and E. Krebbers, in preparation) indicates that steps in the proteolytic processing of 2S albumin precursors can be carried out in vitro. This may prove to be useful in evaluating the potential of modified 2S albumins, which are discussed in more detail later.

It is more difficult to imagine a pure in vitro system to follow the targeting of modified seed storage proteins, but some groups have used biological systems to study targeting and other aspects of seed storage protein expression, as well as to test modified storage proteins. Chrispeels and colleagues (4,49) have used yeast to look for vacuolar targeting determinants in phytohemagglutinin (PHA), a seed lectin. The results obtained in yeast did not correlate exactly with those obtained in transgenic plants. This suggests that yeast may not always be a good model for targeting of seed storage proteins. Saalbach et al. (39,50) used yeast both to study vacuolar targeting of a *Vicia faba* 11S globulin and to express modified forms of the protein. The targeting studies (39) gave comparable results in yeast and in transgenic plants, but the modified protein was accumulated in yeast and not in transgenic tobacco seeds (50). Similarly, the relative stability of different modified 2S albumin precursors is not the same in yeast and transgenic tobacco plants (K. D'Hondt, E. Krebbers, and J. Vandekerckhove, unpublished), again casting doubt on yeast as a suitable model for seed storage protein expression. Coraggio et al. (51) showed that a modified 23-kDa zein containing an extra lysine residue is accumulated in yeast, as well as an unmodified version. Protein was detected only in the endoplasmic reticulum, however.

Wallace et al. (46) made more significant changes in the structure of a 19-kDa zein and tested these in a biological system not usually associated with plant biologists, *Xenopus* oocytes. Hurkman et al. (52) had shown that when zein mRNA is injected into such oocytes, zeins are synthesized and, after passing into the endoplasmic reticulum, are assembled into membrane-bound structures similar to maize protein storage vacuoles. The zeins can be detected if the proteins are labeled with ^{3}H-leucine, and zeins can be specifically extracted with ethanol. Wallace et al. used this system to test 13 different modifications, including single and double substitutions involving lysine additions, short (eight-amino-acid) insertions, and one long (17-kDa) insertion. Only the long insertion construct failed to result in the formation of PSV-like structures. These results are promising, but they remain to be confirmed in maize itself. A later study (53) showed that it was not possible to express the modified zeins in tobacco by using a dicot promoter; this is discussed further in Section V. It is unfortunate that similar transient expression experiments using electroporation or polyethylene glycol–mediated transfer of DNA into cotyledon or endosperm protoplasts cannot be used to study modified seed storage accumulation, but this has not yet proved to be possible.

Kim et al. (43) have tested the expression of modified soybean glycinin genes in *E. coli.* As criteria for successful expression they used high-level expression, solubility, and self-assembly into trimers, all of which were obtained for unmodified glycinin. A series of modifications were made, including deletions of the variable regions defined by Wright (42) and insertion of methionine residues into some of these regions. Both positive and negative results were obtained, depending on the construction. On the basis of the results obtained in these experiments, as well as those of the in vitro experiments described earlier (48), it would seem that small deletions and insertions are best accommodated in the extreme N-terminal and C-terminal regions of proglycinin, as well as in the hypervariable region defined by Argos et al. (41). However, the evidence available indicates that deletions that include small portions outside of the hypervariable region behave abnormally, at least according to the test criteria that were used. More recent work (S. Utsumi, personal communication) has shown that in the *E. coli* system, some mutants that either remove the disulfide bridge linking the basic and acidic subunits or change the net charge of the molecule will stably assemble. These mutations are of potential interest because disulfide bonds affect the gel-forming ability of glycinins, whereas the net charge of unmodified glycinin endows it with an isoelectric point of pH 4.5, limiting its use in acidic foods. It remains to be seen if the various modifications tested in the *E. coli* system will be reflected in plant systems.

In summary, a variety of in vitro and in vivo test systems have been used to try to speed up the evaluation of modified storage proteins before initiating the long plant transformation process. Each has its advantages and disadvantages, and none has yet been shown to consistently predict correctly whether or not a modified protein will be stably accumulated in plant seeds.

V. EXPRESSION OF MODIFIED STORAGE PROTEINS IN TRANSGENIC PLANTS

It is hoped that modeling and expression in other systems will allow rapid testing of different modifications, but the ultimate test will always be expression in the species of choice. The experience here is thus far more limited, particularly in the monocots, for which transformation has until recently not been generally possible. Monocot storage proteins have been expressed in dicots, however. Robert et al. (54) expressed an unmodified wheat high-molecular-weight glutenin in tobacco using its own promoter, while Hoffman et al. (16), expressing an unmodified 15-kDa zein in tobacco seeds using the phaseolin promoter, were able to reach 1.6% of total protein. However, when Ohtani et al. (53) attempted to express the modified 19-kDa zeins previously tested in the *Xenopus* system (see Section IV) in tobacco seeds using the same promoter, they found that these were degraded, as was the unmodified version of the protein. The different result obtained with the 19-kDa zein may be due either to structural differences between the 19- and 15-kDa zeins (there is little structural similarity between the two classes) or to the inability of the 19-kDa zeins to form protein storage vacuoles on their own (53). Work by the same group (26) has shown that zeins are deposited in the PSV in an orderly pattern.

The modified seed storage proteins expressed in tobacco by Voelker et al. (35), working with PHA, Bustos et al. (37), working with phaseolin, and Wilkins et al. (36), working with barley lectin, all involve single or double amino acid changes to alter glycosylation sites. These have already been discussed in the context of glycosylation, but they do represent examples of modified seed storage proteins successfully expressed in transgenic plants, albeit in leaves in the third case. Although there was wide variation between individual transgenic plants, Bustos et al. reported expression levels of 0.2–0.9% of total seed protein for the glycoform mutants, the average being lower for the double mutant. Voelker et al. reported levels of 0.01–1.7% of total seed protein.

Saalbach et al. (55) have reported the construction of a modified legumin (11S globulin) gene from *Vicia faba* in which, through a frameshift, the final exon was modified to contain four methionine residues. This exon was then substituted for the corresponding exon in the 11S globulin from soybean, resulting in a hybrid glycinin-modified legumin gene. The modified gene was transformed into both yeast and tobacco plants. It was accumulated in the vacuoles of the former, but no modified protein could be detected in transgenic tobacco seeds (50). Because the mRNA was detectable, the defect appears to be at the translational or posttranslational level; it has been suggested that the problem may be in intracellular transport (K. Müntz, personal communication).

The work of Hoffman et al. (24) with modified phaseolin has already been referred to. A 45-nucleotide sequence containing six methionine codons was inserted in the third exon of the phaseolin gene, encoding a 7S globulin from *Phaseolus*, and the modified gene was expressed in tobacco. Analysis showed that the gene was transcribed and the mRNA translated, but the protein was (relatively slowly) degraded during the intracellular transport process. As related earlier, the elucidation of the complete structure of the protein after the experiments described revealed that the modification was made in what is probably a structurally sensitive region. Sonnewald (56) reported similar results when modifications were made in the tuber storage protein palatin that, although in a region thought to be on the outside of the protein, changed its pI.

The 2S albumins have proved to be more tolerant of changes, and work in our laboratories has led to the expression of six different modified 2S albumins in the model plant *Arabidopsis thaliana* and, in some cases, the crop plant *Brassica napus* (canola). This work has recently been reviewed (10,57) and will only be described briefly here. The 2S albumins are among the smallest of seed storage proteins, consisting of two subunits of approximately 9 and 3 kDa linked by disulfide bridges. As discussed in the section on protein transport and compartmentalization, they are synthesized as precursors that undergo significant posttranslational processing, and a priority in the design of the modifications was to avoid regions that might be important in this process. This precluded making fusions at the carboxyl terminus, which also undergoes proteolytic processing. Fortunately, the small size of the protein, the availability of sequences from several species (summarized in Ref. 31), and the presence of eight conserved cysteine residues in the all these sequences made it possible to construct a hypothetical structural model (Fig. 1). The model predicts that the region between the sixth and seventh cysteine residues forms a flexible loop on the outside of the protein. The model was further supported by computer analysis of several of the available 2S albumin sequences, which showed that in all cases this region was hydrophilic, and by the determination of some of the disulfide bridge positions in the *Arabidopsis* 2S albumin (J. Van Damme and J. Vandekerckhove, unpublished). While the initial work was being done, another group (44) independently suggested, on the basis of circular dichroism studies combined with acetylation and succinylation studies, that a large part of the corresponding sequence of napin (the 2S albumin of *Brassica*) forms a flexible region between two alpha helices. It is interesting that Radke et al. (58) made a nine-amino-acid insertion, including five methionines, adjacent to the fourth cysteine residue. Although the modified gene was transcribed, no evidence for modified storage protein accumulations was found in transformants harboring this construct (M. Moloney, personal communication). The distance between the fourth and fifth cysteine residues, the region where the insertion was made, is conserved across several species.

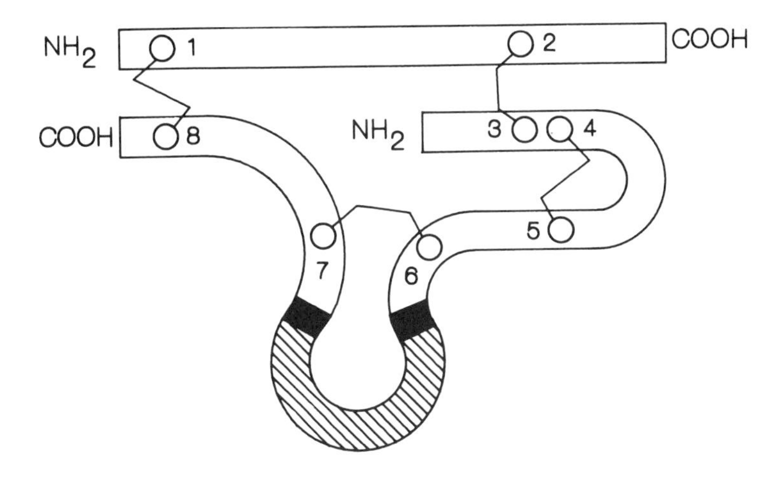

CVCPTLKQAAKAVRLQGQHQPMQVRKIYQTAK	HLPNVC	at2S1
CVCPTLKQAAK**YGGFLK**QHQPMQVRKIYQTAK	HLPNVC	Leu-enkephalin
CVCPTLK**G**AAKAV**KMEK**Q**KK**P**E**QV**K**K**M**Y**K**TAK	HLPNVC	Lysine rich
CVCPTLK**GIMMMRM**	**I**HLPNVC	Methionine rich
CVCPTLK**GIMMMM**R**M**Q**PRGDQ**M**RRMM**	**I**HLPNVC	Methionine rich
CVCPTLK**GIMMMQPRGDMMMI**M**MMQPRGDMMM**	**I**HLPNVC	Methionine rich
CVCPTLK**GIRGIGKFLHSAGKFGKAFVGEIM**K**SRI**HLPNVC		Magainin

Figure 1 Top panel: Hypothetical model of 2S albumins showing the position of the cysteine residues and the region into which modifications were introduced (hatched region). The solid sections represent possible protease cleavage sites used to excise biological peptides. Bottom panel: Summary of the modifications introduced into the variable region. The top line shows the unmodified sequence between the fifth and seventh cysteine residues. Amino acid differences relative to this sequence are shown in bold, and protease cleavage sites used to excise biological peptides are underlined. Gaps have been introduced as appropriate to allow alignment of the sequences. See Note Added in Proof.

On the basis of this model six different modifications (Fig. 1) of at2S1, an *Arabidopsis* 2S albumin (31), have been expressed in transgenic plants. The modifications range in size from the seven-amino-acid substitution consisting of leu-enkephalin (a neuropeptide) and flanking protease cleavage sites to a 28-amino-acid sequence containing magainin (an antibacterial peptide from *Xenopus*) along with cleavage sites. These two constructions were made as models for the production of peptides in plants. The other four substitutions are directed toward altering seed amino acid composition, three containing methionine and one containing lysine residues. In all cases the presence of the modified 2S albumin amid the endogenous 2S albumins present in both species was confirmed by the isolation and sequencing of peptides unique to the modified isoform after digestion of a mixture of 2S albumins with an appropriate protease, as previously described (18,19). An extra advantage of this method of detection is that it allows more accurate quantitation of the protein level than is possible by using Western blots. As usual, expression levels varied between transgenic plants, but in both *Brassica* and *Arabidopsis* the highest levels of expression were in the range of 100–200 nmol of peptide per gram of seed. The yields, which reflect actual isolation yields uncorrected for losses, translate to approximately 1 kg of a 25 amino acid peptide per hectare of *Brassica* plants, which produce 2000–3000 kg/ha of seed. However, these levels of expression were insufficient to make measurable increases in the methionine content of *Arabidopsis* seeds, possible reasons for which were discussed (19). The expression levels reported were obtained by using the at2S1 promoter, which is expressed only in the axis of *Arabidopsis* embryos (59). These low expression levels, combined with the instability of methionine during amino acid determination procedures, were probably responsible for the lack of a detectable change in methionine content. Preliminary data suggest that, at least in transgenic *Arabidopsis* plants, the use of the at2S2 promoter results in significantly higher levels of expression of chimeric 2S albumin genes. Experiments are currently underway to determine if this is the case in *Brassica* as well.

The results just described demonstrate that modified 2S albumins can be stably expressed in transgenic plants. Current work is directed at raising expression levels and streamlining the procedures required to test new modifications. So long as a crystal structure of the 2S albumin is not available, each new modification, either for peptide production or amino acid composition modification, must be tested empirically. Experiments are underway to further explore the limits of the flexibility of 2S albumins in terms of the size of the substitution/insertion and the sequence composition of the alteration. There are some other interesting issues specific to the use of modified 2S albumins for the production of biological peptides. For instance, yields can probably be further optimized by using varieties of *Brassica* that have a higher protein content (and thus usually a lower oil content, making them less useful for agricultural purposes). Increases in extraction efficiencies will in any case lead to enhanced yields. A particular problem is the low number of specific proteases available, a problem common to all systems that rely on fusion proteins. The identification of restriction proteases (e.g., factor Xa; see also Ref. 60) may lessen this problem. All biological peptide production systems face the problem of peptides with modified amino acids. The establishment of an economic method of carboxyl terminal amidation will be particularly important; fortunately, progress is being made in this area (61). As long as the cost of growing plants on a large scale remains low, the concept holds the promise of being economically feasible for peptides longer than a few amino acids that are required in medium-to-large amounts, although much work remains to be done.

VI. TRANSGENIC PLANTS IN THE FIELD

A. Regulatory Issues

The regulatory situation regarding the growth of transgenic plants in the field is gradually being resolved, and field trials are now commonplace. The production of peptides in plants involves some regulatory issues aside from those related to the growth of the plants in the field. In particular, it is not yet clear if a pharmaceutical product produced in plants but already on the market through a different production system will have to go through a new set of clinical trials or whether it will be sufficient to demonstrate that the peptide is chemically identical to a product already approved (10). There may also be some sensitivity about the growth of plants expressing a biologically active peptide, even one embedded in another protein and thus inactive. As suggested previously (10), precautions to prevent the unwanted spread of genes encoding storage proteins modified in this way, such as the use of male sterile plants (62) in the production phase, may be necessary.

B. Field Performance

Farmers are unlikely to grow plants producing seeds with an altered amino acid composition if those plants are not agronomically acceptable. A list of characteristics of an acceptable new line would include good agronomic performance (normal growth rate, flowering, lodging, etc.), normal yields of seed, normal yields of oil from the seed in cases where this is relevant, normal oil quality, equal protein content, and digestibility of the protein by livestock. Some seeds contain antinutritional factors, and the levels of these must be as low as possible. As described previously, the original transgenic plants produced in our laboratories did not show any changes in methionine content because of low expression levels. However, it was of interest to determine the effects of the transformation and regeneration process, as well as the expression (at whatever level) of simply chimeric as well as modified storage proteins, on the agronomic performance of the plants and the quality of the seed produced. In 1989 and 1990 multiple-site field trials were conducted (63) using *Brassica napus* plants expressing such constructs. Establishment, date of flowering, height of the plants at flowering, and yield of seed were measured. The results of one such trial, carried out in Canada in 1990, are shown in Table 1. There was a slight delay in the development of the transgenic plants (days to full flower) which was coupled with a reduction in yield. These effects could be traced back to poor seed quality resulting from an accelerated production procedure in the greenhouse. (The seed for the control plots in the field trials was not produced in the greenhouse, but for reasons of scheduling this was necessary for the transgenic seed.) This was confirmed by additional germination tests and comparisons with control material, and a repetition of the trial using field produced trransgenic seed the following season. All other parameters were normal. The quality characteristics of the field harvested seed (fatty acid composition, including erucic acid content, glucosinolate content, and amino acid composition of amino acids other than methionine, which as

Table 1 Results of Controlled Field Trial of Transgenic *Brassica napus*[a]

Characteristic	Control 1	Control 2	Transgenic
Yield (kg/ha)	2346	2352	1946*
Days to full flower	54	53	56*
Days to maturity	95	95	96
Plant height (cm)	106	98	106
Lodging score (1–5)	1.5	1.5	1.6
Seed protein content (%)	19.2	18.9	19.7
Seed oil content (%)	41.1	40.5	42.5

[a]Two lots of control seeds (untransformed *Brassica napus* Drakkar) were compared to a transgenic line expressing a chimeric seed storage protein gene. The values marked with an asterisk (*) are significant deviations from the control values. See text for further details.

expected also showed no change) showed no statistically significant variations from the controls. Although these results may at first glance appear trivial, they are encouraging because they suggest that the transformation and regeneration procedures combined with low level expression of a foreign gene in developing seed tissues do not result in poor agronomic performance or changes in seed quality relative to normal commercial lines.

VII. SUMMARY

Expression of modified storage proteins in transgenic plants is clearly still in its infancy. Most of the efforts discussed are directed toward the protein engineering and gene expression side of the challenge. As reliable systems are developed for testing modified storage proteins quickly and as more structural data become available, the field will advance more quickly. However, once stable modified proteins are designed and expression levels optimized, the effects on the whole plant and, in particular, the quality of the seed will become important. There are as yet no data on what effect of overexpressing proteins rich in a particular amino acid will have on amino acid pools or other physiological factors. As suggested by de Lumen (3), protein quality is not the only parameter of importance, and the interaction of changes in protein quality or quantity with changes in oil or starch will have to be closely monitored. These must be balanced in such a way as to provide a useful product accepted by the market. Different considerations apply to the use of seeds as production systems for peptides, but, as previously discussed, the process must be shown to be economically as well as technically feasible. Finally, the regulatory issues related to the growth in the field of transgenic plants must be dealt with, and this will require communication with public and governmental bodies.

ACKNOWLEDGMENTS

We thank Shigeru Utsumi of Kyoto University, Klaus Müntz of the Institute of Genetics and Crop Plant Research in Gatersleben, Maurice Moloney of the University of Calgary, and Mauricio Bustos of the University of Maryland for permission to cite data unpublished or still in press. The contributions of our collaborators at Plant Genetic Systems and the University of Ghent to the work described here are warmly acknowledged. Work at the University of Ghent was partially funded by a grant to Joël Vandekerckhove from the NFWO (Belgian National Fund for Scientific Research).

NOTE ADDED IN PROOF

Recent work has suggested that the model for the structure of 2S albumins shown in Fig. 1 is not the most likely one. Various alternative models exist, but all are consistent with the notion that the region between the sixth and seventh cysteine residues is on the outside of the protein.

REFERENCES

1. Higgins, T. J. V. Synthesis and regulation of major proteins in seeds, *Annu. Rev. Plant Physiol.*, *35*: 191–221, 1984.
2. Osborne, T. B. *The Vegetable Proteins*. Longmans, Green, London, 1924.
3. de Lumen, B. O. Molecular approaches to improving the nutritional and functional properties of plant seeds as food sources: developments and comments, *J. Agric. Food Chem.*, *38*: 1779–1788, 1990.
4. Goldberg, R. B., Barker, S. J., and Perez-Grau, L. Regulation of gene expression during plant embryogenesis, *Cell*, *56*: 149–160, 1989.
5. Chrispeels, M. J. Sorting of proteins in the secretory system. *Annu. Rev. Plant. Physiol. Plant Mol. Biol.*, *42*: 21–53, 1991.
6. Clandinin, D. R., ed. *Canola Meal for Livestock and Poultry*, Canola Council, Winnepeg, Man., Canada, 1986.
7. Huyghebaert, G., Fontaine, G., and De Groote, G. The feeding value of rapeseed meals with different glucosinolate contents (*Brassica* sp.), as measured by digestibility experiments with broiler chicks and adult roosters, *Arch. Gegflügelk.*, *47*: 50–60, 1983.
8. Altenbach, S. B., and Simpson, R. B. Manipulation of methionine-rich protein genes in plant seeds, *Trends Biotechnol.*, *8*: 156–160, 1990.
9. Payne, P. I., Holt, L. M., Jackson, E. A., and Law, C. N. Wheat storage proteins: their genetics and their potential for manipulation by plant breeding, *Philos. Trans. R. Soc. Lond [Biol.]*, *304*: 359–371, 1984.
10. Krebbers, E., Bosch, D., and Vandekerckhove, J. Prospects and progress in the production of foreign proteins and peptides in plants, *Plant Protein Engineering* (P. R. Shewry, and S. Gutteridge, eds.), Cambridge University Press, England, U.K., in press.
11. Casey, R., and Domoney, C. The structure of plant storage protein genes, *Plant Mol. Biol. Reporter*, *5*: 261–281, 1987.
12. Okamuro, J. K., and Goldberg, R. B. Regulation of plant gene expression: general principles, *Biochem. Plants*, *15*: 1–82, 1989.
13. Gordon-Kamm, W. J., Spencer, T. M., Mangano, M. L., Adams, T. R., Daines, R. J., Start, W. G., O'Brien, J. V., Chambers, S. A., Adams, W. R., Willetts, N. G., Rice, T. B., Mackey, C. J., Krueger, R. W., Kausch, A. P., and Lemaux, P. G. Transformation of maize cells and regeneration of fertile transgenic plants, *Plant Cell*, *2*: 603–618, 1990.

14. Datta, S. K., Peterhans, A., Datta, K., and Potrykus, I. Genetically engineered fertile indica-rice recovered from protoplasts, *Biotechnology, 8*: 736–740, 1990.
15. Ueng, P., Galili, G., Sapanara, V., Goldsbrough, P. B., Dube, P., Beachy, R. N., and Larkins, B. A. Expression of a maize storage protein gene in petunia plants is not restricted to seeds, *Plant Physiol., 86*: 1281–1285, 1988.
16. Hoffman, L. M., Donaldson, D. D., Bookland, R., Rashka, K., and Herman, E. M. Synthesis and protein body deposition of maize 15-kd zein in transgenic tobacco seeds, *EMBO J., 6*: 3213–3221, 1987.
17. Guerche, P., De Almeida, E. R. P., Schwarztein, M. A., Gander, E., Krebbers, E., and Pelletier, G. Expression of the 2S albumin from *Bertholletia excelsa* in *Brassica napus*, *Mol. Gen. Genet., 221*: 306–314, 1990.
18. Vandekerckhove, J., Van Damme, J., Van Lijsebettens, M., Botterman, J., De Block, M., Vandewiele, M., De Clercq, A., Leemans, J., Van Montagu, M., and Krebbers, E. Enkephalins produced in transgenic plants using modified 2S seed storage proteins. *Biotechnology, 7*: 929–932, 1989.
19. De Clercq, A., Vandewiele, M., Van Damme, J., Guerche, P., Van Montagu, M., Vandekerckhove, J., and Krebbers, E. Stable accumulation of modified 2S albumin seed storage proteins with higher methionine contents in transgenic plants. *Plant Physiol., 94*: 970–979, 1990.
20. Chee, P. P., Fober, K. A., and Slightom, J. L. Transformation of soybean (*Glycine max*) by infecting germinating seed with *Agrobacterium tumefaciens*, *Plant Physiol., 91*: 1212–1218, 1989.
21. Okamuro, J. K., Jofuku, K. D., and Goldberg, R. B. Soybean seed lectin gene and flanking nonseed protein genes are developmentally regulated in transformed tobacco plants, *Proc. Natl. Acad. Sci. USA, 83*: 8240–8244, 1986.
22. Williamson, J. D., Galili, G., Larkins, B. A., and Gelvin, S. B. The synthesis of a 19 kilodalton zein protein in transgenic Petunia plants, *Plant Physiol., 88*: 1002–1007, 1988.
23. Altenbach, S. B., Pearson, K. W., Meecker, G., Staraci, L. C., and Sun, S. S. M. Enhancement of the methionine content of seed proteins by the expression of a chimeric gene encoding a methionine-rich protein in transgenic plants, *Plant Mol. Biol., 13*: 513–522, 1989.
24. Hoffman, L. M., Donaldson, D. D., and Herman, E. M. A modified storage protein is synthesized, processed, and degraded in the seeds of transgenic plants, *Plant Mol. Biol., 11*: 717–729, 1988.
25. Lawrence, M. C., Suzuki, E., Varghese, J. N., Davis, P. C., Van Donkelaar, A., Tulloch, P. A., and Colman, P. M. The three-dimensional structure of the seed storage protein phaseolin at 3 Å resolution, *EMBO J., 9*: 9–15, 1990.
26. Lending, C. R., and Larkins, B. A. Changes in the zein composition of protein bodies during maize endosperm development, *Plant Cell, 1*: 1011–1023, 1989.
27. Chrispeels, M. J., Higgins, T. J., and Spencer, D. Assembly of storage protein oligomers in the endoplasmic reticulum and processing of the polypeptides in the protein bodies of developing pea cotyledons, *J. Cell Biol., 93*: 306–313, 1982.

28. Dickinson, C. D., Hussein, E. H. A., and Nielsen, N. C. Role of posttranslational cleavage in glycinin assembly, *Plant Cell., 1*: 459–469, 1989.
29. Chrispeels, M. J. The Golgi apparatus mediates the transport of phytohemagglutinin to the protein bodies in bean cotyledons, *Planta, 158*: 140–151, 1983.
30. Dorel, C., Voelker, T. A., Herman, E. M., and Chrispeels, M. J. Transport of proteins to the plant vacuole is not by bulk flow through the secretory system, and requires positive sorting information, *J. Cell Biol., 108*: 327–337, 1989.
31. Krebbers, E., Herdies, L., De Clercq, A., Seurinck, J., Leemans, J., Van Damme, J., Segura, M., Gheysen, G., Van Montagu, M., and Vandekerckhove, J. Determination of the processing sites of an *Arabidopsis* 2S albumin and characterization of the complete gene family. *Plant Physiol., 87*: 859–866, 1988.
32. Lott, J. N. A. Protein bodies, *Biochem. Plants, 1*: 589–623, 1980.
33. Krishnan, H. B., Franceschi, V. R., and Okita, T. W. Immunochemical studies on the role of the Golgi complex in protein body formation in rice seeds, *Planta, 169*: 471–480, 1986.
34. Bollini, R., Ceriotti, A., Daminati, M. G., and Vitale, A. Glycosylation is not needed for the intracellular transport of phytohemagglutinin in developing *Phaseolus vulgaris* cotyledons and for the maintenance of its biological activities, *Physiol. Plant, 65*: 15–22, 1985.
35. Voelker, T. A., Herman, E. M., and Chrispeels, M. J. In vitro mutated phytohemagglutinin genes expressed in tobacco seeds: role of glycans in protein targeting and stability, *Plant Cell, 1*: 95–104, 1989.
36. Wilkins, T. A., Bednarek, S. Y., and Raikhel, N. V. Role of propeptide glycan in post-translational processing and transport of barley lectin to vacuoles in transgenic tobacco, *Plant Cell, 2*: 301–313, 1990.
37. Bustos, M. M., Kalkan, F. A., Vandenbosch, K. A., and Hall, T. C. Differential accumulation of four phaseolin glycoforms in transgenic tobacco, *Plant Mol. Biol., 16*: 381–395, 1991.
38. Pelham, H. R. B. Control of protein exit form the endoplasmic reticulum, *Annu. Rev. Cell. Biol., 5*: 1–23, 1989.
39. Saalbach, G., Jung, R., Kunze, G., Saalbach, I., Adler, K., and Müntz, K. Different legumin propolypeptide domains act as vacuolar targeting signals, *Plant Cell*, in press.
40. Colman, P. M., Lawrence, M. C., Varghese, J., Hall, T., and Bustos, M. M. PCT Patent Publication WO 91/04270, 1991.
41. Argos, P., Narayana, S. V. L., and Nielsen, N. C. Structural similarity between legumin and vicilin storage proteins from legumes, *EMBO J., 4*: 1111–1117, 1985.
42. Wright, D. J. The seed globulins II, *Developments in Food Proteins*, Vol. 6 (B. J. F. Hudson, ed.), Elsevier Applied Science, London, 1988, pp. 119–177.
43. Kim, C.-S., Kamiya, S., Sato, T., Utsumi, S., and Kito, M. Improvement of nutritional value and functional properties of soybean glycinin by protein engineering, *Protein Eng., 3*: 725–731, 1990.

44. Schwenke, K. D., Drescher, B., Zirwer, D., and Raab, B. Structural studies on the native and chemically modified low-molecular mass basic storage protein (napin) from rapeseed (*Brassica napus* L.), *Biochem. Physiol. Pflanzen, 183*: 219–224, 1988.
45. Argos, P., Pedersen, K., Marks, M. D., and Larkins, B. A. A structural model for maize zein proteins, *J. Biol. Chem., 257*: 9984–9990, 1982.
46. Wallace, J. C., Galili, G., Kawata, E. E., Cuellar, R. E., Shotwell, M. A., and Larkins, B. A. Aggregation of lysine-containing zeins into protein bodies in *Xenopus* oocytes, *Science, 240*: 662–664, 1988.
47. Dickinson, C. D., Floener, L. A., Lilley, G. G., and Nielsen, N. C. Self-assembly of proglycinin and hybrid proglycinin synthesized in vitro from cDNA, *Proc. Natl. Acad. Sci. USA, 84*: 5525–5529, 1987.
48. Dickinson, C. D., Scott, M. P., Hussein, E. H. A., Argos, P., and Nielsen, N. C. Effect of structural modifications on the assembly of a glycinin subunit, *Plant Cell, 2*: 403–413, 1990.
49. Tague, B. W., Dickinson, C. D., and Chrispeels, M. J. A short domain of the plant vacuolar protein phytohemagglutinin targets invertase to the yeast vacuole, *Plant Cell, 2*: 533–546, 1990.
50. Saalbach, G., Jung, R., Kunze, G., Manteuffel, R., Saalbach, I., and Müntz, K. Expression of modified legume storage protein genes in different systems and studies on intracellular targeting of *Vicia faba* legumin in yeast, *Proceedings of Genetic Engineering of Crop Plants*, 49th Nottingham Easter School, Sutton Bonington, Butterworth, London, 1990, pp. 1151–1158.
51. Coraggio, I., Martegani, E., Compagno, C., Porro, D., Alberghina, L., Bernard, L., Faoro, F., and Viotti, A. Differential targeting and accumulation of normal and modified zein polypeptides in transformed yeast, *Eur. J. Cell Biol., 47*: 165–172, 1988.
52. Hurkman, W. J., Smith, L. D., Richter, J., and Larkins, B. A. Subcellular compartmentalization of maize storage proteins in *Xenopus* oocytes injected with zein messenger RNAs, *J. Cell. Biol., 89*: 292–299, 1981.
53. Ohtani, T., Galili, G., Wallace, J. C., Thompson, G. A., and Larkins, B. A. Normal and lysine-containing zeins are unstable in transgenic tobacco seeds, *Plant Mol. Biol., 16*: 117–128, 1991.
54. Robert, L. S., Thompson, R. D., and Flavell, R. B. Tissue-specific expression of a wheat high molecular weight glutenin gene in transgenic tobacco, *Plant Cell, 1*: 569–578, 1989.
55. Saalbach, G., Jung, R., Saalbach, I., and Müntz, K. Construction of storage protein genes with increased number of methionine codons and their use in transformation experiments, *Biochem. Physiol. Pflanzen, 183*: 211–218, 1988.
56. Sonnewald, U. Konstruktion und Analyse von Chimären Patatin-Genen unter Besonderer Berücksichtigung der Stabilitat und Subzellulären Kompartmentierung der Patatin Proteine, Thesis, University of Cologne, Germany, 1989.
57. Krebbers, E., and Vandekerckhove, J. Production of peptides in plant seeds, *Trends in Biotechnol., 8*: 1–3, 1990.

58. Radke, S. E., Andrews, B. M., Moloney, M. M., Crouch, M. L., Kridl, J. C., and Knauf, V. C. Transformation of *Brassica napus* L. using *Agrobacterium tumefaciens*: developmentally regulated expression of a reintroduced napin gene, *Theor. Appl. Genet., 75*: 685–694, 1988.
59. Guerche, P., Tire, C., Grossi De Sa, F., De Clercq, A., Van Montagu, M., and Krebbers, E. Differential expression of the Arabidopsis 2S albumin genes and the effect of increasing gene family size, *Plant Cell, 2*: 469–478, 1990.
60. Dougherty, W. G., Cary, S. M., and Parks, T. D. Molecular genetic analysis of a plant virus polyprotein cleavage site: a model, *Virology, 171*: 356–364, 1989.
61. Beaudry, G. A., Mehta, N. M., Ray, M. L., and Bertelsen, A. H. Purification and characterization of functional recombinant alpha-amidating enzyme secreted from mammalian cells, *J. Biol. Chem., 265*: 17694–17699, 1990.
62. Mariani, C., De Beuckeleer, M., Truettner, J., Leemans, J., and Goldberg, R. B. Induction of male sterility in plants by a chimeric ribonuclease gene, *Nature, 347*: 737–741, 1990.
63. Krebbers, E., Rudelsheim, P., De Greef, W., and Vandekerckhove, J. Laboratory and field performance of transgenic *Brassica* plants expressing chimeric 2S albumin genes, *Proceedings of GCIRC 1991*, Saskatoon, Sask., Canada, in press.
64. Bednarek, S. Y., and Raikhel, N. V. The barley lectin carboxyl terminal propeptide is a vacuolar protein sorting determinant in plants. *Plant Cell 3*: 1195–1206, 1991.
65. Matsuoka, K., and Nakamura, K. Propeptide of a precursor to a plant vacuolar protein required for vacuolar targeting. *Proc. Natl. Acad. Sci. USA 88*: 834–838, 1991.
66. Chrispeels, M. J,. and Tague, B. W. Protein sorting in the secretory system of plant cells. *International Review of Cytology, 125*: 1–45, 1991.

4

Transgenic Plants: A Tool to Study Intercellular Transport

Moshe Lapidot and Roger N. Beachy

The Scripps Research Institute, La Jolla, California

I. INTRODUCTION

The movement of molecules between plant cells involves both transmembrane events and channel transport. The latter encompasses the function(s) of the cytoplasmic connections that form a network between adjacent cells through channels known as "plasmodesmata." Plasmodesmata, with a capacity to exclude from passage molecules with a molecular mass greater than 1 kDa, effectively control the flow of materials from cell to cell. Although these structures were first described more than 100 years ago, only during the past 10 years or so have experimental tools been developed sufficiently to permit in-depth studies of their function. The details are, however, only now beginning to unfold. Although a few high-resolution electron-microscopic studies have exposed the structural details of plasmodesmata, investigations of more detailed structures of plasmodesmata in different plants and different tissues remain to be done. As diagrammed in Fig. 1 and discussed by Robards and Lucas (2), there is evidence that plasmodesmata include one or more types of protein molecules that are probably associated with one or more types of cellular membranes, most likely the plasmalemma and endoplasmic reticulum. However, no one has yet isolated and characterized a known component of plasmodesmata.

Several experimental approaches are being used to identify the structural components and biological function(s) of plasmodesmata. The first is to

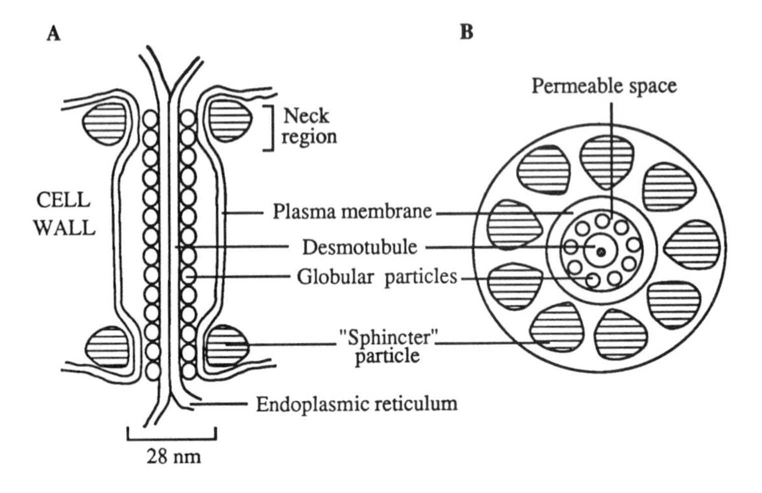

Figure 1 Structure of a simple plasmodesmata: (A) longitudinal section; (B) transverse section. Adapted from Ref. 1 with permission.

biochemically isolate their components by subcellular fractionation, with the ultimate intention of isolating the genes that encode the components. The second is to characterize changes that occur during infection and cell-to-cell spread of plant viruses, some of which are known to move to adjacent cells after modifying the structure or function of the plasmodesmata. There is growing evidence that some, if not most, plant viruses encode a protein that by itself, or in combination with one or more host factors, enables the virus or its nucleic acid to spread between adjacent cells. To better characterize the role of these proteins in modifying plasmodesmata in the absence of virus infection, virus transport (movement) proteins have been expressed in transgenic plants, and the nature of the functional modifications and the structural changes associated with such modifications have been studied. In this chapter we describe one such system, the movement protein of tobacco mosaic virus and its host, tobacco.

Multicellular and systemic virus infections in plants require (a) that the virus replicate in the initially infected cells and (b) that the virus spread from the initially infected cells into neighboring healthy cells and, eventually, throughout the plant. Interference with one of these processes inhibits virus infection and renders the plant a nonhost to the virus, i.e., the plant reacts as if it were resistant to the virus.

The identification of a virus-encoded protein that facilitates the spread of virus from cell to cell was first completed for tobacco mosaic virus (TMV), (3,4). Although dispensable for virus replication, the TMV 30-kDa movement protein (MP) is essential for cell-to-cell movement of the infection and is an important component in determining virulence and host range of TMV. From a number of studies that examined virus replication in protoplasts, it is clear that some plant viruses are able to replicate in protoplasts of a nonhost (5,6). This leads one to conclude that a major barrier to multicellular infection of a nonhost plant is the spread of the infectious agent, either from the initially infected cells into noninfected ones or from one leaf to another.

The tobacco mosaic virus, a member of the tobamovirus group, is the most extensively studied plant virus. The virus is rod shaped (18 × 300 nm), and its genome is composed of a single-stranded RNA of positive polarity encapsidated by a single type of capsid protein. As predicated from the complete nucleotide sequence of the TMV U1 (common) strain determined by Goelet et al. (7), the virus genome codes for at least four proteins, with molecular masses of 183, 126, 30, and 17.5 kDa, each of which has been detected in vivo and in vitro (5). There is an open reading frame for a fifth protein of 54 kDa, but the protein has yet to be found in TMV-infected plants (8). The 126- and 183-kDa proteins are translated directly from the genomic RNA, with the 183-kDa protein being the readthrough product of the amber termination codon of the 126-kDa cistron (9). Both proteins are apparently subunits of the viral replicase (10–12). Translation of the two other open reading frames to produce the MP (268 amino acids) and the 17.5-kDa protein (coat protein; CP) requires the formation of two subgenomic mRNAs. The MP mRNA was designated I_2 RNA (13,14), and the CP mRNA was called the low-molecular-weight-component (LMC) mRNA (11,15). Contained within the 3′ region of the MP gene are two cis-acting sequences: the origin of assembly (OAS), which is necessary for virion assembly, and the promoter responsible for transcription of the CP mRNA (16,7).

II. DEMONSTRATION THAT TMV 30-kDa PROTEIN IS A MOVEMENT PROTEIN

Early indirect evidence suggesting that plant viruses may encode a movement protein was provided by temperature-sensitive mutants of two closely related tobamoviruses, TMV (17) and tomato mosaic virus (ToMV) (18). The best studied temperature-sensitive mutant is the Ls1 strain of ToMV, which is temperature sensitive in its capacity to spread throughout the plant. At nonpermissive temperatures (32°C), the Ls1 strain replicates and assembles normally in inoculated leaves and infected protoplasts but is not capable of cell-to-cell movement in the inoculated leaves (19). Peptide maps indicated that the Ls1 30-kDa protein was different from that of the parental (L) strain (20). A comparison of the nucleotide sequence of the L and Ls1 strains revealed that the Ls1 strain had a single base change in the 30-kDa protein gene, which replaced a proline by a serine residue at amino acid position 154 (21). Taliansky et al. (22) showed that the temperature-sensitive defect in virus movement of Ls1 could be complemented in tobacco plants infected by related strains of TMV.

Direct evidence that the 30-kDa protein is essential for viral movement was provided by two different and complementary approaches. Meshi et al. (4) introduced in vitro a single base change into the MP gene of the L strain to produce the Ls1 virus. They also showed that the MP does not participate in viral replication. The second approach was taken by Deom et al. (3), who used transgenic plants expressing the wild-type 30-kDa movement protein of TMV to directly define the role of the protein in viral movement. A chimeric gene containing the 35S promoter of cauliflower mosaic virus (CaMV) and the polyadenylation signal of the nopaline synthase gene (NOS) flanking a cloned cDNA of the 30-kDa protein was constructed. Transgenic plants (*Nicotiana tabacum* cv. Xanthi) regenerated from transformed tobacco cells expressed the new gene, accumulated the MP, and complemented the temperature-sensitive defect of the Ls1 virus. That is, transgenic plants expressing the TMV MP [MP(+) plants], infected with Ls1 and maintained at the nonpermissive temperature, complemented cell-to-cell movement of the Ls1 virus. Complementation occurred in both the inoculated leaves and systemically infected upper leaves. More recently, it was shown by Holt and Beachy (23) that a TMV mutant containing a frame shift in the MP gene that abolishes its function is complemented by the MP(+) transgenic plant line and that the mutant virus spreads both locally and systemically. These results demonstrated that the TMV 30-kDa protein is a movement protein and that it can mimic the role of the ToMV movement protein.

III. MODIFICATION OF PLASMODESMATA SIZE EXCLUSION LIMITS BY THE MP

As indicated earlier, the most obvious, and perhaps the only, route for a plant virus to spread from cell to cell is via the cytoplasmic connections between plant cells, the plasmodesmata. Electron-microscopic studies have localized the MP from TMV-infected plants (24), as well as from MP(+) transgenic plants (25,26), to the plasmodesmata. Plasmodesmata appear to be membrane-lined pores or channels 20–80 nm in diameter with a thin strand of modified endoplasmic reticulum, called the desmotubule, that runs through the channel (Fig. 1) (for review see Ref. 2). Most plasmodesmata have a neck region that constricts the plasmodesmata at the entrance to each cell. A ring of particles, thought to act as a sphincter to regulate the size exclusion limits of the plasmodesmatum, can be found just below the plasma membrane at the neck region (2,27,28). A cross section of the plasmodesmatal pore reveals nine closely packed globular particles (thought to be protein subunits) that surround the desmotubule (27). It appears that whereas the desmotubule is impermeable, intercellular transport occurs through the spaces between the protein subunits (29). These spaces have an effective diameter of 1.5–3.0 nm (29,30). Microinjection of fluorescent glycans reveals that the molecular exclusion limit of plasmodesmata is 700–1000 Da (29,31). This limit is much too low to permit the passage of any known plant virus or its nucleic acid (32). This leads to the question of whether plant viruses alter the structure or function of plasmodesmata or both.

To address this question, Wolf et al. (33) developed a novel technique to deliver probes to the cytosol of leaf mesophyll cells. With this method a fluorescent probe, such as fluorescein isothiocyanate–labeled dextrans (F-dextran), was pre-encapsulated in liposomes. After they were injected into the vacuole, the liposomes fused with the tonoplast and released the fluorescent dye into the cytoplasm. The movement of the fluorescent probes to adjacent calls was then followed in living tissue by microscopy. This technique was used to compare the plasmodesmatal size exclusion limits of control and transgenic tobacco plants that express the MP. F-dextrans with an average molecular mass of 9400 Da and an approximate Stokes radius of 2.4 nm moved between cells of MP(+) transgenic plants, while the size exclusion limit of the control plants was 700–800 Da (Stokes radius ≈0.7 nm). Thus, the TMV MP clearly directly affects a plasmodesmatal function by altering the size exclusion limits.

The question of whether the infection moves from cell to cell via virus particles or RNA molecules or as an undescribed ribonucleoprotein (RNP) has been raised during the last several years. Particles of TMV are probably not necessary for cell-to-cell spread of infection; it has been shown that coat protein production and encapsidation are not required for local spread of TMV (34–36), although effective long-distance movement requires coat protein (37,38). Since it has been suggested (39) that the molecular dimensions of

TMV RNA are approximately 10 nm, it is clear that the MP-mediated alteration of the plasmodesmata by itself is not sufficient to facilitate cell-to-cell spread of infection. Additional information is required before we can account for the role of the MP in mediating viral spread. Citovsky et al. (40) have shown that MP produced in and purified from *Escherichia coli* binds in vitro to small, single-stranded nucleic acids (ssDNA and RNA), and this may indicate that the MP has multiple functions. These researchers proposed that the MP binds to TMV RNA and forms an unfolded elongated protein–RNA structure that moves the RNA from cell to cell. Citovsky et al. (41) have also shown that two domains in the MP, designated A (amino acids 112 to 184) and B (amino acids 185 to 268), are apparently involved in binding to nucleic acids to form the elongated protein–RNA complex (41). It is possible that the MP functions by two different but complementary mechanisms, i.e., through interactions with the plasmodesmata and through RNA binding.

IV. FURTHER ANALYSIS OF MP IN TRANSGENIC PLANTS

The preceding discussion documents that transgenic plants provide a good opportunity to study the MP in the absence of expression of other viral genes. Therefore, more extensive analyses of the MP(+) transgenic plants were performed. Deom et al. (42) demonstrated that the MP gene, under the control of the CaMV 35S promoter, is expressed in leaf, stem, and root tissues. The highest levels of MP were detected in older leaves. When subcellular fractions from transgenic leaf tissue were analyzed, the MP was most abundant in a detergent-washed cell wall fraction of older leaves and remained at high levels as the leaves aged. Significant levels of the MP were detected in a crude membrane/organelle fraction (P30) and in a soluble fraction (S30) of younger leaves, but the levels were lower in older leaves. The highest levels of MP mRNA were detected in the youngest leaves, and significantly less MP mRNA was detected in older leaves, which parallels a decrease in total RNA from young to old leaves. These results are consistent with the observation that decreases of MP in the P30 and S30 fractions of older leaves are due to decreases in MP mRNA. The accumulation of the MP in the cell wall fraction of older leaves, concomitant with a decrease of MP in the P30 and S30 fractions of older leaves, suggests that the MP has an affinity for, or is very stable in, the cell wall fraction in both transgenic and TMV-infected leaf tissue.

To determine whether leaf age or maturity has an influence on the modification of the function of plasmodesmata, liposomes containing the 9.4-kDa fluorescent probe were injected into mesophyll cells of leaves of MP(+) tobacco plants at different stages of maturity (42). It was found that the effect of the MP on the molecular exclusion limits of plasmodesmata depends upon leaf maturity. Movement of the 9.4-kDa probe in leaves 5 and 6 (fully expanded leaves) was detected in less than a minute after injection. In contrast, in leaves 1 through 4 (young leaves) the 9.4-kDa probe was restricted to the injected cell at 10 min after injection (42).

The finding that the plasmodesmata in young leaves of MP(+) plants are not functionally altered indicates that the capacity of the MP to alter plasmodesmatal size exclusion limits is influenced by leaf age or maturity in transgenic MP(+) plants. It is important to note that TMV moves into, and replicates in, very young leaves of infected plants. A possible explanation for the differences observed in young tissue between MP(+) plants and TMV-infected plants is that in addition to the MP one or more other components of the infection process are required for virus movement. Alternatively, movement of virus between mesophyll cells of young leaf tissue may require the highly localized concentration of MP that occurs during virus infection but is not attained in transgenic plants.

V. ALTERATION OF PLASMODESMATAL STRUCTURE BY THE MP

Moore et al. (26) examined the ultrastructural characteristics of plasmodesmata in plants that express the MP gene and those that do not [MP(-) plants]. They found that the structure of plasmodesmata changes as the leaf ages in both MP(+) and MP(-) plants. The structural features of plasmodesmata in mesophyll cells of young leaves (<7 cm long) of MP(+) plants were essentially identical to those observed in MP(-) plants. The plasmodesmata between mesophyll cells of a young leaf are relatively simple tubular structures, whereas in a mature, fully expanded leaf of an MP(-) plant the central cavity of the plasmodesmata swells in the region of the middle lamella. Closely adjacent plasmodesmata appear to undergo anastomosis; several plasmodesmata were observed to be connected to a single central cavity in both MP(+) and MP(-) plants. However, in mature leaves of MP(+) plants, but not MP(-) plants, a fibrous electron-dense material was observed in the enlarged central cavities. Most, but not all, of the anti-MP antibody labeling was localized to the plasmodesmata and to some extent to the fibrous material itself. The identity of the fibrous structures is not known, but they are clearly not viral particles. The fibrous structures may be composed of MP in association with membranes or other cellular components, or they may be a cellular response to the MP. Most of the anti-MP antibody labeling was associated with the margins of the plasmodesmata, with much less over the fibers themselves, suggesting that the fibers might be a cellular response to the MP rather than the MP itself.

In the report of Moore et al. (26) there was a more complete investigation of the subcellular fractionation of the MP in transgenic plants. It was found that significant amounts of the MP are associated with plasma membranes, with washed cell walls, and with a soluble fraction. The MP that was associated with the plasma membrane had the characteristics of a hydrophobic integral membrane protein. It is possible that the MP associated with the cell wall is actually associated with the plasma membrane that is within the plasmodesmata. The MP that lacks a classical signal peptide was largely soluble in extracts of young leaves and was mostly absent from a fraction enriched for intracellular membranes. On the basis of these findings, it was proposed that the pathway of synthesis and transport of the MP is not dependent on the endomembrane system but moves through the fraction to the plasma membrane and thus to the plasmodesmata (24).

VI. FUNCTION OF THE MP IN A HYPERSENSITIVE HOST

As implied in the Introduction, one of the defense mechanisms employed by plants that are nonhosts for TMV is interference with functions of the MP. Support for this concept derives from Meshi et al. (43), who analyzed a strain of tomato mosaic tobamovirus that is able to spread in tomato plants carrying the Tm-2 gene for resistance to TMV. It was shown that the capacity to overcome the resistance conferred by the Tm-2 gene results from two amino acid changes within the MP.

Resistance to plant viruses can take many forms (44,45). In some cases the host–defense response is a necrotic, hypersensitive resistance reaction (HR) in which the plant specifically recognizes the pathogen and mounts an active defense response that restricts movement of the virus.

A second gene for resistance to TMV, the N gene, was originally recognized in *N. glutinosa*. The N gene confers hypersensitive resistance in response to infection by most tobamoviruses (46). Plants that express the N gene are resistant to all strains of TMV and respond to infection by forming small necrotic lesions at the site of infection. This localizes the virus to cells within and immediately surrounding the necrotic lesion. In *N. tabacum* cv. Xanthi NN (homozygote in the N gene) necrotic lesions form on TMV-infected leaves approximately 2 days after inoculation and increase in size for about 8 days. The N gene–mediated HR is temperature sensitive (47), and at temperatures above 28°C lesions do not develop and the virus spreads systemically. Because the N gene–mediated HR limits the spread of TMV in infected Xanthi NN plants, it was suggested that the MP is involved in some way with the local lesion reaction. To determine whether that is the case, transgenic Xanthi NN plants that express the MP gene were produced and the plasmodesmatal size exclusion limits in leaf mesophyll cells were determined (48). At 24°C the 3.9-kDa probe moved from cell to cell, whereas the 9.4-kDa probe did not move. At 33°C the 9.4-kDa probe moved readily from cell to cell. As already described, MP(+) Xanthi plants permit the movement of the 9.4-kDa probe from cell to cell at both 24°C and 33°C. These results may indicate that one or more products of the N gene modify the ability of the MP to alter plasmodesmatal size exclusion limits, although expression of the MP by itself did not induce the HR. Furthermore, when MP(+) Xanthi NN plants were inoculated with a mutant of TMV that lacks the MP gene, the HR was induced and the size of the local lesions was correlated with the amount of MP produced by the transgenic plant line. Unfortunately, little is known about the N gene or how it functions, and no direct evidence is yet available on how the MP and N gene might interact. However, it is expected that transgenic plants will help to clarify the interactions between viral and host-resistance genes.

Recently, another approach was taken to study the TMV–N gene interactions in transgenic plants. Turpen et al. (49) introduced into systemic (Xanthi) tobacco plants a chimeric gene that encodes a TMV "replicon," i.e., a gene that encodes the 126- and 183-kDa subunits of the viral replicase and also encodes the MP. During the regeneration of the transgenic plants, mild mosaic symptoms were observed. When these plants were crossed with the Xanthi nc cultivar (which contains the N gene) and the R1 progeny were maintained at a temperature above 30°C, the plants grew normally. When the temperature was lowered below 28°C the plants became necrotic. Turpen et al. (49) proposed to use these plants to find mutations in the N gene, or another host gene, that suppress the necrotic reaction in the R1 progeny.

VII. EFFECT OF CALLOSE DEPOSITION ON PLASMODESMATAL EXCLUSION LIMITS

To better understand the MP and its role in virus movement, Wolf et al. (50) further explored the underlying mechanism(s) responsible for the temperature-sensitive nature of the LS1 virus. A temperature-sensitive mutant of the TMV MP (referred to as ts-MP) was generated in vitro by changing the proline residue at amino acid 154 to alanine. The Pro-to-Ala mutation was chosen because alanine is not commonly found in turn structures (51) and contains an uncharged side chain. Transgenic tobacco plants that express the ts-MP gene were produced. Dye-coupling experiments using F-dextrans were employed to study the effects of permissive versus nonpermissive temperatures on plasmodesmatal function in the plants that contained ts-MP.

At 24°C, 90% of the ts-MP(+) plants complemented the movement of MP(-) TMV. However, when the ts-MP(+) plants were held at 32°C following inoculation with MP(-) virus, no disease symptoms developed on the inoculated or systemically infected leaves, indicating that the ts-MP is temperature sensitive at 32°C. By contrast, symptoms on inoculated as well as upper leaves developed at both temperatures in transgenic plants that express the wild-type MP.

The amounts of the ts-MP were the same at 24°C and 32°C, suggesting that the failure of the ts-MP to complement movement of the MP(-) TMV was not due to lower levels of the mutant protein at 32°C. Furthermore, the subcellular localization of ts-MP at both temperatures was similar to that of the wild-type MP (50).

Microinjection of fluorescently labeled dextrans into the transgenic plants that express the ts-MP gene were performed to determine its effect on the size exclusion limits of the plasmodesmata at both 24°C and 32°C. At the permissive temperature (24°C), the 9.4-kDa F-dextran moved between mesophyll cells of ts-MP(+) plants. After the plants were shifted to the nonpermissive temperature (32°C), movement of both the 3.9-kDa and 9.4-kDa probes was inhibited. After 48 hr at 32°C, plants were shifted back to 24°C, and within 6 hr at 24°C, both the 3.9-kDa and the 9.4-kDa probes exhibited cell-to-cell movement in the ts-MP(+) plants. Thus, the temperature-related inhibition of movement is reversible.

Surprisingly, a transient inhibition of probe movement was observed in wild-type MP(+) plants that had been held at 32°C. However, after 48 hr at 32°C this inhibition was no longer detected and the 9.4-kDa probe moved in all MP(+) plants tested.

Callose ([1-3]-β-D-glucan) synthesis and accumulation has long been implicated as a mechanism for limiting the spread of virus in hypersensitive tissues (52). Furthermore, temperature-associated changes in callose formation have been observed in several plant systems (53–55). A recent model for plasmodesmatal function suggests that (1-3)-β-D-glucan synthase may be involved in the rapid deposition of callose, which, by compressing against the cell wall, would close an otherwise open cytoplasmic annulus (56). To determine whether callose deposition is involved in the transient inhibition of cell-to-cell movement of the glycan probes in MP(+) plants at 32°C, microinjection experiments were performed in the presence of Sirofluor, an inhibitor of (1-3)-β-D-glucan synthase (57). There was no cell-to-cell movement of the 9.4-kDa probe in MP(+) plants (at 32°C) that were pretreated with distilled water, whereas the 9.4-kDa probe moved in MP(+) plants in which the tissue was pretreated with 5 m*M* Sirofluor solution. In parallel experiments performed on ts-MP(+) plants at 32°C, the 9.4-kDa probe did not move out of the injected cell in tissue pretreated with either distilled water or Sirofluor (50). These types of data support the hypothesis that at 32°C callose synthesis and accumulation affect movement of the fluorescent dye.

Nishiguchi et al. (19) suggested two explanations for the inability of the Ls1 virus (Ls1 has a similar amino acid mutation in its MP as was reconstructed in the ts-MP) to move from cell to cell under restrictive temperature: (a) The host plant may be induced to produce a material that blocks viral movement, or (b) the inhibition of viral movement may be caused by a functional defect in a viral protein. The decrease in movement of fluorescent probes in MP(+) plants at 32°C is in agreement with the hypothesis that a blocking agent is being deposited at the elevated temperature that might modulate viral cell-to-cell movement. Also, the observation that the reduction in plasmodesmatal size exclusion limits in MP(+) plants at 32°C is transient is consistent with temperature-induced synthesis of callose. The involvement of callose synthesis in the temperature-sensitive response of the plasmodesmatal size exclusion limits was evident from the results showing that Sirofluor-treated tissue maintained a higher exclusion limit. However, down-regulation of plasmodesmata by callose synthesis is not sufficient to account for the restriction of plasmodesmatal size exclusion limits in ts-MP(+) plants maintained at 32°C. The restriction at 32°C was not transient, and Sirofluor did not affect the plasmodesmatal size exclusion limits at the restrictive temperature. Moreover, the total amount and subcellular distribution of the ts-MP were the same as those of wild-type MP. Hence, the temperature-dependent reduction in plasmodesmatal exclusion limits in the ts-MP(+) plants may result from a change in the ability of the mutant MP to interact with a structural component of the plasmodesmata, whose function is to regulate the physical pore size within the cytoplasmic annulus.

VIII. TRANSGENIC PLANTS EXPRESSING THE MP GENE AS A LABORATORY TOOL

In addition to being used for studies of the plasmodesmata, the MP(+) tobacco plants are also used to maintain viruses that lack cell-to-cell movement functions. Since MP(-) TMV can be complemented by MP(+) plants, these plants can serve as hosts for proliferation of mutant viruses and maintenance of viral stocks. Gafny et al. (58) created in vitro deletions in various regions of the MP gene of TMV and inoculated the viruses to MP(+) as well as to wild-type plants. Those mutants that eliminated or reduced movement function did not induce symptoms on the wild-type plants but did induce symptoms on the MP(+) plants, showing that other vital viral functions, such as replication, were unaffected by the mutation.

Berna et al. (59) produced plants that express truncated MP genes and demonstrated that it is possible to delete 55 amino acids from the carboxy terminus of the MP and retain the capacity to complement movement of MP(-) viruses. The plasmodesmatal size exclusion limits in leaves of the truncated-MP(+) plants were also studied and were found to be essentially the same as in wild-type MP(+) plants (59).

Holt and Beachy (unpublished) introduced a number of reporter and other genes into a full-length cDNA clone of TMV in place of the MP coding region. The local and systemic movement of these viruses was complemented by the MP(+) plants. Marker genes expressed by viral-based clones will aid in studying very early events in viral infection and monitoring viral movement and replication. Furthermore, transgenic plants that express the TMV MP gene will likely expand the use of TMV-based expression vectors, examples of which are described in another chapter of this volume.

IX. CONCLUSION

The purpose of this chapter was to demonstrate the use of transgenic plants in studies of the function of plasmodesmata and the movement protein of TMV. The MP(+) transgenic plants are extremely useful tools in the study of the MP mode of action, as well as in understanding some of the mechanisms regulating plasmodesmatal function.

Unfortunately, the mechanisms of viral cell-to-cell movement are not fully understood; likewise, the mechanisms governing plasmodesmatal functions are poorly understood. Undoubtedly, transgenic plants will be important in elucidating how viruses move from cell to cell in the infected plant. These results will likely influence our understanding of how plant cells communicate and may eventually help us to devise ways to inhibit viruses spread within the infected plant.

REFERENCES

1. Citovsky, V., and Zambryski, P. How do plant virus nucleic acids move through intercellular connections? *Bioessays, 13*: 373–379, 1991.
2. Robards, A. W., and Lucas, W. J. Plasmodesmata, *Annu. Rev. Plant Physiol. Plant Mol. Biol., 41*: 369–419, 1990.
3. Deom, C. M., Oliver, M. J., and Beachy, R. N. The 30-kilodalton gene product of tobacco mosaic virus potentiates virus movement, *Science, 237*: 384–389, 1987.
4. Meshi, T., Watanabe, Y., Saito, T., Sugimoto, A., Maeda, T., and Okada, Y. Function of the 30kD protein of tobacco mosaic virus: involvement in cell-to-cell movement and dispensibility for replication, EMBO J., 6: 2557–2563, 1987.
5. Zaitlin, M., and Hull, R. Plant virus-host interactions, *Annu. Rev. Plant Physiol., 38*: 291–315, 1987.
6. Hull, R. The movement protein of viruses in plants, *Annu. Rev. Phytopathol., 27*: 241–245, 1989.
7. Goelet, P., Lomonossoff, G. P., Butler, P. J. G., Akam, M. E., Gait, M. J., and Karn, J. Nucleotide sequence of tobacco mosaic virus RNA, *Proc. Natl. Acad. Sci. USA 79*: 5818–5822, 1982.
8. Palukitis, P., and Zaitlin, M. Tobacco mosaic virus: infectivity and replication, *The Plant Viruses, Vol. 2, The Rod-Shaped Plant Viruses* (M. H. V. Regenmortel and H. Fraenkel-Conrat, eds.), Plenum, New York, 1986, pp. 105–131.
9. Pelham, H. R. B. Leaky AUG termination codon in tobacco mosaic virus RNA, *Nature, 272*: 469–471, 1978.
10. Zaitlin, M., Duda, C. T., and Petti, M. A., Replication of tobacco mosaic virus, V: properties of the bound an solubilized replicase, *Virology, 53*: 300–311, 1973.
11. Hunter, T. R., Hunt, T., Knowland, J., and Zimmern, D. Messenger RNA for the coat protein of tobacco mosaic virus, *Nature, 260*: 759–764, 1976.
12. Scalla, R., Romaine, P., Aselin, A., Rigaud, I., and Zaitlin, M. An *in vivo* study of a nonstructural polypeptide synthesized upon TMV infection and its identification with a polypeptide synthesized *in vitro* from TMV RNA, *Virology, 91*: 182–193, 1987.
13. Bruening, G., Beachy, R. N., Scalla, R., and Zaitlin, M. *In vitro* and *in vivo* translation of ribonucleic acids of cowpea strain of tobacco mosaic virus, *Virology, 17*: 498–517, 1976.
14. Beachy, R. N., and Zaitlin, M. Characterization and in vitro translation of the RNAs from less-than-full-length, virus related, nucleoprotein rods present in tobacco mosaic virus preparations, *Virology, 81*: 160–169, 1977.
15. Beachy, R. N., Zaitlin, M., Bruening, G., and Israel, H. W. A genetic map for the cowpea strain of TMV, *Virology, 73*: 498–507, 1976.

16. Zimmern, D. The nucleotide sequence at the origin for assembly of tobacco mosaic virus RNA, *Cell, 11*: 463–482, 1977.
17. Jockusch, H. Two mutants of tobacco mosaic virus temperature sensitive in two different functions, *Virology, 35*: 94–101, 1968.
18. Nishiguchi, M., Motoyoshi, F., and Oshima, N. Behavior of a temperature sensitive strain of tobacco mosaic virus in tomato leaves and protoplasts, *J. Gen. Virology, 39*: 53–61, 1978.
19. Nishiguchi, M., Motoyoshi, F., and Oshima, N. Further investigation of a temperature-sensitive strain of tobacco mosaic virus in tomato leaves and protoplasts, *J. Gen. Virology, 46*: 497–500, 1980.
20. Leonard, D. A., and Zaitlin, M. A temperature-sensitive strain of tobacco mosaic virus defective in cell to cell movement generates an altered virus-coded protein, *Virology, 117*: 416–424, 1982.
21. Ohno, T., Takamatsu, N., Meshi, T., Okada, Y., Nishiguchi, M., and Kiho, Y. Single amino acid substitution in 30K protein of TMV defective in virus transport function, *Virology, 131*: 255–258, 1983.
22. Taliansky, M. E., Malyshenko, S. I., Pshennikova, E. S., Kaplan, I. B., Ulanova, E. F., and Atabekov, J. G. Plant virus-specific transport function, 1: virus genetic control required for systemic spread, *Virology, 122*: 318–326, 1982.
23. Holt, C. A., and Beachy, R. N. *In vivo* complementation of infectious transcripts from mutant tobacco mosaic virus cDNAs in transgenic plants, *Virology, 181*: 109–117, 1991.
24. Tomenius, K., Clapham, D., and Meshi, T. Localization by immunogold cytochemistry of the virus-coded 30K protein in plasmodesmata of leaves infected with tobacco mosaic virus, *Virology, 160*: 363–371, 1987.
25. Atkins, D., Hull, R., Wells, B., Roberts, K., Moore, P., and Beachy, R. N. The tobacco mosaic virus 30K movement protein in transgenic tobacco plants is localized to plasmodesmata, *J. Gen. Virol., 72*: 209–211, 1991.
26. Moore, P. J., Fenczik, C. A., Deom, C. M., and Beachy, R. N. Developmental changes in the structure of plasmodesmata in transgenic tobacco expressing the movement protein of tobacco mosaic virus (submitted for publication).
27. Robards, A. W. Plasmodesmata, *Annu. Rev. Plant Physiol., 26*: 13–29, 1975.
28. Olsen, P. The neck constriction in plasmodesmata: evidence for a peripheral sphincter-like structure revealed by fixation with tannic acid, *Planta, 144*: 349–358, 1979.
29. Terry, B. R., and Robards, A. W. Hydrodynamic radius alone governs the mobility of molecules through plasmodesmata, *Planta, 171*: 145–157, 1987.
30. Lucas, W. J., Wolf, S., Deom, C. M., Kishore, G. M., and Beachy, R. N. Plasmodesmata-virus interaction, *Parallels in Cell-to-Cell Junction in Plants and Animals* (A. W. Robards, H. Jongsma, W. J. Lucas, J. Pitts, and D. Spray, eds.), Springer-Verlag, Berlin, 1990, pp. 261-274.
31. Baron-Eppel, O., Hernandes, D., Jiang, L.-W., Meiners, S., and Schindler, M. Dynamic continuity of cytoplasmic and membrane compartments between plant cells, *J. Cell Biol., 106*: 715–721, 1988.

32. Matthews, R. E. F. *Plant Virology, 3rd ed.*, Academic Press, New York, 1991.
33. Wolf, S., Deom, C. M., Beachy, R. N., and Lucas, W. J. Movement protein of tobacco mosaic virus modifies plasmodesmatal size exclusion limit, *Science, 246*: 377–379, 1989.
34. Dorokhov, Y. L., Alexandrov, N. M., Miroshnichenko, N. A., and Atabekov, J. G. Isolation and analysis of virus-specific ribonucleoprotein of tobacco mosaic virus-infected tobacco, *Virology, 127*: 237–252, 1983.
35. Takamatsu, N., Ishikawa, M., and Okada, Y. Expression of bacterial chloramphenicol acetyltransferase gene in tobacco plants mediated by TMV-RNA, *EMBO J., 6*: 307–311, 1987.
36. Dawson, W. O., Bubric, P., and Grantham, G. L. Modifications of the tobacco mosaic virus coat protein gene affecting replication, movement, and symptomatology, *Phytopathology, 78*: 783–789, 1988.
37. Dorokhov, Y. L., Alexandrov, N. M., Miroshnichenko, N. A., and Atabekov, J. G. The informosome-like virus-specific ribonucleoprotein (vRNP) may be involved in the transport of tobacco mosaic virus infection, *Virology, 137*: 127–134, 1984.
38. Saito, T., Yamanaka, K., and Okada, Y. Long-distance movement and viral assembly of TMV mutants, *Virology, 176*: 329–336, 1990.
39. Gibbs, A. J. Viruses and plasmodesmata, *Intercellular Studies in Plants: Studies on Plasmodesmata* (B. E. S. Gunning and A. W. Robards, eds.), Springer-Verlag, Berlin, 1976, pp. 149-164.
40. Citovsky, V., Knorr, D., Schuster, G., and Zambryski, P. The P30 movement protein of tobacco mosaic virus is a single-strand nucleic acid binding protein, *Cell, 60*: 637–647, 1990.
41. Citovsky, V., Wong, M. L., Knorr, D., and Zambryski, P. How do nucleic acids move across membrane channels? International Society for Plant Molecular Biology 3rd International Congress, Abstract 1067, 1991.
42. Deom, C. M., Schubert, K., Wolf, S., Holt, C. A., Lucas, W. J., and Beachy, R. N. Molecular characterization and biological function of the movement protein of tobacco mosaic virus in transgenic plants, *Proc. Natl. Acad. Sci. USA, 87*: 3284–3288, 1990.
43. Meshi, T., Motoyoshi, F., Maeda, T., Yoshiwoka, S., Watanabe, H., and Okada, Y. Mutations in the tobacco mosaic virus 30K protein gene overcome Tm-2 resistance in tomato, *Plant Cell, 1*: 515–522, 1989.
44. Ponz, F., and Bruening, G. Mechanisms of resistance to plant viruses, *Annu. Rev. Phytopathol., 24*: 355–381, 1986.
45. Goodman, R. N., Kiraly, Z., and Wood, K. R. *The Biochemistry and Physiology of Plant Diseases*, University of Missouri Press, Colombia, 1986.
46. Holmes, F. O. Inheritance of resistance to tobacco mosaic disease in tobacco, *Phytopathology, 28*: 553–561, 1938.
47. Weststeijin, E. A. Lesion growth and virus localization in leaves of *Nicotiana tabacum* cv. Xanthi nc after inoculation with tobacco mosaic virus and incubation alternately at 22°C and 32°C, *Physiol. Plant Pathol., 18*: 357–368, 1981.

48. Deom, C. M., Wolf, S., Holt, C. A., Lucas, W. J., and Beachy, R. N. Altered function of the tobacco mosaic virus movement protein in a hypersensitive host, *Virology, 180*: 251–256, 1991.
49. Turpen, T. H., Turpen, A. M., and Dawson, W. O. A conditional lethal phenotype caused in N gene tobacco (*Nicotiana tabacum* cv. Xanthi nc) by transfection from chromosomally integrated tobacco mosaic virus cDNA, International Society for Plant Molecular Biology 3rd International Congress, Abstract 1166, 1991.
50. Wolf, S., Deom, C. M., Beachy, R. N., and Lucas, W. J. Plasmodesmatal function is probed using transgenic tobacco plants that express a virus movement protein, *Plant Cell, 3*: 593–604, 1991.
51. Chou, P. Y., and Fasman, G. D. Empirical predictions of protein conformation, *Annu. Rev. Biochem., 47*: 251–276, 1978.
52. Allison, A. V., and Shalla, T. A. The ultrastructure of local lesions induced by potato virus X: a sequence of cytological events in the course of infection, *Phytopathology, 64*: 784–793, 1974.
53. Webster, D. H., and Currier, H. B. Heat-induced callose and lateral movement of assimilates from phloem, *Can. J. Bot., 46*: 1215–1220, 1968.
54. Eschrich, W. Sealing system in phloem, *Encyclopedia of Plant Physiology*, New Series, Vol. 1: Transport in Plants (M. H. Zimmermann and J. A. Milburn, eds.), Springer-Verlag, Berlin, 1975, pp. 39–56.
55. Dinar, M., Rudich, J., and Zamski, E. Effect of heat stress on carbon transport from tomato leaves, *Ann. Bot., 51*: 97–103, 1983.
56. Olsen, P., and Robards, A. W. The neck region of plasmodesmata: general architecture and some functional aspects, *Parallels in Cell-to-Cell Junction in Plants and Animals* (A. W. Robards, H. Jongsma, W. J. Lucas, J. Pitts, and D. Spray, eds.), Springer-Verlag, Berlin, 1990, pp. 145–170.
57. Morrow, D. L., and Lucas, W. J. 1-3-β-D-Glucan synthase from sugar beet, II: product inhibition by UDP, *Plant Physiol., 84*: 565–567, 1987.
58. Gafny, R., Lapidot, M., Berna, A., Holt, C. A., Deom, C. M., and Beachy, R. N. Effects of terminal deletion mutations on function of the movement protein of tobacco mosaic virus, *Virology, 187*: 499–507, 1992.
59. Berna, A., Gafny, R., Wolf, S., Lucas, W. J., Holt, C. A., and Beachy, R. N. The TMV movement protein: role of the C-terminal 73 amino acids in subcellular localization and function, *Virology, 182*: 682–689, 1991.

II

VIRAL PATHOGEN RESISTANCE

5

Virus Resistance Mediated by a Nonstructural Viral Gene Sequence

George P. Lomonossoff

John Innes Institute, John Innes Centre
Norwich, England

I. INTRODUCTION

Since the demonstration that tobacco plants transformed with and expressing the coat protein gene of tobacco mosaic virus (TMV) are resistant to infection by the virus (1), there have been numerous attempts to produce virus-resistant plants by using the tools of plant transformation. Such attempts have included the transformation of a variety of different plant species with the coat protein genes from a number of different plant viruses (for a review, see Ref. 2), with cDNA sequences corresponding to small satellite RNAs (3,4), or with cDNA sequences that are transcribed into virus-specific antisense RNAs in the transformed plants (5–8). Although the results have been somewhat variable, all three methods have led to the production of plants that are at least partially resistant to virus infection.

This chapter describes a novel approach to the production of virus-resistant plants. This new approach, which was discovered fortuitously, involves transforming plants with a sequence of a nonstructural virus gene. Although it has, to date, been shown to work only in the TMV/tobacco system (9), the method appears to have considerable potential for use with other virus–plant combinations.

II. STRUCTURE OF THE GENOME OF TOBACCO MOSAIC VIRUS

The genome of the common (U1 or *vulgare*) strain of TMV consists of a single strand of positive-sense RNA, that is 6395 residues long (10). It is generally accepted that the RNA encodes at least four proteins, two of which, the 126K and 183K proteins (with molecular weights of 126 and 183 kDa, respectively), are directly translated from the genomic RNA. The 126K and 183K proteins are amino coterminal and arise as a result of initiation at the AUG at position 69 on the TMV sequence (Fig. 1). The 126K protein terminates at the UAG codon at position 3417, the 183K protein being generated by readthrough of this UAG codon into the region marked "54K" (11). The terminator for the 183K protein is at position 4917. The two other proteins whose existence has been unambiguously demonstrated in plants, the 30K protein and the viral coat protein, are synthesized from separate subgenomic mRNAs.

The one controversial aspect of the genome organization of TMV is whether there is a separate protein (termed the 54K protein) that corresponds to the readthrough portion of the 183K protein. Such a protein has not, as yet, been detected in TMV-infected plants, but there is circumstantial evidence for its existence. For example, an RNA termed I_1 (12), whose 3′ end is coterminal with the genomic RNA and whose 5′ end has been mapped to position 3405 of the TMV sequence (13), has been detected in TMV-infected tobacco tissue. The 5′ proximal AUG on this RNA is at position 3495 and is in phase with the readthrough portion of the 183K gene (Fig. 1). Thus, the I_1 RNA has all the

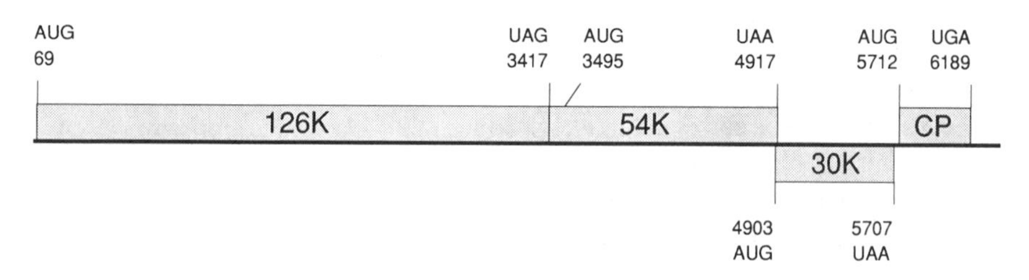

Figure 1 Genome organization of TMV strain U1 (*vulgare*). The open reading frames are indicated by the shaded boxes, and the positions of their initiation and termination codons are indicated. The region designated the 54K protein in the text is encoded between the AUG at 3495 and the UAA at 4917. The UAG at 3417 is the "leaky" terminator readthrough of which gives rise to the 183K protein. CP = viral coat protein gene.

properties necessary for it to act as an mRNA for the synthesis of the 54K protein. Support for the role of the I_1 RNA as an mRNA has also been provided by the observation that it is found on polyribosomes (13). Regardless of whether the 54K protein has an independent existence, throughout this chapter I shall refer to the region of the TMV genome between the AUG at position 3495 and the UAA at 4917 as the 54K gene and the protein encoded by this region as the 54K protein.

Irrespective of whether it is expressed as a separate protein or is always fused to the C terminus of the 126K protein, the 54K polypeptide sequence appears to be vitally concerned with viral replication. It contains the Gly-Asp-Asp (GDD) sequence characteristic of RNA polymerases of positive-strand RNA viruses (14) and shows amino acid sequence homology with replication-associated proteins from a number of RNA viruses from both plants and animals (for a recent review, see Ref. 15). Furthermore, both the 183K and 126K proteins have been found to be associated with the membrane-bound complexes in which viral RNA replication is believed to occur (16). In fact, the transformation of tobacco with the 54K sequence was originally undertaken in an attempt to clarify the role of the 54K polypeptide sequence in the viral replication cycle.

III. CLONING THE TMV 54K GENE

To obtain a clone containing the sequence of the readthrough portion of the 183K gene (the 54K gene), cDNA was synthesized with Moloney murine leukemia virus reverse transcriptase using a primer that consisted of the sequence of a BamH1 site linked to a sequence complementary to nucleotides 4906–4923 of TMV RNA. The resulting cDNA was rendered double stranded by using loop-back synthesis, digested with BamH1 (which cleaves once within the TMV RNA sequence at position 3332) and ligated into the BamH1 site of M13mp18. Analysis of the resulting clones showed that they all contained the sequence from the BamH1 site at position 3332 of the TMV sequence but, curiously, did not contain the BamH1 site provided by the primer. As a result, the termination codon for the 54K gene was deleted, leading to the sequence of the 54K protein being fused at its C terminus to five amino acids encoded by the vector. To remove sequences upstream of the AUG codon at position 3495, the replicative form DNA of one of the clones was digested with Hae11 (which cuts at position 3467 in the TMV RNA sequence), the 3′ overhang was "polished" with DNA polymerase (Klenow fragment), and the DNA was further digested with Pst1 (which cuts within the M13 polylinker). The resulting 1.5-kb fragment, containing the sequence of TMV RNA from positions 3472–4916 linked to part of the M13 polylinker, was isolated and ligated into Smal/Pst1-digested pBS(-) (Stratagene) to yield the clone pRTT-1.

The structure of the TMV-specific insert in pRTT-1 is shown in Fig. 2. The insert consists of the entire 54K open reading frame downstream of the T7 promoter of pBS(-) and includes 23 TMV-specific residues upstream of the AUG at 3495. The insert was designed so that the AUG at position 3495, believed to be the initiator for the synthesis of the 54K protein (13), is the first AUG encountered in T7 transcripts derived from pRTT-1. The presence of an intact open reading frame for the 54K protein in pRTT-1 was verified by synthesizing T7 transcripts and translating them in rabbit reticulocyte lysate. This resulted in the synthesis of a product that migrated with a size of 54 kDa on polyacrylamide/SDS gels. This product could be immunoprecipitated with an antiserum raised against a synthetic peptide whose sequence corresponded to residues 243–257 of the 54K sequence. These results confirmed that pRTT-1 contains a functional 54K open reading frame and strongly suggested that the AUG at position 3495 could, indeed, act as an initiation codon, at least in vitro. Confirmation that initiation of 54K synthesis does indeed occur from the AUG at position 3495, at least in vitro, has been provided by constructing a site-directed mutant of pRTT-1 in which the AUG at position 3495 has been

(A)

```
EcoR1  Sac1  Kpn1                         M  Q  F  Y
GAATTCGAGCTCGGTACCCCAAAGACTGGTGATATTTCTGATATGCAGTTTTAC.......
                   :                      :
                3472                   3495
```

(B)

```
      D  G  S  S  C [V  E  I  V  L  *]  Sal1  Pst1  Sph1 Hind111
.....GATGGCTCTAGTTGTGTGGAAATTGTGCTCTAGAGTCGACCTGCAGGCATGCAAGCTT
                   :
                4916
```

Figure 2 Sequences at the 5′ (A) and 3′ (B) ends of the insert in pRTT-1. The positions of the beginning (3472) and end (4916) of the TMV-specific portion of the insert are marked, as is the position of the AUG believed to be the initiator for the synthesis of the 54K protein (13). The amino acid sequence of the protein encoded by the insert is shown using the standard one-letter code. The five nonviral amino acids fused to the C terminus of the 54K protein are shown in brackets.

changed to an AUC codon. Translation of T7 transcripts derived from this mutant yields a product that is approximately 2 kDa smaller than the 54-kDa product derived from transcripts of pRTT-1 (G. P. Lomonossoff, unpublished observations). This smaller product presumably arises from initiation at the next in-phase AUG (at position 3537) in the 54K sequence.

IV. PRODUCTION OF TRANSGENIC TOBACCO PLANTS HARBORING THE TMV 54K GENE

To introduce the 54K gene into tobacco plants, the insert from pRTT-1 was excised by digestion with Hind111 and Sac1, made blunt-ended by treatment with DNA polymerase (Klenow fragment), and ligated into either Xhol- or Smal-digested pMON316 (17). The unique Xho1 site of pMON316 is located in the polylinker between the cauliflower mosaic virus (CaMV) 35S promoter and the nopaline synthase 3′ untranslated region. Insertion of the 54K sequence in the sense orientation at this site gave rise to plasmid pTS541 (Fig. 3).

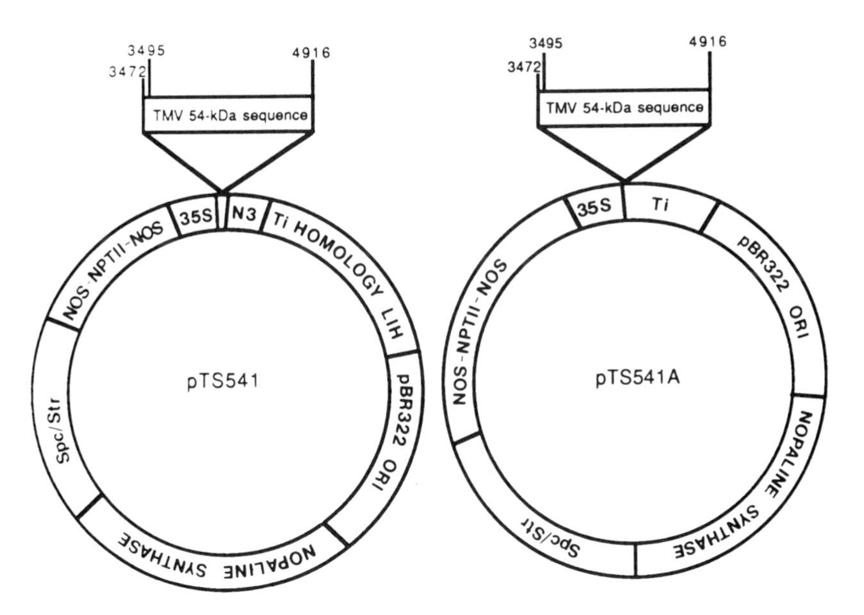

Figure 3 Structure of the plasmids used to transform tobacco plants with the TMV 54K sequence. The sequence of the TMV 54K gene was inserted downstream of the cauliflower mosaic virus 35S promoter in the vector pMON316 by using either the Xhol site (pTS541) or the Smal site (pTS541A). Reproduced from Ref. 9 with permission.

Because pMON316 contains two Sma1 sites, one in the polylinker region and the other in the Ti homology region, insertion of the 54K sequence into Sma1-cut pMON316 resulted in the construction of a clone, pTS541A, in which the nopaline synthase 3′ untranslated region and part of the Ti plasmid homology region was deleted (Fig. 3). Both constructs were transferred to *Agrobacterium tumefaciens* strain GV3111 carrying plasmid pTiB6S3-SE by means of triparental mating (18). Transconjugants were selected by their resistance to the antibiotics kanamycin and streptomycin.

Leaf disks of *Nicotiana tabacum* cv. Xanthi nn were transformed with cultures of *Agrobacterium tumefaciens* that had acquired either pTS541 or pTS541A by using standard techniques (19). Transformed calli were selected on a regeneration medium containing 300 μg/ml kanamycin, and resistant calli were induced to regenerate shoots and roots. The regenerated plantlets were finally transferred to soil and maintained in a greenhouse. Altogether, eight independently transformed plants were generated by this method, four of which were transformed with pTS541 (plants 5411, 5412, 5413, and 5414) and four with pTS541A (plants 541A11, 541A12, 541A20, and 541A21). All eight plants tested positive for the production of nopaline synthase.

V. EXPRESSION OF THE 54K SEQUENCE IN THE TRANSGENIC PLANTS

Six of the independently transformed plants (5411, 5413, 541A11, 541A12, 541A20, and 541A21) were analyzed for the 54K sequence in their nuclear DNA by Southern blotting. All six plants gave the expected pattern of bands, although the copy number of the 54K sequence appeared to vary from one to five copies per diploid genome. No copies of the 54K sequence could be detected in the nuclear DNA from untransformed plants or from plants transformed with pMON316.

The presence of 54K-specific transcripts in all eight transformed plants was confirmed by Northern blot analysis of RNA extracted from leaf tissue. In the case of plants transformed with pTS541, a single major 54K-specific transcript that had the expected size of 1.6 kb could be detected. In the case of the plants transformed with pTS541A, a more heterogeneous pattern was observed; several of the 54K-specific transcripts were significantly longer than 1.6 kb. This result was not surprising because the nopaline synthase 3′ region, which contains a transcription termination sequence, was missing in pTS541A (Fig. 3).

To date, attempts to detect the 54K protein in tissue from transformed plants have proved unsuccessful. Two methods have so far been used—Western blot analysis of proteins extracted from the leaves of transformed plants and immunoprecipitation of ^{35}S-labeled extracts from either transformed plants or protoplasts prepared from them. In each case two different antisera were used: One was raised against a synthetic peptide corresponding to residues 243–257 of the 54K sequence; the other was raised against a fusion protein containing 432 residues of the readthrough portion of the 183K protein linked to β-galactosidase (20). The failure to detect the 54K protein in the transgenic plants was unexpected in view of the fact that 54K-specific mRNA could be detected. The failure to detect the protein indicates that it is either synthesized at a very low level or that it is a very labile protein that is rapidly turned over within the cell and so does not accumulate to a significant extent. In regard to these two possibilities it should be noted that the AUG at position 3495 is in a distinctly suboptimal "context" to serve as an efficient initiator of protein synthesis in plants (21). Thus, the 54K mRNA present in the transgenic plants may not be translated very efficiently.

VI. RESISTANCE OF TRANSFORMED PLANTS TO VIRUS INFECTION

To examine the effect of the presence of the 54K sequence on the ability of TMV to multiply in the transgenic plants, four rooted cuttings from each of the eight primary transformants were inoculated with the U1 strain of TMV at a concentration of 50 μg/ml. Surprisingly, none of the transgenic plants harboring the 54K sequence developed any symptoms up to 48 days after infection. Nontransformed Xanthi nn plants and Xanthi nn plants transformed with pMON316 all developed characteristic mosaic symptoms by 5 days after inoculation (Fig. 4). To examine whether the apparent resistance manifested by the 54K transgenic plants was due to an inability of TMV to multiply in them or was simply due to a failure of the plants to develop symptoms, homogenates of the inoculated and upper leaves of the plants were applied to the local lesion host *N. tabacum* cv. Xanthi nc. No local lesions developed in response to any of the homogenates, indicating the absence of detectable virus in the 54K transgenic plants.

To investigate the heritability of the 54K-mediated resistance phenomenon, progeny seedlings from self-fertilized transgenic plants were scored for their resistance to TMV infection. In each case examined, the ratio of resistant to susceptible R1 plants was approximately 3:1. This was somewhat surprising in

Figure 4 Resistance of transformed plants to infection with TMV. The plant on the left was transformed with the vector pMON316 alone, and the plant on the right was transformed with pMON316 harboring the sequence of the TMV 54K protein. The plants are shown 15 days after they were inoculated with 100 μg/ml of TMV strain YSI/1 [mutant b6 of Garcia-Arenal et al. (22)]. Symptoms are visible only on the vector-transformed plant on the left. Reproduced from Ref. 9 with permission.

view of the fact that a number of the primary transformants clearly contained multiple copies of the 54K sequence. The segregation of resistance implies that all copies of the 54K sequence in plants with a multiple insertion are on the same chromosome.

The properties of the 54K-mediated resistance phenomenon were investigated using a segregating population of R1 seedlings derived from plant 541A11. These investigations demonstrated that resistance could still be observed when extremely high levels of TMV (up to 0.5 mg/ml) were applied to the plants. Furthermore, the resistance could not be overcome by using TMV RNA (rather than virions) as an inoculum. In these properties, the 54K-mediated resistance to TMV differs from that of the coat protein–induced resistance which has previously been reported (1). The ability of the TMV 54K sequence to protect transgenic plants against infection by other strains of TMV and other plant viruses has also been tested. Resistance could still be observed in plants inoculated with a mutant derived from U1 (Fig. 4) but not when plants were inoculated with the more distantly related TMV strains U2 and L.

Similarly, the 54K transgenic plants were not resistant to infection with cucumber mosaic virus (CMV).

VII. FURTHER CHARACTERIZATION OF THE 54K-INDUCED RESISTANCE

Since the discovery of the 54K-mediated resistance (9), a number of experiments have been carried out in an attempt to further characterize the phenomenon. It has recently been shown, for example, that the resistance operates at the level of single cells, since protoplasts isolated from 54K transgenic plants are also resistant to infection with the U1 strain of TMV (J. P. Carr and M. Zaitlin, unpublished observations). The results of the protoplast experiments also suggested, however, that the presence of the 54K transcripts or protein does not completely abolish the ability of TMV RNA to replicate in tobacco cells. Although no infectious virus or coat protein could be detected in 54K transgenic protoplasts that had been electroporated with TMV RNA, small quantities of both plus- and minus-sense viral RNA appeared to be made.

To eliminate the possibility that the 54K transgenic plants have been transformed with a copy of the 54K gene that is in some way aberrant, the complete sequence of the TMV-specific insert of pRTT-1 has recently been determined (G. P. Lomonossoff, unpublished observations). The analysis revealed that the insert contains eight nucleotide changes from the sequence originally published for the corresponding region of TMV RNA (10). The changes are shown in Table 1. All are transitions, and none affects the amino acid sequence of the encoded protein. Four of the changes were, in fact, picked as heterogeneities in the TMV RNA sequence by Goelet et al. (10). It therefore seems highly probable that the differences in sequence between the insert in pRTT-1 and the previously determined sequence of TMV RNA arise through natural variations in the sequence of the TMV RNA population. It therefore appears that the only difference between the putative 54K protein produced in the transgenic plants and that proposed to be produced during virus infection (13) is the presence of the extra five amino acids at the C terminus of the former.

VIII. HOW DOES 54K-MEDIATED RESISTANCE WORK?

The point in the TMV replication cycle that is inhibited in the 54K transgenic plants has been clarified somewhat by the discovery that protoplasts derived from 54K transgenic plants are also resistant to TMV infection. This indicates that the resistance mechanism involves an inhibition of virus replication in

Table 1 Nucleotide Sequence Changes in pRTT-1

Position[a]	Change	
	From[b]	To[c]
3791[d]	A	G
3941[d]	A	G
3962	G	A
3998[d]	A	G
4076[d]	U	C
4091	U	C
4499	C	U
4820	G	A

[a]Nucleotide positions are from the published sequence of TMV (10).
[b]From = nucleotide found in published sequence of TMV strain *vulgare* (10).
[c]To = nucleotide found in sequence of pRTT-1.
[d]Heterogeneities reported in Ref. 10.

individual cells rather than an interference with the ability to spread from cell to cell in the intact plant. Although it appears that individual 54K transgenic tobacco cells are not completely immune to TMV infection, replication in such cells is so poor that it prevents detectable virus infection of whole plants. However, the precise point (or points) in the TMV replication cycle inhibited by the presence of the 54K gene is at present unknown.

How the 54K gene interferes with viral replication in tobacco cells is obscure. At present it is not even known whether the resistance is mediated by the expression of the 54K protein or by the mRNA that potentially encodes it. Indeed, it is, in theory, even possible that the 54K-specific DNA sequence is directly responsible for the resistance phenomenon. Despite this lack of knowledge, it is possible to construct a number of reasonably plausible hypotheses to explain 54K-mediated resistance. For example, a separate 54K protein, which is a truncated version of the 183K protein, may not normally be synthesized during virus infection. If such a truncated version of the 183K protein is present in a cell, it could act as a competitive inhibitor of the 183K protein, interfering with its binding either to other components of the viral replication complex or to the viral RNA. If this proves to be the mechanism of 54K-mediated resistance, it implies that it may be necessary to make plants transgenic for a deleted form of a replication-associated protein to achieve resistance. Some support for this notion is provided by recent experiments in which tobacco plants transformed with complete copies of the replication-associated proteins (the P1 and P2 proteins) of alfalfa mosaic virus (AlMV) were found to be susceptible to infection by AlMV (23).

There are several ways in which the synthesis of 54K-specific transcripts in the cells of transgenic plants might give rise to resistance. For example, it is possible that the transcripts might inhibit viral replication by binding nonproductively to the replication complex. In this regard, it should be noted that the 54K-specific transcripts produced in the transgenic plants do not resemble natural viral subgenomic RNA in that they are not 3′ coterminal with the TMV genomic RNA. Alternatively, the 54K transcripts might hybridize to minus strands of TMV RNA and inhibit replication by the minus-strand capture mechanism originally put forward to explain cross protection (24). However, until the basic fact about the level—DNA, RNA, or protein—at which the 54K sequence induces resistance is known, all theories about its exact mode of operation are highly speculative.

IX. FUTURE PROSPECTS

To fully exploit the newly discovered 54K-mediated resistance phenomenon, it is necessary not only to gain further understanding of how the resistance operates but also to determine how widely applicable the results obtained with the U1 strain of TMV in tobacco are to other virus–plant combinations. The fact that the presence of the 54K gene from the U1 strain of TMV protects tobacco plants from infection with that strain and mutants derived from it but not from infection with more distantly related strains, such as U2 and L, suggests that the protection against virus infection afforded by a 54K-like gene is likely to be specific for the virus from which the gene was isolated. In this specificity, the 54K-mediated resistance is similar to that of coat protein–mediated protection.

As a first step towards determining the general applicability of 54K-mediated resistance, it will be of considerable interest to transform plants with the 54K gene from other TMV strains and with the equivalent protein from other members of the tobamovirus group. However, for substantial progress to be made, it will be necessary to demonstrate that 54K-mediated resistance can also be effective against plant viruses from other groups. As mentioned previously, the 54K protein (or readthrough region of the 183K protein) appears to have a vital role in viral RNA replication and, as such, must have counterparts in all positive-strand RNA viruses. Sequence comparisons have clearly identified proteins equivalent to the TMV 54K protein in many groups of plant viruses, including the tobraviruses, bromoviruses, cucumoviruses, and AlMV (15). Clearly, transformation of a variety of plant species with the sequences encoding such proteins is an urgent priority.

ACKNOWLEDGMENTS

I would like to thank Dr. J. Carr and Prof. M. Zaitlin for communicating their results before publication and Dr. J. Stanley for critically reading the manuscript. Much of the experimental detail that appears in this chapter regarding the production and characterization of the 54K transgenic tobacco plants were taken from Ref. 9.

REFERENCES

1. Powell-Abel, P. A., Nelson, R. S., De, B., Hoffman, N., Rogers, S. G., Fraley, R. T., and Beachy, R. N. Delay of disease development in transgenic plants that express the tobacco mosaic virus coat protein, *Science, 232*: 738, 1986.
2. Beachy, R. N., Loesch-Fries, S., and Tumer, N. E. Coat protein-mediated resistance against virus infection. *Annu. Rev. Phytopathol., 28*: 451, 1990.
3. Gerlach, W. L., Llewellyn, D., and Haseloff, J. Construction of a plant disease resistance gene from the satellite RNA of tobacco ringspot virus, *Nature, 328*: 802, 1987.
4. Harrison, B. D., Mayo, M. A., and Baulcombe, D. C. Virus resistance in transgenic plants that express cucumber mosaic virus satellite RNA, *Nature, 328*: 799, 1987.
5. Cuozzo, M., O'Connell, K. M., Kaniewski, W., Fang, R. X., Chua, N.-H., and Tumer, N. E. Viral protection in transgenic tobacco plants expressing the cucumber mosaic virus coat protein or its antisense RNA, *Biotechnology, 6*: 549, 1988.
6. Hemenway, C., Fang, R. X., Kaniewski, W., Chua, N.-H., and Tumer, N. E. Analysis of the mechanism of protection in transgenic plants expressing the potato virus X coat protein or its antisense RNA, *EMBO J., 7*: 1273, 1988.
7. Powell, P. A., Stark, D. M., Sanders, P. R., and Beachy, R. N. Protection against tobacco mosaic virus in transgenic plants that express tobacco mosaic virus antisense RNA, *Proc. Natl. Acad. Sci. USA, 86*: 6949, 1989.
8. Kawchuk, L. M., Martin, R. R., and McPherson, J. Sense and antisense RNA-mediated resistance to potato leafroll virus in Russet Burbank potato plants, *Mol. Plant–Microbe Interact., 4*: 247, 1991.
9. Golemboski, D. B., Lomonossoff, G. P., and Zaitlin, M. Plants transformed with a tobacco mosaic virus nonstructural gene are resistant to the virus, *Proc. Natl. Acad. Sci. USA, 87*: 6311, 1990.

10. Goelet, P., Lomonossoff, G. P., Butler, P. J. G., Akam, M. E., Gait, M. J., and Karn, J. Nucleotide sequence of tobacco mosaic virus RNA, *Proc. Natl. Acad. Sci. USA, 79*: 5818, 1982.
11. Pelham, H. R. B. Leaky UAG termination codon in tobacco mosaic virus RNA, *Nature, 272*: 469, 1978.
12. Beachy, R. N., and Zaitlin, M. Characterization and *in vitro* translation of the RNAs from less-than-full-length, virus-related nucleoprotein rods present in tobacco mosaic virus preparations, *Virology, 81*: 160, 1977.
13. Sulzinski, M. A., Gabard, K., Palukaitis, P., and Zaitlin, M. Replication of tobacco mosaic virus, VIII: characterization of a third subgenomic TMV RNA, *Virology, 145*: 132, 1985.
14. Kamer, G., and Argos, P. Primary structural comparison of RNA-dependent polymerases from plant, animal and bacterial viruses, *Nucleic Acids Res., 12*: 7269, 1984.
15. Goldbach, R., Le Gall, O., and Wellink, J. Alpha-like viruses in plants, *Semin. Virol., 2*: 19, 1991.
16. Young, N., Forney, J., and Zaitlin, M. Tobacco mosaic virus replicase and replicative structures, *J. Cell Sci. Suppl., 7*: 277, 1987.
17. Rogers, S. G., Klee, H. J., Horsch, R. B., and Fraley, R. T. Improved vectors for plant transformation: expression cassette vectors and new selectable markers, *Methods Enzymol., 153*: 253, 1987.
18. Rogers, S. G., Horsch, R. B., and Fraley, R. T. Gene transfer in plants: production of transformed plants using Ti plasmid vectors, *Methods Enzymol., 118*: 627, 1986.
19. Horsch, R. B., Fry, J. E., Hoffman, N. L., Eichholtz, D., Rogers, S. G., and Fraley, R. T. A simple and general method for transferring genes into plants, *Science, 227*: 1229, 1985.
20. Saito, T., Watanabe, Y., Meshi, T., and Okada, Y. Preparation of antibodies that react with the large non-structural proteins of tobacco mosaic virus by using *Escherichia coli* expressed fragments, *Mol. Gen. Genet., 205*: 82, 1986.
21. Lütcke, H. A., Chow, K. C., Mickel, F. S., Moss, K. A., Kern, H. F., and Scheele, G. A. Selection of AUG codons differs in plants and animals, *EMBO J., 6*: 43, 1987.
22. Garcia-Arenal, F., Palukaitis, P., and Zaitlin, M. Strains and mutants of tobacco mosaic virus are both found in virus derived from single-lesion-passaged inoculum, *Virology, 132*: 131, 1984.
23. Taschner, P. E. M., van der Kuyl, A. C., Neeleman, L., and Bol, J. F. Replication of an incomplete alfalfa mosaic virus genome in plants transformed with viral replicase genes, *Virology, 181*: 445, 1991.
24. Palukaitis, P., and Zaitlin, M. A model to explain the 'cross-protection' phenomenon shown by plant viruses and viroids, *Plant–Microbe Interactions: Molecular and Genetic Perspectives, Vol. 1* (T. Kosuge and E. W. Nester, eds.), Macmillan, New York, 1984, p. 420.

6

Virus Resistance in Transgenic Plants: Coat Protein-Mediated Resistance

Ann P. Sturtevant and Roger N. Beachy

The Scripps Research Institute, La Jolla, California

I. INTRODUCTION

Transgenic plants that express a viral coat protein gene are often resistant to infection by the virus from which the coat protein sequence was derived. This phenomenon is referred to as coat protein–mediated resistance or genetically engineered cross protection. Viral coat proteins (CPs) from a number of different viruses have been cloned into plant shuttle vectors, usually under the control of a strong constitutive plant promoter, such as the cauliflower mosaic virus (CaMV) 35S promoter, and used to transform plants. Expression of the viral CP confers resistance to viral infection, which is demonstrated by a delay or absence of systemic disease symptoms.

Before the application of genetic engineering to provide protection against virus infections, and in the absence of other genes to protect plants, classical cross protection was sometimes used to provide increased resistance to severe viral infections. In classical cross protection, plants are infected with a mild strain of virus to provide protection against subsequent infection by a more severe, related strain of the virus. Although these plants exhibit the mild disease symptoms characteristic of the "protecting" strain of virus, they do not develop the severe disease symptoms of the challenge virus that result in reduced crop yield.

Although classical cross protection can be beneficial in cases where no other effective means of disease control exists, it has many disadvantages. Unfortunately, not all agriculturally important viruses have a mild strain that can be used as the protecting virus; therefore, cross protection cannot be used in most cases in which protection is required. Furthermore, whereas infecting crops with a protecting strain of virus prevents severe viral infection, it may still produce mild disease symptoms that can suppress crop yield. Also, it is not always possible to predict whether one virus will protect against subsequent infection by another virus. In some cases the two viruses act synergistically to produce a much more severe infection than either virus would cause on its own. Furthermore, even in instances in which classical cross protection can be useful in disease control, the possibility of infection of neighboring fields with the protecting strain of virus must be considered. Because of this, farmers are often reluctant to use classical cross protection to control virus disease.

Coat protein–mediated resistance is the expression of a viral CP gene in transgenic plants to develop resistance to the virus from which the gene was derived. In most cases the symptoms that develop in plants expressing the CP genes [CP(+) plants] are less severe than symptoms in control plants [CP(-) plants]. In many cases of CP-mediated resistance, less virus accumulates in infected leaves of CP(+) plants than in those of CP(-) plants. In plant genotypes that exhibit a localized response, CP(+) plants have fewer local lesions after infection than do CP(-) plants. Furthermore, the presence of viral CP in transgenic plants does not produce disease symptoms.

Coat protein–mediated resistance was first demonstrated in transgenic tobacco plants that express the CP gene of tobacco mosaic virus (TMV) (1). Plant lines that express the CP gene are resistant to infection by TMV and also to certain related tobamoviruses. Furthermore, transgenic tomato plants that express the TMV CP gene are resistant to virus infection under field conditions (2). Since these first demonstrations of CP-mediated resistance, similar transgenic approaches have been successfully used to produce transgenic plants resistant to many viruses, including alfalfa mosaic virus, several members of the potyvirus group, tobravirus, ilarvirus, cucumovirus, potexvirus, carlavirus, luteovirus, and tospovirus groups.

II. COAT PROTEIN–MEDIATED RESISTANCE TO TMV INFECTION

Tobacco mosaic virus, the type member of the tobamovirus group, is a rod-shaped virus possessing a (+)-stranded (sense) genomic RNA helically encapsidated by coat protein. On entering the plant cell, capsid proteins are removed from the virion, starting at the 5′ end of the virus. This process is called disassembly or uncoating. After coat protein is removed from the first few helical turns of the virus, the ribosome binding site of TMV RNA is exposed. Translation of RNA into viral proteins displaces additional capsid proteins until the entire RNA is uncoated and available for transcription, translation, and replication. For further information on TMV infection and replication see the reviews by Dawson et al. (3), and Matthews (4).

Coat protein–mediated resistance, first described by Powell Abel et al. (1), resulted when the CP gene from the U1 strain of TMV was cloned into a plant shuttle vector under the control of the CaMV 35S promoter and nopaline synthase terminator. The CP gene was introduced into a tobacco cultivar (*Nicotiana tabacum* cv. Xanthi) that is a systemic host for TMV infection. A high level of coat protein expression was detected in some of the transgenic tobacco lines, equal to as much as 0.1% of total extractable leaf protein. Leaves from the top, middle, and bottom of transgenic plants showed the same level of CP expression.

When transgenic plants were inoculated with TMV, CP(+) plants showed later development of disease symptoms than did CP(-) plants, and many plants escaped infection or did not develop symptoms. An absence of disease symptoms indicated a lack of viral replication (5). When CP(+) plants were inoculated with the U1 strain of TMV, the inoculated leaves contained less than 30% of the virus found in the inoculated leaves of control plants (6). As in classical cross protection, increasing the concentration of the inoculum decreased the length of time between infection and symptom development. When CP(+) plants were infected with high concentrations of TMV, a greater percentage of plants developed disease symptoms and the lag time between inoculation and symptom development was decreased. To explain the resistance, Powell Abel et al. theorized that there were fewer sites susceptible to TMV infection in CP(+) plants than in CP(-) plants (1). They also reported that the degree of resistance exhibited by CP(+) plants is directly related to the level of CP expression (7).

Coat protein–mediated resistance conferred by the CP of the common (U1) strain of TMV is also active against a severe strain of TMV, PV230 (6). Infection by PV230 is characterized by the appearance of localized chlorotic lesions before the development of systemic symptoms. The number of chlorotic lesions on plants expressing the TMV CP was only 10–30% the number of lesions on control plants. There was a strong correlation between the number of chlorotic lesions and the rapidity of systemic symptom development. The CP(+) plants that did not develop chlorotic lesions (71%) did not develop systemic symptoms or accumulate virus. The delay in the development of systemic symptoms can be accounted for, in part, by the 70% reduction in lesion development and virus accumulation. There may also be less movement of virus from the infected leaf to upper leaves, because virus accumulation in the first systemic leaf of CP(+) plants infected with U1 TMV is inhibited to an even greater extent than is virus accumulation in the inoculated leaf (6).

The effect of intracellular levels of CP on the development of necrotic local lesions was studied by introducing the TMV CP gene into a cultivar of tobacco that is a local lesion host for TMV (*N. tabacum* cv. Xanthi NN) (6). When CP(+) plants were inoculated with 1 μg/ml TMV virus, the number of necrotic local lesions that developed was less than or equal to 5% of the number that developed on control plants (6). However, in contrast to plants infected by virions, CP(+) plants showed little resistance to infection with TMV RNA (6). The ability of TMV RNA to overcome resistance in CP(+) plants has important implications for the mechanism of CP-mediated resistance, as will be shown in the discussion that follows.

A. Protection Against Infection by Other Tobamoviruses

Because tobacco plants expressing the CP of the U1 strain of TMV were resistant to infection by the PV230 strain of TMV (6), CP(+) tobacco and tomato plants were challenged with viruses other than TMV to determine the breadth of protection conferred by expression of the TMV CP. The CP levels in transgenic tomato lines were approximately 0.05% of total leaf protein (2), comparable to CP levels found in the transgenic tobacco lines. Transgenic CP(+) tomato plants were infected with TMV U1, TMV PV230, and tomato mosaic virus strain L. (ToMV). The CP(+) tomato plants were resistant to high concentrations (20.0 μg/ml) of TMV U1. Only 9% of the CP(+) plants exhibited disease symptoms 30 days after inoculation, compared with 100% of the control plants. Coat protein–mediated resistance was less effective against the PV230 strain of TMV and ToMV, with 54% and 62% of the CP(+) tomato plants showing disease symptoms 29 days after inoculation when infected with 20.0 μg/ml PV230 and ToMV, respectively. However, the CP(+) tomato plants developed symptoms significantly later than did the control plants; 100% of the CP(-) tomato plants developed disease symptoms 9–12 days after inoculation when infected with PV230 or ToMV.

Transgenic tomato plants tested under field conditions were resistant to infection by TMV (U1 strain) (2). Even when plants were inoculated with high levels of TMV (40 μg/ml), less than 10% of the CP(+) tomato plants showed disease symptoms 61 days after inoculation, compared with 100% of the control plants. Fruit yields were decreased by 26–35% in virus-infected control plants as compared with uninfected control plants. The fruit yield of uninoculated CP(+) plants was the same as that of uninfected control plants and did not decrease when the CP(+) tomato plants were inoculated with TMV. These results show that CP-mediated resistance effectively controls viral disease under field conditions (2).

Nejidat et al. challenged transgenic CP(+) tobacco plants with five other tobamoviruses to determine if expression of the CP gene from TMV, the type member of the tobamovirus family, conferred protection against these viruses (8). The viruses varied in their degree of relatedness to TMV, as determined by percent amino acid homology to the TMV CP, and included tomato mosaic virus (ToMV, 82%), tobacco mild green mosaic virus (TMGMV, 72%), pepper mild mosaic virus (PMMV), odontoglossom ringspot virus (ORSV, 62%), and ribgrass mosaic virus (RMV, 45%) (8). The viruses tested differed in their ability to infect systemic and local lesion varieties of tobacco.

Transgenic CP(+) systemic host tobacco plants inoculated with high concentrations (up to 100 μg/ml) of either ToMV or TMGMV remained free of disease symptoms or demonstrated a significant delay in symptom development as compared with control plants (8). This protection was overcome by inoculation with viral RNA, as was previously described for infection by TMV RNA (6). Coat protein–mediated resistance was not effective against infection with RMV. Thus, expression of the TMV CP gene in transgenic tobacco plants provides a significant degree of protection against infection by closely related viruses, such as ToMV and TMGMV, but little if any resistance to RMV, a more distantly related tobamovirus (8).

These experiments were repeated with transgenic CP(+) local lesion tobacco host plants inoculated with PMMV and ORSV in addition to the viruses just described, and the results were similar (8). Infection of CP(+) plants with the viruses most closely related to TMV (ToMV, TMGMV, PMMV, and ORSV) produced <20% the number of necrotic lesions observed on control plants. The CP(+) plants were not resistant to infection by RMV, which is the least related to TMV (45% amino acid homology) (8). There was little or no resistance against another distantly related tobamovirus, sunn hemp mosaic virus (Cc TMV), or against the unrelated viruses potato virus X, potato virus Y, cucumber mosaic virus, and alfalfa mosaic virus (9).

B. Resistance in Protoplasts

Studies of TMV infection in protoplasts provide further insight into the mechanisms of CP–mediated resistance (10). Protoplasts from leaves of CP(+) and CP(-) plants were inoculated with TMV by electroporation; increase in CP accumulation was used to determine infection. Only 5–9% of the CP(+) protoplasts became infected, compared with 40–55% of the CP(-) protoplasts. In the CP(+) protoplasts that became infected, virus accumulated to approximately the same level as in control protoplasts.

To determine the events in the TMV infection cycle that were inhibited, Register et al. monitored levels of CP and TMV (+)-strand RNA after infection with either TMV or TMV RNA (10). When TMV was used as the inoculum, CP(+) protoplasts accumulated virtually no CP or viral RNA, whereas CP(-) protoplasts accumulated both. Production of TMV (-)-strand RNA was also inhibited in CP(+) protoplasts. When protoplasts were inoculated with TMV RNA, both CP(+) and CP(-) protoplasts became infected and accumulated similar levels of CP and (+)-strand RNA. These results show that the step in the TMV infection cycle that is inhibited in CP(+) protoplasts occurs at or before RNA replication (10).

Since CP(+) protoplasts were resistant to infection with TMV but susceptible to infection with TMV RNA, the effectiveness of CP-mediated protection against partially uncoated TMV was investigated (10). Wilson showed that briefly treating TMV at pH 8.0 increased its translatability in vitro, probably by removing a few CP subunits from the 5′ end of the virion (11). Although the number of lesions produced by infection with partially uncoated TMV (pH 8.0) was generally less than the number of lesions following inoculation with TMV RNA, the number was 4 to 15 times as high as the number of lesions produced by infection with TMV virus under standard inoculation conditions (i.e., pH 7.5). These results indicate that CP-mediated protection occurs at or before uncoating of viral RNA (10).

C. Proposed Mechanisms

Coat protein–mediated resistance has several characteristics in common with classical cross protection. The presence of CP, or a "protecting" strain of virus, confers resistance to infection by closely related viruses but not to more distantly related or unrelated viruses. The protection provided by either CP-mediated resistance or cross protection can be overcome by high concentrations of virus. Both methods of protection are active against inoculated virus but are much less effective when viral RNA is used as the infectious agent. These results suggest that CP-mediated protection and classical cross protection may have similar mechanisms, at least one of which acts at or before uncoating of viral RNA. In TMV CP-mediated resistance the degree of resistance is related to the level of CP expression. Two models for the mechanism of CP-mediated protection have been proposed, re-encapsidation and prevention of viral entry and/or viral uncoating.

In the re-encapsidation theory, endogenous CP subunits prevent viral uncoating by binding to viral RNA as soon as viral CP subunits are removed. Viral disassembly will be absent or delayed if the levels of CP in the cell are above a minimum level. This theory is consistent with the observation that plants expressing higher levels of CP are more resistant to viral infection than are plants that express lower levels. The specificity of CP-mediated protection is also explained; if the CP cannot bind to the RNA of the infecting virus, no resistance is possible. Re-encapsidation was proposed by de Zoeten et al. (12) to explain classical cross protection, but it is equally applicable to CP-mediated resistance.

In the second theory, endogenous CP binds to putative cellular receptors for viral entry or disassembly. The prior binding of CP to the cellular receptor prevents the incoming virus from entering the cell or forming disassembly complexes. This explains how, in some cases, relatively low levels of CP can provide viral protection in transgenic plants. In this case the specificity of CP-mediated protection depends on the specificity of the viral receptor.

III. COAT PROTEIN–MEDIATED RESISTANCE TO OTHER VIRUSES

Genes encoding CPs of a number of different viruses have been expressed in transgenic plants, and CP-mediated resistance has been confirmed as a common means for providing viral resistance. In the 6 years since genetically engineered resistance was described in plants expressing the CP of TMV. CP-mediated resistance has been achieved in transgenic plants expressing the CP of a number of different viruses of various virus groups. The organization of the viral genome and mode of replication and infection of these viruses are, in some cases, very different from those of TMV. Resistance conferred by CP gene expression has many similarities to TMV CP-mediated protection, but in some cases there are important differences. Whether the mechanism of CP-mediated resistance is the same for each virus is currently unknown.

A. Alfalfa Mosaic Virus

The genome organization of alfalfa mosaic virus (AlMV) is very different from that of TMV. Alfalfa mosaic virus is tripartite, and each of the three genomic RNA molecules is separately encapsidated. RNAs 1 and 2 encode proteins involved in viral replication, and RNA 3 codes for a putative viral movement protein. A subgenomic mRNA, RNA 4, is transcribed from RNA 3 and encodes the CP. Viral infection by AlMV RNA requires RNAs 1 and 2 as well as RNA 4 or a few subunits of CP. Despite these differences from other viruses, expression of the AlMV CP in transgenic plants confers resistance to infection by AlMV.

Tumer et al. (13,14), Loesch-Fries et al. (15), and Van Dun et al. (16) developed transgenic plants that contained a gene encoding the AlMV CP and demonstrated that CP(+) plants were protected against viral infection by AlMV. Transgenic tobacco and tomato plants expressed high levels (0.1–0.8% extractable leaf protein) of the AlMV CP (13). The CP(+) tobacco plants exhibited an absence or delay of symptom development when inoculated with AlMV, and increasing the concentration of the infecting virus decreased the degree of AlMV CP-mediated protection. The CP(+) tobacco plants produced fewer chlorotic lesions and accumulated less virus in the inoculated and upper leaves than did CP(-) plants. The CP(+) tomato plants showed fewer necrotic local lesions when infected with AlMV and likewise accumulated lower levels of virus in inoculated and systemically infected leaves than did control plants. The CP(+) plants that expressed higher levels of CP showed a greater delay in disease symptom development and greater viral resistance than did plants with lower levels of CP. This relationship between level of CP expression and degree of disease resistance was also observed in transgenic plants expressing the TMV CP and suggests that there is a minimum level of expression necessary to confer CP-mediated protection in transgenic plants (13).

Van Dun et al. transformed plants with a gene encoding the AlMV CP and with a similar gene that contained a frame-shift mutation (16). Although plants expressing the AlMV CP gene were resistant to infection by AlMV, those expressing the frame-shift mutation were not protected, indicating that CP protein and not CP mRNA is responsible for resistance (16). The CP(+) plants were not resistant to infection by AlMV RNAs or to infection by TMV (15,17). The CP(+) plants were not protected against infection by a more closely related virus, tobacco streak virus (16).

Tumer et al. (14) transformed tobacco plants with either a gene encoding the wild-type AlMV CP or a gene encoding the AlMV CP with a change in the second amino acid (Ser to Gly). Plants accumulating the wild-type AlMV CP were resistant to infection by AlMV, whereas plants accumulating the mutant CP were not protected. The mutant CP binds AlMV RNA and participates in AlMV replication; therefore, the lack of protection in plants expressing the mutant CP is not due to loss of activity. In contrast to results from other groups (16), transgenic plants expressing the wild-type AlMV CP were resistant to infection by AlMV RNA, possibly because of higher levels of CP expression (14). Although CP-mediated protection in AlMV CP(+) plants may act before the virus is uncoated, protection against AlMV RNA implies an additional level of resistance that acts after virus disassembly, possibly by CP regulation of translation, replication, or both (14).

B. Tobacco Streak Virus

Tobacco streak virus (TSV), an ilarvirus, is structurally related to AlMV and has a similar genome organization. Like AlMV, TSV has three (+)-stranded genomic RNAs that are separately encapsidated; the coat protein is encoded by a subgenomic mRNA of RNA 3. Also like AlMV, TSV requires the presence of CP for infection. Although there is no amino acid similarity between the AlMV and TSV CP, the AlMV CP may substitute for the TSV CP in viral replication. Because of these similarities, Van Dun et al. investigated CP-mediated cross protection against AlMV and TSV (16).

Transgenic plants expressing the TSV CP gene were challenged with TSV and AlMV virus (16). The TSV CP(+) plants were resistant to infection by TSV virions but not to TSV RNA. Although the TSV and AlMV CPs are interchangeable in their role in viral replication, CP(+) plants were not protected against infection by AlMV. These results suggest that CP-mediated resistance to TSV acts before viral replication (16).

C. Tobacco Rattle Virus

Tobacco rattle virus (TRV), a tobravirus, has a bipartite genome consisting of two (+)-stranded RNAs; RNA 1 can replicate and move throughout the plant independent of RNA 2, which encodes the CP. The fact that TRV is not dependent on the presence of CP for replication or for systemic movement led Van Dun et al. to investigate the role of the TRV CP in CP-mediated resistance (18).

Transgenic plants expressing the CP of TRV strain TCM were produced and assayed for viral resistance (17,18). The CP(+) plants were resistant to infection by the TCM strain of TRV but not to infection by a distantly related strain, PLB. The CP of strain PLB can, however, encapsidate RNA from strain TCM;

presumably, the CP of strain TCM may also encapsidate RNA from strain PLB. This suggests that the mechanism behind TRV CP-mediated protection is not re-encapsidation but may involve a cellular receptor (18). The CP(+) plants were also protected against infection by a closely related virus, pea early browning virus (PEBV) (18).

D. Beet Necrotic Yellow Vein Virus

Beet necrotic yellow vein virus (BNYVV) is a member of the furovirus group and possesses four separately encapsidated genomic RNAs (4). Although little is known about its replication, BNYVV causes rhizomania, a severe disease in sugar beets. The CP(+) sugar beet protoplasts derived from transgenic suspension cultures demonstrated resistance to BNYVV inoculated by electroporation (19). It is not yet known whether BNYVV CP-mediated protection is active in transformed plants, although such studies are in progress (19).

E. Cucumber Mosaic Virus

Cucumber mosaic virus (CMV), a member of the cucumovirus group, is perhaps the single most important plant virus, with a wide host range and a large number of identified strains. Like AlMV, CMV has a multipartite genome, but it differs in that its particles are icosahedral in structure. Unlike AlMV, CMV does not require the presence of either RNA 4 or CP for viral replication. Cucumber mosaic virus is divided into two subgroups, group I (strains C, D, and Fny) and group II (strains Q and WL). Within subgroups the amino acid homology of the CPs is approximately 95%, whereas between subgroups the homology is approximately 80% (20). Although CMV can be mechanically inoculated, it is normally transmitted by aphids.

Transgenic plants expressing the CPs of various strains of CMV have been tested for resistance to infection by viruses of both subgroups (20–22). Expression of the CP of strain D in transgenic tobacco plants conferred a high degree of protection against mechanical infection by strain C (21). Transgenic plants expressing the CP gene of strain C, which is a group I virus, were protected against infection by strains C and Chi and by strain WL, which is a group II virus (20). Some CP(+) lines were more resistant to infection by a virus of the heterologous group (strain WL) than to infection with group I viruses (20). The CP(+) plants were also resistant to infection caused by aphid transmission of the virus (20). Namba et al. developed transgenic plants that express the CP of a group II virus, WL, and reported that they were protected almost equally against infection by CMV strains WL, Chi, and C (22). Furthermore, the transgenic plant lines expressing the highest levels of WL CP were not the lines that demonstrated the highest degree of virus resistance (22).

F. Potato Virus X

Potato virus X (PVX), like TMV, is a rod-shaped virus containing a single (+)-sense, ssRNA genome. Hemenway et al. expressed the PVX CP, or its antisense RNA, in transgenic tobacco plants (23). The CP(+) plants were protected against infection by PVX, and the level of protection was correlated with the level of CP accumulation, as has been previously observed with plants expressing the TMV CP. Plants expressing CP antisense RNA were resistant to only low concentrations of PVX. Unlike certain other types of CP-mediated resistance using CPs of other viruses, PVX CP(+) plants were also resistant to infection by PVX RNA (23). This suggests that the PVX CP inhibits some early step in the infection process other than uncoating of viral RNA (23).

Coat protein–mediated resistance to PVX has been applied in several commercially important potato cultivars. Potato infection by PVX generally causes mild symptoms. However, mixed infection with PVX and potato virus Y (PVY), a potyvirus, results in a synergistic increase in the severity of disease symptoms and can severely depress crop yield. The severity of symptoms is directly related to the level of PVX replication. Hoekema et al. successfully transformed two potato cultivars, Bintje and Escort, with the PVX CP gene (24). The CP(+) potato plants of both cultivars were protected against infection by PVX and showed a direct correlation between CP levels and the degree of resistance (24). Lawson et al. produced lines of the potato cultivar Russet Burbank that expressed the CP genes of both PVX and PVY (25). Plants that expressed both CP genes were protected against infection by both viruses. However, the degree of resistance to PVY infection was not related to the level of PVY CP (25,26). The plant line that exhibited the greatest degree of viral resistance contained the lowest levels of both the PVX and PVY CP (25). Under field conditions, transgenic potato plants of this line were protected against dual infections and did not show a reduction in yield, as did infected CP(-) plants (26).

G. Potyviruses

The potyviruses are a large and agriculturally important group of viruses. Potyviruses are rod shaped and possess a single (+)-stranded RNA genome that contains a covalently attached protein at the 5′ end and a polyadenylated 3′ end. The RNA encodes a polyprotein that is posttranslationally processed to yield a number of proteins. Because of the large number of potyviruses that infect crops, it may be desirable to achieve broad-spectrum resistance to potyviral infection through the expression of a single CP gene.

Stark et al. introduced the CP of soybean mosaic virus (SMV) into tobacco, a nonhost for SMV (27). The CP(+) plants were challenged with the potyviruses PVY and tobacco etch virus (TEV). Although the CPs of PVY and TEV have limited amino acid similarity to the SMV CP (61% and 58%, respectively), expression of the SMV CP in tobacco plants conferred a considerable degree of protection against infection by both PVY and TEV. The CP(+) plants demonstrated a significant delay in symptom development and a lower level of virus accumulation than did the control plants. Furthermore, the symptoms that eventually developed on CP(+) plants were much less severe than those that developed on control plants (27). The SMV CP(+) plants were not resistant to infection by TMV (27).

The expression of the papaya ringspot virus (PRV) CP in transgenic plants also conferred resistance to a variety of potyviruses (28). Ling et al. introduced the CP of PRV into tobacco, a nonhost for PRV infection, and found protection against infection by the potyviruses TEV, pepper mosaic virus (PeMV), and PVY but no resistance to the unrelated CMV (28). The CP(+) plants either escaped infection or demonstrated a delay in symptom development and a decrease in symptom severity when infected with TEV, PeMV, or PVY.

Regner et al. expressed the CP gene of plum pox virus (PPV) in *N. clevelandii* and *N. benthamiana* (29). The CP(+) plants demonstrated a lack of symptoms or delay in symptom development when infected with PPV as compared with control plants. In PPV CP(+) plants, as in all other instances of potyviral CP-mediated viral protection studied to date, the most resistant transgenic plant lines were not the lines that expressed the highest level of CP.

H. Potato Virus S

Potato virus S (PVS), a carlavirus, contains a single genomic RNA encapsidated to form a rod-shaped particle. Transgenic *N. debneyii* plants expressing the PVS CP were resistant to infection by PVS virus as well as to infection by PVS RNA (30). The CP(+) plants did not develop disease symptoms nor did they accumulate virus (30). MacKenzie et al. introduced the PVS CP gene into the commercial potato cultivar Russet Burbank and demonstrated resistance to infection by PVS and to the related virus potato virus M (PVM) (31). Like transgenic plants expressing the PVX CP, transgenic potato plants expressing the PVS CP were protected against infection by PVS RNA (31). The CP(+)

plants were not resistant to infection by the heterologous PVX, despite the similarities between PVS CP-mediated protection and that conferred by the PVX CP (31).

I. Potato Leafroll Virus

Potato leafroll virus (PLRV), a member of the luteovirus group, is a phloem-limited virus possessing a single (+)-sense RNA genome and is transmitted only by aphids. Kawchuk et al. introduced a chimeric PLRV CP gene into two cultivars of potato, Desiree and Russet Burbank (32,33). The PLRV CP cDNA sequence was also introduced in antisense orientation in the cultivar Russet Burbank (33). Although CP(+) potato plants expressed high levels of CP mRNA, only low levels of CP accumulated. Transgenic CP(+) potato plants were resistant to PLRV infection when challenged by aphids despite the low endogenous level of CP, and they accumulated lower levels of virus than did control plants (32). In contrast to the results of other studies using antisense genes, potato plants expressing an antisense gene to the PLRV CP were as resistant to infection by PLRV as plants that expressed the (+)-sense gene (33).

J. Tomato Spotted Wilt Virus

Tomato spotted wilt virus (TSWV), a member of the tospovirus genus of the Bunyaviridae, is an enveloped virus with three single-stranded genomic RNAs. The RNAs M and L are of antisense polarity, whereas RNA S is ambisense, encoding the nucleoprotein (NP) in antisense orientation and a nonstructural protein in the sense orientation. Tomato spotted wilt virus has an extremely wide host range and causes significant yield loss in a number of crops because it is transmitted by thrips, which are resistant to all or most chemical pesticides. Gielen et al. (34) and MacKenzie et al. (35) expressed the TSWV NP gene in tobacco plants. The NP(+) plants were resistant to mechanical infection with TSWV, demonstrating a lack or delay of symptom development. The degree of resistance in NP(+) plants showed no correlation with the level of NP accumulation. Gielen et al. proposed that TSWV NP-mediated resistance may be due to the putative role of the NP in regulating the shift of the viral polymerase from transcription to replication mode (34). According to this hypothesis, high endogenous levels of NP could result in a premature switch to the replication mode, disrupting the normal infection cycle.

Table 1 Examples of Coat Protein-Mediated Protection

Viral CP	Group	Challenge virus	Group	Protection	Other characteristics	References
TMV	Tobamovirus	TMV U1	Tobamovirus	Yes	Not resistant to infection by RNA; degree of protection directly correlated with CP level	1,2,6,8,10
		TMV P230	Tobamovirus	Yes		
		ToMV	Tobamovirus	Yes		
		TMGMV	Tobamovirus	Yes		
		PPMV	Tobamovirus	Yes		
		ORSV	Tobamovirus	Yes		
		RMV	Tobamovirus	No		
		SHMV	Tobamovirus	No		
		AlMV	Alfalfa mosaic	No		
AlMV	Alfalfa mosaic	AlMV	Alfalfa mosaic	Yes	CP protection active against RNA in some plant lines	13–17
		TRV	Tobravirus	No		
		TSV	Ilarvirus	No		
		TMV	Tobamovirus	No		
TSV	Ilarvirus	TSV	Ilarvirus	Yes	Same as TMV	16
		AlMV	Alfalfa mosaic	No		
		TMV	Tobamovirus	No		
TRV-TCM	Tobravirus	TRV TCM	Tobravirus	Yes	Same as TMV	18
		TRV PLB	Tobravirus	No		
		PEBV	Tobravirus	Yes		
BNYVV	Furovirus	BNYVV	Furovirus	Yes	Protection observed in protoplasts	19

CMV	Cucumovirus	CMV	Cucumovirus	Yes	Degree of protection not correlated with CP level	20–22
PVX	Potexvirus	PVX	Potexvirus	Yes	CP protection active against RNA	23,24
PVX+PVY	Potexvirus/ Potyvirus	PVX+PVY	Potexvirus/ Potyvirus	Yes		25,26
SMV	Potyvirus	PVY	Potyvirus	Yes	Degree of protection not correlated with CP level	27
		TEV	Potyvirus	Yes		
		TMV	Tobamovirus	No		
PRV	Potyvirus	TEV	Potyvirus	Yes	Degree of protection not correlated with CP level	28
		PeMV	Potyvirus	Yes		
		PVY	Potyvirus	Yes		
		CMV C	Cucumovirus	No		
PPV	Potyvirus	PPV	Potyvirus	Yes	Degree of protection not correlated with CP level	29
PVS	Carlavirus	PVS	Carlavirus	Yes	CP protection active against RNA	30,31
		PVM	Carlavirus	Yes		
		PVX	Potexvirus	No		
PLRV	Luteovirus	PLRV	Luteovirus	Yes		32,33
TSWV	Tospovirus	TSWV	Tospovirus	Yes	Degree of protection not correlated with CP level	34,35

IV. SUMMARY

Virus resistance in transgenic plants expressing the CPs of the viruses AlMV, TSV, TRV, and PLRV shares many of the characteristics of TMV CP-mediated resistance, even though the genome organization and mode of replication of these viruses are different from those of TMV. Plants that are TMV-CP(+) are not resistant to infection by viral RNA, and the degree of protection in TMV-CP(+) plants is directly correlated with the level of CP expression. The CP(+) plants are resistant to infection by closely related members of the homologous virus group but not to more distantly related members or unrelated viruses. In general, protection conferred by expression of CP antisense RNA is not as effective as CP-mediated resistance. These similarities suggest that CP-mediated protection may function by a common mechanism, despite differences in viral replication and infection. However, not all CP-mediated resistance follows the TMV model (see Table 1). Transgenic plants expressing the CPs of members of the cucumovirus, potyvirus, and tospovirus groups do not show a correlation between the level of viral resistance and level of CP accumulation. The CP-mediated resistance in transgenic plants expressing the CP of PVX (a potexvirus) or PVS (a carlavirus) is effective against infection by viral RNA as well as by virions. Protection against PLRV is equally effective in transgenic plants that express the PLRV CP and those that express PLRV antisense RNA. Whether the mechanism behind CP-mediated protection in these plants is the same as in plants expressing the TMV CP has yet to be determined.

Coat protein–mediated resistance has proven to be effective under field conditions, as well as in greenhouses. Transgenic tomato plants expressing the TMV CP were resistant to infection by TMV and ToMV in the field (2). The fruit yield of CP(+) tomato plants did not decrease when plants were inoculated with virus and was equivalent to that of uninoculated control tomato plants; in virus-infected control plants the fruit yield decreased by as much as 35% (2). In the field, transgenic potato plants that express the CP of PVX, or both PVX and PVY, were resistant to infection by PVX or by both viruses (24,26).

The resistance in the field of transgenic CMV CP(+) cucumber plants was compared with the resistance of the cultivar Marketmore, a resistant cultivar of cucumber developed through plant breeding (36,37). Field trials were performed in an isolated field planted with cucumber plants of the susceptible cultivar Poinsett, transgenic Poinsett CP(+) lines 47 and 60, and the CMV-resistant cultivar Marketmore. Poinsett cucumber plants infected with CMV were transplanted to the field and served as a source of virus; CMV spread relied on aphid transmission. In the 1989 trial 70% of the susceptible Poinsett cucumber plants became infected, compared with 10% of the Marketmore plants and 23% and 24% of transgenic R_1 CP(+) lines 47 and 60, respectively (36). Because the R_1 CP(+) cucumbers are heterozygous, one would expect these rates of CMV infection. The 1990 field trial was performed similarly but with R_2 seeds of the transgenic CP(+) lines. Although protection against CMV infection was not as effective in the transgenic CP(+) cucumber plants as in Marketmore (8% versus 1% infected plants, respectively), there was a significant degree of protection as compared with the susceptible Poinsett control plants (98% infected plants) (37).

Genetic engineering shows great promise for protection of crops against virus infections. Besides the approach of CP-mediated resistance, additional strategies to obtain increased protection may be investigated. For example, viral CP genes may be introduced into resistant crop cultivars to provide a greater degree of protection against viral infection. The expression of CP genes from tissue-specific promoters may confer greater resistance in transgenic plants to infection by certain viruses than does expression from nominally constitutive promoters. Plants may be transformed with a chimeric gene encoding a polyprotein consisting of multiple CP genes from unrelated viruses to confer broad-spectrum resistance to viral disease. Also, a combination of different strategies may be employed in the same transgenic plant, such as expression of both CP genes and antisense RNAs or other viral sequences. The performance of these plants under large-scale field conditions will, in the end, determine their commercial value.

REFERENCES

1. Powell Abel, P., Nelson, R. S., De, B., Hoffmann, N., Rogers, S. G., Fraley, R. T., and Beachy, R. N. Delay of disease development in transgenic plants that express the tobacco mosaic virus coat protein gene, *Science, 232*: 738–743, 1986.
2. Nelson, R. S., McCormick, S. M., Delannay, X., Dube, P., Layton, J., Anderson, E. J., Kaniewska, M., Proksch, R. K., Horsch, R. B., Rogers, S. G., Fraley, R. T., and Beachy, R. N. Virus tolerance, plant growth, and field performance of transgenic tomato plants expressing coat protein from tobacco mosaic virus, *Biotechnology, 6*: 403–409, 1988.
3. Dawson, W. O., and Lehto, K. M. Regulation of tobamovirus gene expression, *Adv. Virus Res., 38*: 307–342, 1990.
4. Matthews, R. E. F. *Plant Virology*, 3rd ed., Academic Press, San Diego, 1991.

5. Cassells, A. C., and Herrick, C. C. The identification of mild and severe strains of tobacco mosaic virus in doubly inoculated tomato plants, *Ann. Appl. Biol., 86*: 37–46, 1977.
6. Nelson, R. S., Powell Abel, P., and Beachy, R. N. Lesions and virus accumulation in inoculated transgenic tobacco plants expressing the coat protein gene of tobacco mosaic virus, *Virology, 158*: 126–132, 1987.
7. Powell Abel, P., Sanders, P. R., Tumer, N., Fraley, R. T., and Beachy, R. N. Protection against tobacco mosaic virus infection in transgenic plants requires accumulation of coat protein rather than coat protein RNA sequences, *Virology, 175*: 124–130, 1990.
8. Nejidat, A., and Beachy, R. N. Transgenic tobacco plants expressing a coat protein gene of tobacco mosaic virus are resistant to some other tobamoviruses, *Mol. Plant Microbe Interact., 3*: 247–251, 1990.
9. Anderson, E. J., Stark, D. M., Nelson, R. S., Powell, P. A., Tumer, N. E., and Beachy, R. N. Transgenic plants that express the coat protein genes of tobacco mosaic virus or alfalfa mosaic virus interfere with disease development of some nonrelated viruses, *Phytopathology, 79*: 1284–1290, 1989.
10. Register, J. C., III, and Beachy, R. N. Resistance to TMV in transgenic plants results from interference with an early event in infection, *Virology, 166*: 524–532, 1988.
11. Wilson, T. M. A. Cotranslational disassembly increases the efficiency of expression of TMV RNA in wheat germ cell-free extracts, *Virology, 138*: 353–356, 1984.
12. de Zoeten, G. A., and Fulton, R. W. Understanding generates possibilities, *Phytopathology, 65*: 221–222, 1975.
13. Tumer, N. E., O'Connell, K. M., Nelson, R. S., Sanders, P. R., Beachy, R. N., Fraley, R. T., and Shah, D. M. Expression of the alfalfa mosaic virus coat protein gene confers crossprotection in transgenic tobacco and tomato plants, *EMBO J., 6*: 1181–1188, 1987.
14. Tumer, N. E., Kaniewski, W., Haley, L., Gehrke, L., Lodge, J. K., and Sanders, P. The second amino acid of alfalfa mosaic virus coat protein is critical for coat protein-mediated protection, *Proc. Natl. Acad. Sci., USA, 88*: 2331–2335, 1991.
15. Loesch-Fries, L. S., Merlo, D., Zinnen, T., Burhop, L., Hill, K., Krahn, K., Jarvis, N., Nelson, S., and Halk, E. Expression of alfalfa mosaic virus RNA 4 in transgenic plant confers virus resistance, *EMBO J., 6*: 1845–1851, 1987.
16. Van Dun, C. M. P., Overduin, B., Van Vloten-Doting, L., and Bol, J. F. Transgenic tobacco expressing tobacco streak virus or mutated alfalfa mosaic virus coat protein does not cross-protect against alfalfa mosaic virus infection, *Virology, 164*:383–389, 1988.
17. Van Dun, C. M. P., Bol, J. F., and Van Vloten-Doting, L. Expression of alfalfa mosaic virus and tobacco rattle virus coat protein genes in transgenic tobacco plants, *Virology, 159*: 299–305, 1987.
18. Van Dun, C. M. P., and Bol, J. F. Transgenic tobacco plants accumulating tobacco rattle coat protein resist infection with tobacco rattle virus and pea early browning virus, *Virology, 167*: 649–652, 1988.

19. Kallerhoff, J., Perez, P., Bouzoubaa, S., Ben Tahar, S., and Perret, J. Beet necrotic yellow vein virus coat protein-mediated protection in sugarbeet (*Beta vulgaris* L.) protoplasts, *Plant Cell Rep., 9*: 224–228, 1990.
20. Quemada, H. D., Gonsalves, D., and Slightom, J. L. Expression of coat protein from cucumber mosaic virus strain C in tobacco: protection against infection by CMV strains transmitted mechanically or by aphids, *Phytopathology, 81*: 794–802, 1991.
21. Cuozzo, M., O'Connell, K. M., Kaniewski, W., Fang, R. X., Chua, N. H., and Tumer, N. E. Viral protection in transgenic tobacco plants expressing the cucumber mosaic virus coat protein or its antisense RNA, *Biotechnology, 6*: 549–557, 1988.
22. Namba, S., Ling, K., Gonsalves, C., Gonsalves, D., and Slightom, J. L. Expression of the gene encoding the coat protein of cucumber mosaic virus (CMV) strain-WL appears to provide protection to tobacco plants against infection by several different CMV strains, *Gene, 107*: 181–188, 1991.
23. Hemenway, C., Fang, R. X., Kaniewski, W. K., Chua, N. H., and Tumer, N. E. Analysis of the mechanism of protection in transgenic plants expressing the potato virus X coat protein or its antisense RNA, *EMBO J., 7*: 1273–1280, 1988.
24. Hoekema, A., Huisman, M. J., Molendijk, L., Van Den Elzen, P. J. M., and Cornelissen, B. J. C. The genetic engineering of two commercial potato cultivars for resistance to potato virus X, *Biotechnology 7*: 273–278, 1989.
25. Lawson, C., Kaniewski, W., Haley, L., Rozman, R., Newell, C., Sanders, P., Tumer, N. E. Engineering resistance to mixed virus infection in a commercial potato cultivar: resistance to potato virus X and potato virus Y in transgenic Russet Burbank, *Biotechnology, 8*: 127–134, 1990.
26. Kaniewski, W., Lawson, C., Sammons, B., Haley, L., Hart, J., Delannay, X., and Tumer, N. E. Field resistance of transgenic Russet Burbank potato to effects of infection by potato virus X and potato virus Y, *Biotechnology, 8*: 750–754, 1990.
27. Stark, D. M., and Beachy, R. N. Protection against potyvirus infection in transgenic plants: evidence for broad spectrum resistance, *Biotechnology, 7*: 1257–1262, 1989.
28. Ling, K., Namba, S., Gonsalves, C., Slightom, J. L., and Gonsalves, D. Protection against detrimental effects of potyvirus infection in transgenic tobacco plants expressing the papaya ringspot virus coat protein gene, *Biotechnology, 9*: 752–758, 1991.
29. Regner, F., da Camara Machado, A., da Camara Machado, M. L., Steinkellner, H., Mattanovich, D., Hanzer, V., Weiss, H., and Katinger, H. Coat protein mediated resistance to plum pox virus in *Nicotiana clevelandii* and *benthamiana, Plant Cell Rep., 11*: 30–33, 1992.
30. MacKenzie, D. J., and Tremaine, J. H. Transgenic *Nicotiana debneyii* expressing viral coat protein are resistant to potato virus S infection, *J. Gen. Virol., 71*: 2167–2170, 1990.
31. MacKenzie, D. J., Tremaine, J. H., and McPherson, J. Genetically engineered resistance to potato virus S in potato cultivar Russet Burbank, *Mol. Plant Microbe Interact., 4*: 95–102, 1991.

32. Kawchuk, L. M., Martin, R. R., and McPherson, J. Resistance in transgenic potato expressing the potato leafroll virus coat protein gene, *Mol. Plant Microbe Interact.*, *3*: 301–307, 1990.
33. Kawchuk, L. M., Martin, R. R., and McPherson, J. Sense and antisense RNA-mediated resistance to potato leafroll virus in Russet Burbank potato plants, *Mol. Plant Microbe Interact.*, *4*: 247–253, 1991.
34. Gielen, J. J. L., de Haan, P., Kool, A. J., Peters, D., van Grinsven, M. Q. J. M., and Goldbach, R. W. Engineered resistance to tomato spotted wilt virus, a negative-strand RNA virus, *Biotechnology*, *9*: 1363–1367, 1991.
35. MacKenzie, D. J., and Ellis, P. J. Resistance to tomato spotted wilt virus infection in transgenic tobacco expressing the viral nucleocapsid gene, *Mol. Plant Microbe Interact.*, *5*: 34–40, 1992.
36. Slightom, J. L., Chee, P. P., and Gonsalves, D. Field testing of cucumber plants which express the CMV coat protein gene: field plot design to test natural infection pressures, *Progress in Plant Cellular and Molecular Biology* (H. J. J. Nijkamp, L. H. W. Van Der Plass, and J. Van Aartrijk, eds.), Kluwer, Dordrecht, The Netherlands, 1990, pp. 201–206.
37. Gonsalves, D., Chee, P., Slightom, J. L., and Providentia, R. Field evaluation of transgenic cucumber plants expressing the coat protein gene of cucumber mosaic virus, *Phytopathology Suppl.*, abstract 296, 1991.

III

TECHNIQUES FOR TRANSFORMATION OF PHOTOSYNTHETIC ORGANISMS

7

Expression of Genes Introduced into the Nuclear Genome of *Chlamydomonas reinhardtii*

Stephen P. Mayfield

The Scripps Research Institute, La Jolla, California

I. INTRODUCTION

Chlamydomonas reinhardtii is a eukaryotic green alga that is an excellent organism for the study of the synthesis, assembly, and function of complex cellular components, including those of the flagella or the photosynthetic apparatus. *Chlamydomonas reinhardtii* contains a single large chloroplast that has multiple copies of a 150-kb circular genome. Vegetative cells contain a haploid nuclear genome with a complexity of approximately 100,000 kb, or about four times that of yeast (*Saccharomyces cerevisiae*). The genetic characteristics of *C. reinhardtii* have been well described, and many photosynthetic mutants, both nuclear and chloroplast in origin, have been isolated and described [see Harris (1)]. With the recent success in transformation of both the nuclear and chloroplastic genomes, *C. reinhardtii* appears to be ideally suited for the study of interactions of the nuclear and chloroplast genomes in gene expression. Because of the ease of genetic manipulation, *C. reinhardtii* may also prove to be an ideal system for expression of a variety of proteins from plant and other species. In this chapter the expression of genes introduced into the nuclear genome of *C. reinhardtii* will be examined.

II. SELECTABLE MARKERS USED FOR *C. reinhardtii* NUCLEAR TRANSFORMATION

Earlier reports of transformation of the nuclear genome of *C. reinhardtii*, using either yeast (2,3) or bacterial (4) selectable marker genes, have been unreproducible. In each of these reports the authors could show signs of integration of some foreign DNA but were unable to show, either genetically or by other means, that transformation had occurred. More recently a number of *C. reinhardtii* nuclear genes have been cloned and sequenced, and a surprising codon bias has emerged from analysis of these genes. The nuclear genome of *C. reinhardtii* is relatively G/C rich at about 60%, which would indicate some codon bias toward G and C. However, analysis of the coding region of abundant transcripts has shown that A is almost never used in the third position of any codon except the stop codon (5). This extreme bias in the third position of codons may be fundamental in excluding or reducing the expression of foreign genes in *C. reinhardtii* and may explain why recent successes in transforming the nuclear genome of *C. reinhardtii* have all used homologous (*C. reinhardtii*) genes as the selectable marker. It should be noted that the chloroplastic genome is not G/C rich and does not show this same codon bias (6).

In each of the more recent reports on nuclear transformation, the selectable marker gene used, although different in each case, was a cloned *C. reinhardtii* gene. The selective scheme used was also unique for each gene and included photosynthetic function [*oee1* gene (5)], nitrate utilization [*nit1* gene (7,8)], and arginine autotrophy [*arg7* gene (9)]. In each of these cases an intact *C. reinhardtii* (nonchimeric) gene was used as the selectable marker. There is no reason to suspect that chimeric genes cannot be constructed to function in *C. reinhardtii*, but because so little is known about the *C. reinhardtii* promoter or the 3′ ends it is difficult to determine what the requirements for 5′ and 3′ noncoding sequences may be. As additional constructs are used for transformation we should gain an increased understanding of the role specific sequences (contained within the 5′ and 3′ noncoding regions) have in gene expression and thus a better understanding of what regions can be altered to increase gene expression.

III. CONSTRUCTION OF PLASMID CONTAINING THE SELECTABLE MARKER GENE

Each of the genes used for transformation has been cloned from genomic fragments of wild-type *C. reinhardtii* DNA into *Escherichia coli* plasmids. A typical construct for transformation of the photosynthetic deficient strain Fud44 [*oee1* deficient (10)] contains an 8-kb genomic fragment containing the entire wild-type *oee1* gene as well as 3 kb DNA 5′ and 2 kb of DNA 3′. This fragment was subcloned into plasmid pUC19 at the unique EcoRI and Kpn I sites to form plasmid pSB101. The plasmid was propagated in *Escherichia coli* strain DH5 and double-stranded DNA isolated from a CsCl gradient. No further preparation of the DNA was necessary for transformation, although extraction of the DNA with chloroform before transformation reduced bacterial and fungal contamination on the selection plates. Similar constructs were used for each of the marker genes used in the *C. reinhardtii* transformations reported so far.

IV. METHODS FOR INTRODUCING DNA INTO THE NUCLEAR GENOME

A. Microprojectile Bombardment

Transformation of plant cells by microprojectile bombardment was first demonstrated by Klein et al. (11) and has since been used to transform a variety of plant and algal species. By using a similar technique several laboratories have reported stable transformation of the *C. reinhardtii* chloroplastic genome (12,13). Transformation of the nuclear genome by microprojectile bombardment has also been demonstrated by several laboratories using a variety of selectable markers (5,7,9). Transformants recovered following microprojectile bombardment show an apparently random integration of the transforming DNA. Both supercoiled plasmid and linear DNA molecules have been used, and they show similar rates of transformation.

B. Agitation with Glass Beads

A simpler protocol has been identified for the transformation of the *C. reinhardtii* nuclear genome, in which DNA is introduced into the cells by glass-bead disruption of cells in the presence of plasmid DNA (8). Transformants obtained from this type of protocol appear to be indistinguishable from transformants obtained by particle bombardment. There is random integration of transforming DNA into the nuclear genome, with single or, in some cases, multiple copies of the marker DNA present in transformed cells.

C. Electroporation

An electrical discharge to introduce pores into the cell membrane (electroporation) has also been used in two investigations as a means to transform the *C. reinhardtii* nuclear genome (14,15). The selectable markers in both of these reports were the same as those used for microprojectile bombardment, that is, cloned intact *C. reinhardtii* genes. The rate of transformation with this method appears to be greater than the rate with either of the other two methods (14), but the transformants recovered appear otherwise indistinguishable from transformants recovered by either of the earlier protocols.

V. ANALYSIS OF DNA INTRODUCED INTO THE NUCLEAR GENOME

Southern analysis of *nit1* and *arg7* genes introduced into *C. reinhardtii* by particle-gun bombardment (7,9) showed that for both of these constructs multiple copies of the gene were recovered in each of the transformants. Examination of *oee1* transformants obtained by particle-gun bombardment showed only single copies of the *oee1* gene in transformants (5). Transformants obtained by glass-bead agitation with the *nit1* gene also showed multiple copies of the gene in transformants (8). Whether the difference in copy number between the *oee1* and the *nit1* and *arg7* genes in transformants is due to the amount of DNA introduced during transformation or to differences in the genes themselves is not apparent from any of these data. However, as will be shown, there can be quite a range in the number of copies of a gene integrated, even from a single transformation experiment.

In one investigation, DNA was isolated from transformants obtained with uncut, EcoR I–cut, and EcoR I/Kpn I–cut plasmid containing the wild-type *oee1* gene (pSB101). The plasmid was transformed into the *oee1*-deficient mutant strain by electroporation. The DNA was digested with Pst I separated on an agarose gel, blotted to nylon membrane, and probed with a cloned genomic fragment from the 5′ end of the *oee1* gene. The probe was a 700-bp Hind III fragment that hybridizes to a unique Pst I fragment in a wild-type strain, in the *oee1* mutant (Fud44), and in a spontaneous revertant (Fud44-R2). The use of this probe allowed us to differentiate among the three fragments and thus quickly separate transformants from spontaneous revertants.

As shown in Fig. 1, this probe hybridized to a 2.2-kb fragment in the wild-type strain, a 7-kb fragment in Fud44, and a 2.7-kb fragment in Fud44-R2. In each of the transformants the 7-kb Fud44 fragment was still found, suggesting that transformation did not occur by homologous recombination with the endogenous Fud44 *oee1* gene. Homologous recombination is the predominant form of transformation of the chloroplast genome of *C. reinhardtii* (12). Each of the transformants from the uncut and EcoR I–cut plasmid contained only a single new fragment hybridizing to the *oee1* probe. Two of the five transformants from the EcoR I/Kpn I–digested DNA contained more than one new copy of the *oee1* gene, and one of the transformants contained five or six copies (Fig. 1). Most, but not all, of the new fragments hybridizing to the *oee1* probe comigrated with the wild-type *oee1* fragment, suggesting that the introduced *oee1* gene integrated into the genome by recombination in regions of the pSB101 clone outside of the Pst I fragment. The appearance of novel Pst I fragments in several of the transformants shows that integration in which at least one of the Pst I sites is lost can also occur.

Transformation with smaller fragments of the *oee1* gene, obtained by digestion of pSB101 with EcoR I and Xho I or with Mlu I and Xho I, was also attempted. Both of these fragments contain the entire *oee1* coding region but only 700 bp of 3′ flanking DNA. The EcoR I–cut plasmid contains 3 kb of 5′ flanking DNA, as did the DNA used to obtain the transformants shown in Fig. 1. The Mlu I–cut plasmid contained approximately 1 kb of DNA 5′ of the *oee1* coding region. Figure 2 shows the analysis of *oee1* genes in transformants recovered after electroporation with one or the other of these DNAs. Transformants from pSB101 DNA digested with EcoR I and Xho I had Southern blot patterns similar to those of transformants with uncut pSB101 DNA or from pSB101 cut with EcoR I or EcoR I and Kpn I. Transformants recovered after electroporation with DNA cut with Mlu I and Xho I had Southern blot patterns in which the Pst I fragments were all distinct from those of the wild type. This is not surprising because digestion with Mlu I separates the 5′ Pst I site from the *oee1* structural gene; thus, any integration into the *C. reinhardtii* genome should introduce a novel Pst I site 5′ of the *oee1* gene. These data show that as little as 700 bp of 3′ and 1 kb of 5′ of the transcription stop and start respectively are sufficient to allow expression of the *oee1* gene.

VI. INTEGRATION OF VECTOR SEQUENCES ALONG WITH THE MARKER GENE

The *oee1* probe was stripped from the Southern blot shown in Fig. 1, and the blot was reprobed with labeled pUC19. As shown in Fig. 3, pUC19 sequences were present in one transformant from each of the groups. It is somewhat

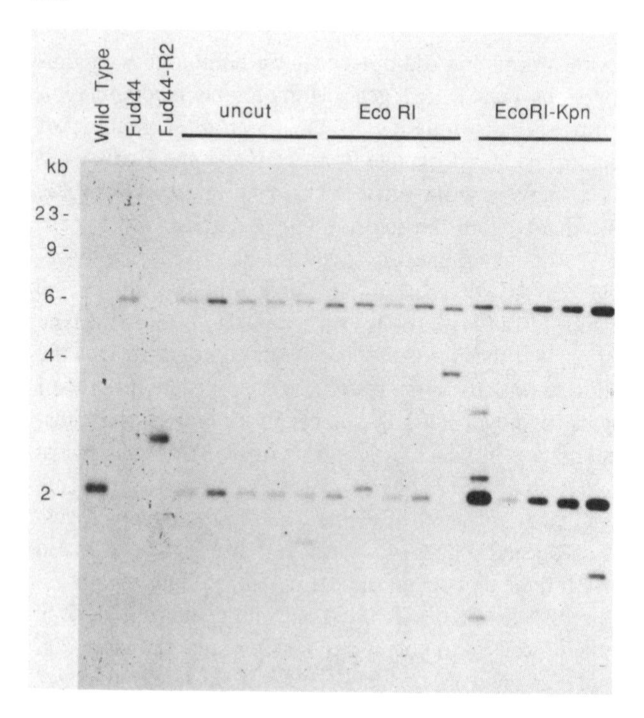

Figure 1 Analysis of *oee1* genes in transformants. The DNA was isolated from a wild-type strain, Fud44, a spontaneous revertant of Fud44 (Fud44-R2), and 15 transformants. Five colonies were selected from the group of transformants obtained with uncut pSB101 plasmid (uncut), five were selected from the group obtained from EcoR I-cut pSB101 plasmid (Eco RI), and five were selected from EcoR I/Kpn I-cut pSB101 plasmid (Eco RI-Kpn). All the DNAs were digested with Pst I and separated on 0.8% agarose gels. After blotting to a nylon membrane, the filter was probed with a 700-bp genomic fragment (5′*oee1* probe), which hybridized to a unique Pst I fragment in the wild-type strain, Fud44, and Fud44-R2. Note that each of the transformants contains a fragment that comigrates with the Fud44 fragment, as well as additional fragments hybridizing to the *oee1* probe. None of the transformants have fragments that comigrate with the revertant band.

surprising that multiple copies of pUC were found in one of the EcoR I–Kpn I groups, because the pUC sequences are unlinked to the *oee1* gene in this DNA. Passage of transformant on nonselective media for extended periods (up to 1 year) had no affect on the stability of the introduced *oee1* genes. However, several, but not all, of the transformants that originally integrated vector

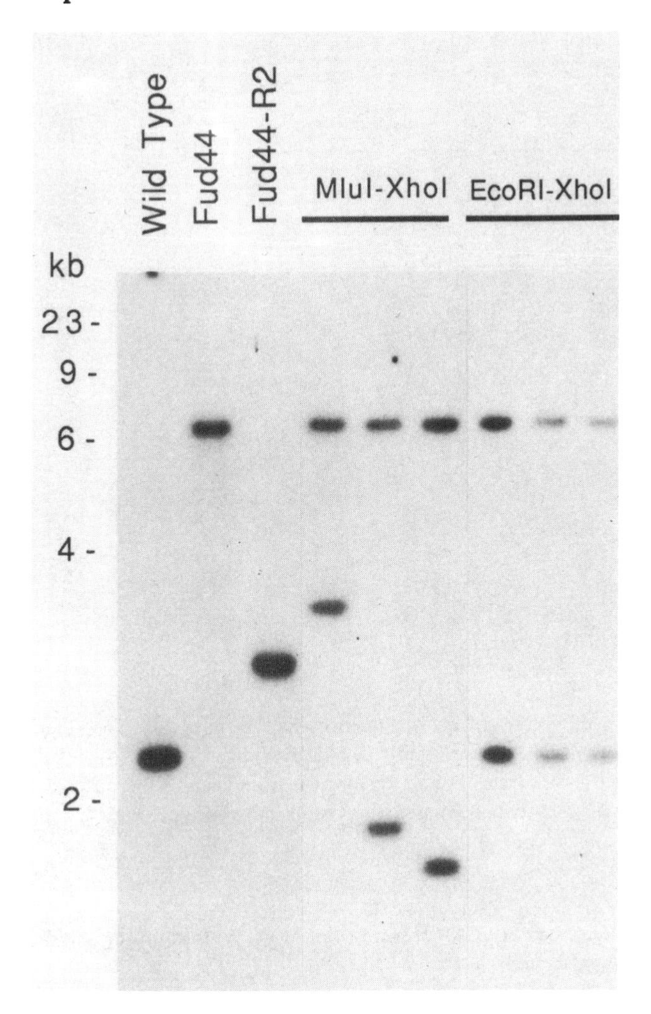

Figure 2 Analysis of *oee1* genes in transformants obtained from pSB101 plasmid cut with EcoR I and Xho I (EcoRI-XhoI) or with Mlu I and Xho I (Mpu I-Xho I). The DNAs from a wild-type strain, Fud44, Fud44-R2, and three transformants, selected at random, were digested with Pst I and separated on agarose gels. After blotting to nylon membrane, the filter was hybridized with the same 5′ *oee1* probe described in Fig. 1. Note that the Pst I fragments for each of the transformants obtained with Mlu I/Xho I-digested pSB101 have sizes different from those of the wild type.

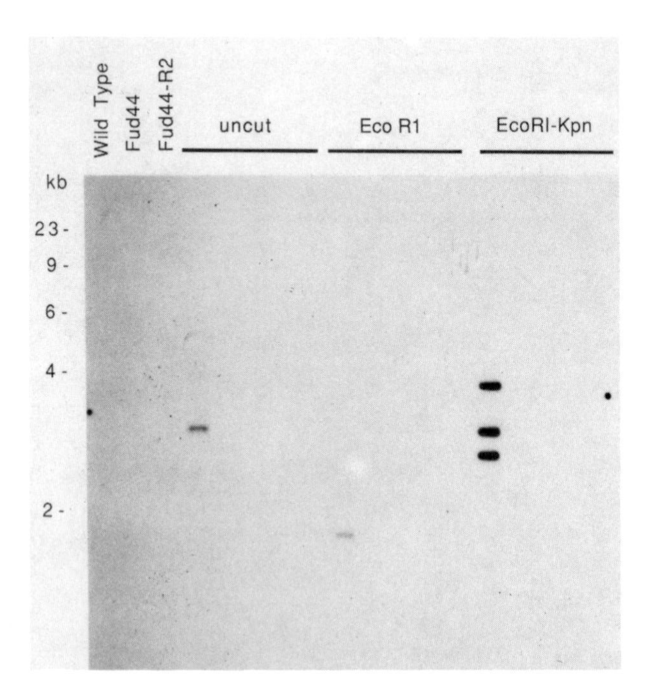

Figure 3 Identification of vector (pUC19) sequences in transformants. The filter shown in Fig. 1 was stripped of the *oee1*, probed, and hybridized with ^{32}P-labeled pUC19. A single fragment hybridized in one transformant from each of the different transformant groups. No pUC19 sequences are observed in the wild-type, Fud44, and Fud44-R2 strains.

sequences lost these sequences after several months of growth on either selective or nonselective plates.

VII. SEGREGATION OF INTRODUCED *oee1* GENES IN A NORMAL MENDELIAN FASHION

A transformant from the EcoR I–cut plasmid group (R1-100-3), was crossed with a wild-type strain, and 16 of the resulting tetrads were dissected and scored for phototrophic growth. Of the 16 tetrads discussed, eight were complete (having four viable daughter cells), and the remaining eight had two or three viable daughters. Of the eight complete tetrads, five had all phototrophic progeny, whereas the other three tetrads had three phototrophic and one nonphototrophic daughter. Southern analyses of a tetrad containing all phototrophic daughters (tetrad 1) and a tetrad with three phototrophic and one nonphototrophic daughters (tetrad 7) are shown in Fig. 4. In each tetrad two daughters contain a copy of the Fud44 *oee1* gene (lanes 1C, 1D, 7B, and 7C), and the other two daughters have a copy of the wild-type *oee1* gene (1A, 1B, 7A, 7D), as expected. The introduced *oee1* gene (4.2-kb fragment) also segregates in a normal Mendelian fashion, with two daughters of each tetrad containing a copy of the gene (1C, 1D, 7C, 7D). Both of the daughters containing the Fud44 gene in tetrad 1 also contain the introduced *oee1* genes (1C, 1D); hence, no nonphotosynthetic colonies are recovered. The segregation of one of the R1-100-3 *oee1* genes with a wild-type *oee1* gene (7D) results in the recovery of a nonphotosynthetic daughter containing only the Fud44 *oee1* gene in that tetrad (7B). The recovery of this progeny also shows that the photosynthetic function of transformant R1-100-3 is not due to a reversion of the Fud44 gene but rather to the presence of the new *oee1* gene.

Genetic analysis of *nit1* transformants obtained from particle bombardment indicated that *nit1* genes also segregated as single Mendelian traits, which is surprising given the high copy number of *nit1* sequences in the transformants (7). However, Southern analysis of the progeny from a cross of a *nit1* transformant and a *nit*(-) parent showed that most of the introduced *nit1* sequences were physically linked to one another in these transformants (7). Close physical linkage of the *nit1* genes would explain this Mendelian segregation. No genetic analysis was performed on the *arg7* transformants or on the *nit1* transformants obtained from glass-bead agitation.

Analysis of cotransformants containing both *nit1* and the flagella protein *RSP3* gene showed that the introduced genes were genetically linked to each other but were not linked to either of the endogenous *nit* or *RSP3* genes (16).

VIII. ANALYSIS OF mRNA ACCUMULATION IN TRANSGENIC STRAINS

RNA was isolated from a wild-type strain, Fud44, and nine transformants shown in Fig. 2. Total RNA was fractionated on denaturing formaldehyde agarose gels, electroblotted to a nylon membrane, and probed with a labeled *oee1* cDNA probe (17). As shown in Fig. 5(A), each of the transformants accumulated *oee1* RNA to levels equal to or greater than those of the wild type. It has been previously shown that Fud44 does not accumulate any detectable *oee1* message (10). One of the transformants (R1-K-50) accumulated appreciably more *oee1* mRNA than did the wild type. This transformant has five or

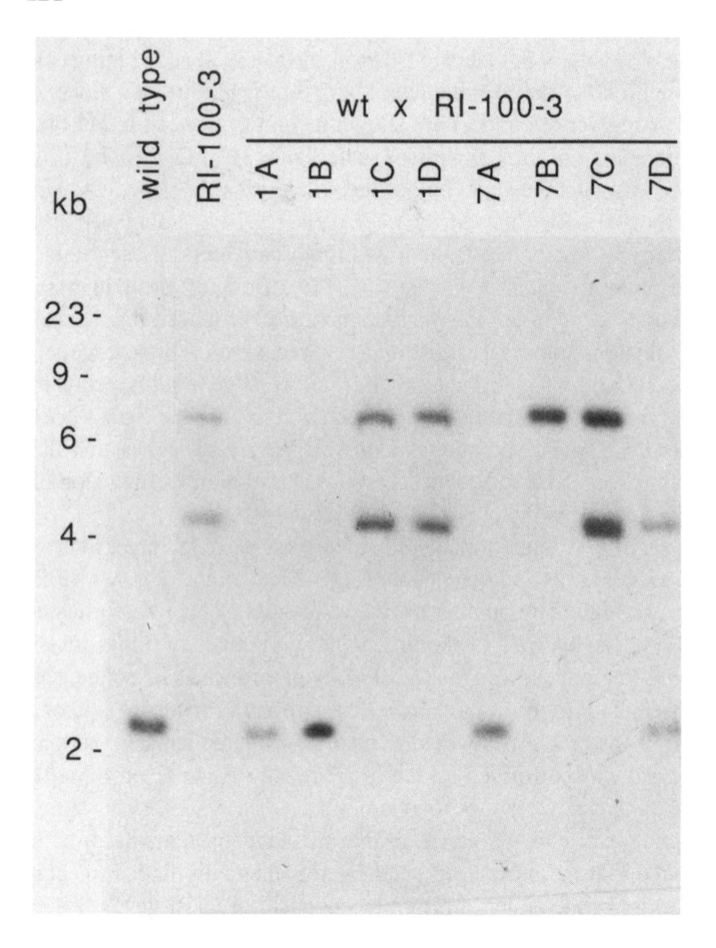

Figure 4 Analysis of *oee1* genes in daughter cells derived from a cross of a wild-type strain and a transformant from the EcoR I group (R1-100-3). The DNA from the wild type, transformant R1-100-3, and each daughter of two tetrads from the cross of the wild type and R1-100-3 were digested with Pst I and separated on an agarose gel. The DNA was blotted to nylon membrane and hybridized with the same 5′ *oee1* probe as in Fig. 2. Each of two daughter cells from both tetrads have a copy of the Fud44 *oee1* gene (1C, 1D, 7B, 7C), and the other two daughters have a copy of the wild-type *oee1* gene (1A, 1B, 7A, 7D), as expected. The R1-100-3 *oee1* genes are also found in each of two daughter cells (1C, 1D, 7C, 7D), showing a normal Mendelian segregation for a single gene. The only daughter to show nonphotosynthetic growth was daughter 7B, which is the only strain that lacks either a wild-type or R1-100-3 *oee1* gene.

six copies of the *oee1* gene, as determined by Southern analysis (first lane of EcoR I–Kpn group in Fig. 1). Two transformants that apparently contain two copies of the *oee1* gene (SC-100 and R1-K-150.2) have RNA levels similar to those of the wild type and of transformants with only single copies of *oee1* genes. To make sure that equal amounts of RNA were loaded onto each lane, the filter was stripped of the *oee1* probe and rehybridized with a probe for the

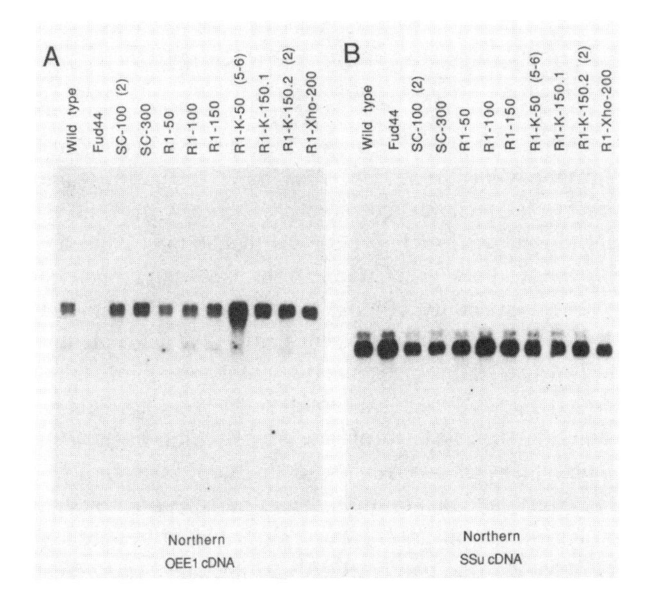

Figure 5 Analysis of *oee1* mRNA in transformants. The RNA was isolated from a wild-type strain, Fud44, and nine transformants selected from the group of transformants shown in Figs. 2 and 4. (A) Total RNA was separated on denaturing agarose gels, blotted to nylon membrane, and hybridized with a ^{32}P-labeled *oee1* cDNA probe. The Fud44 did not accumulate any *oee1* mRNA, but the transformants obtained from uncut (SC), EcoR I-cut (R1), EcoR I/Kpn I-cut (R1-K), and EcoR I/Xho I-cut (R1-Xho) pSB101 plasmid did accumulate *oee1* mRNA. The numbers in parentheses on SC-100, R1-K-50, and R1-K-150.2 are the numbers of *oee1* fragments (genes) that were detected by Southern analysis as shown in Fig. 1. (B) To determine whether other mRNA was expressed at normal levels in the transformants, the accumulation of an unrelated nuclear encoded message (small subunit of RuBPCase, SSu) was measured. The filter shown in panel A was stripped of the *oee1* probe and then hybridized with ^{32}P-labeled cDNA encoding the SSu. Approximately equal amounts of SSu RNA were detected in each of the sample lanes.

small subunit of RuBP Case, SSu (18). As shown in Fig. 5(B), all of the lanes, including Fud44, have similar signals for the SSu RNA.

Examination of RNA in *nit1* transformants showed no correlation between a high number of *nit* gene copies and a high level of *nit* RNA. The *nit* RNA that was present was generally at levels similar to those of wild-type cells and showed regulation by ammonium in a fashion similar to that for wild-type *nit* (7).

IX. EXPRESSION OF PROTEIN IN TRANSGENIC STRAINS

The accumulation of *oee1* protein was determined for several transformants. Proteins were isolated from a wild-type strain, Fud44, and four transformants, including the transformant that accumulated more *oee1* mRNA than did the wild type (R1-K-50). The left panel of Fig. 6 shows the staining pattern of proteins separated on polyacrylamide gels, and the right panel shows an identical set of samples transferred to a nitrocellulose filter, reacted with anti-*oee1* antiserum, incubated in goat anti-rabbit antiserum conjugated with alkaline phosphatase, and subjected to alkaline phosphatase activity staining. As has previously been shown (10), Fud44 does not accumulate *oee1* protein. This can be seen on the stained gel (arrowhead) and on the Western blot. Each of the transformants accumulates at least as much *oee1* protein as does the wild type, whereas the R1-K-50 cells accumulate several times as much *oee1* protein as the wild type. Thus, for *oee1* there appears to be a good correlation between the number of *oee1* gene copies and the *oee1* mRNA and protein levels. When Diener et al. (16) measured protein levels for the *RSP3* gene in cotransformants (which contained multiple copies of the *RSP3* gene), they found levels similar to those of wild-type cells. Protein levels were not determined for either *arg* or *nit* in any of the transformants, but enzyme activity for both of these proteins fluctuated over a large range in transformants (7,9).

X. EXPRESSION OF FOREIGN GENES IN *C. reinhardtii*

Several reports of transformation of *C. reinhardtii* by foreign genes have been published (2–4,19). In only one of these reports (19) has the transformed phenotype been shown to be stable. This single stable transformant was obtained from a chimeric octopine synthetase–neomycin phosphotransferase (*npt-II*) gene that had been used for higher plant transformations. The transformant contained a small amount of *npt-II* RNA and showed some *npt-II* activity,

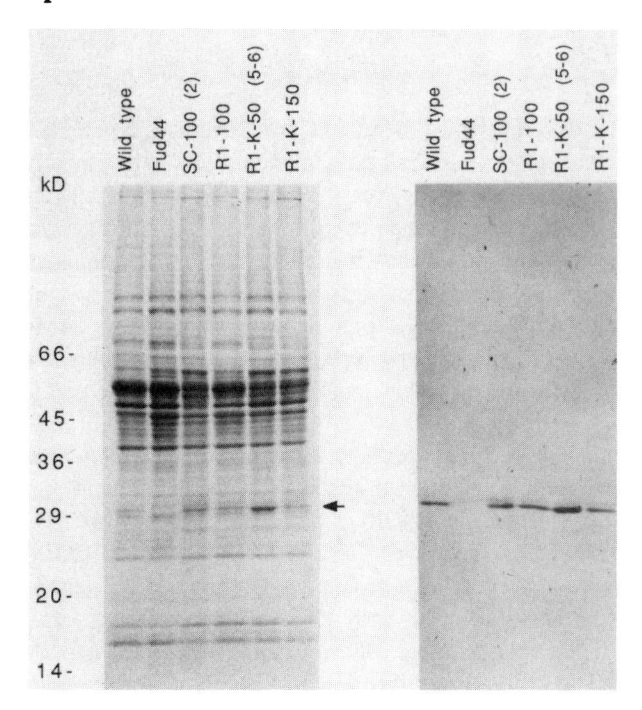

Figure 6 Western blot analysis of *oee1* protein accumulation in transformants. Proteins were isolated from a wild-type strain, Fud44, and four transformants and were then separated on SDS polyacrylamide gels and either stained with Coomassie blue (left panel) or transferred to nitrocellulose membrane (right panel). The nitrocellulose filter was reacted with ant-*oee1* antiserum, incubated with anti-rabbit antiserum conjugated with alkaline phosphatase, and stained with an alkaline phosphatase activity stain. As has been previously shown (10), Fud44 does not accumulate *oee1* protein, whereas each of the transformants accumulates at least as much *oee1* protein as does the wild type. Note that transformant R1-K-50, which accumulates more *oee1* mRNA than does the wild type, also accumulates *oee1* protein to similarly high levels. The arrow indicates the position of the *oee1* protein on the stain gel.

but only slightly more than untransformed cells. The instability and low levels of RNA and protein expression found in these transformants suggest that bacteria and yeast genes may not be suitable for expression in *C. reinhardtii*. It remains to be shown whether foreign genes can be expressed in *C. reinhardtii* without substantial modification to the promoters, terminators, or coding regions.

XI. CONCLUSION

It is clear from the data presented here and from similar experiments using particle-gun technology or glass-bead agitation that transformation of the *C. reinhardtii* nuclear genome is feasible and perhaps trivial. The use of a variety of protocols to produce transformants also indicates that delivery of DNA to the cells is not the limiting factor for *C. reinhardtii* transformation; rather, the limiting factor is identification of a suitable selection scheme and marker gene (homologous gene). These transformation schemes should provide powerful tools for the molecular analysis of genes and their protein products in an already powerful genetic system.

Analysis of genomic DNA from a variety of transformants shows that introduced genes integrate into the nuclear genome at sites unlinked to their endogenous counterparts. Thus, nuclear transformation does not occur by homologous recombination, as is the case for transformation of the chloroplastic genome. More than one copy of a gene may integrate into the genome during a single transformation event. In some cases vector sequences, either linked or unlinked to the transforming DNA, can also integrate into the *C. reinhardtii* genome. After they are integrated, the new genes segregate (in sexual crosses) in a normal Mendelian fashion.

Characterization of the expression of introduced genes indicates that the incorporation of a single gene (or in some cases many copies) results in expression of mRNA at levels similar to those of wild-type cells. This is quite different from higher plants, in which RNA accumulation from genes integrated through transformation can vary dramatically from transformant to transformant and is often much lower than that of wild-type counterparts (20). In some cases integration of multiple copies of a gene (*oee1*) results in the accumulation of mRNA proportional to the number of gene copies, whereas in other cases multiple copies of a gene result in mRNA accumulation no greater than in wild-type cells. Regulation of mRNA accumulation from introduced *nit* genes is similar to that of the endogenous *nit* genes of wild-type cells; this suggests that any sequences necessary for *nit* regulation by ammonium are contained on the *nit1* plasmid.

Protein accumulation in transformants is proportional to mRNA accumulation in those cells for at least the *oee1* gene. Protein accumulation (enzyme activity) for the *nit* and *arg* genes is difficult to determine but appears to be similar to that found in wild-type cells.

Introduced genes segregate in a fashion (Mendelian) that would be expected of any single-copy nuclear gene and appear to be as stable as any other nuclear gene.

Each of these results suggests that transformation of a *C. reinhardtii* gene into the nuclear genome of *C. reinhardtii* results in the formation of a new locus that is expressed and segregates in the same way as any other nuclear gene. The poor expression and unstable phenotype of foreign genes transformed in *C. reinhardtii* suggest that these genes are poor choices as selectable marker genes or for use in expression of foreign proteins in *C. reinhardtii.* If foreign genes are to be expressed in *C. reinhardtii* cells, chimeric genes that contain *C. reinhardtii* promoters and terminators must be constructed, and perhaps coding regions should be modified to optimize codon bias as well. The exact requirements for optimum expression of proteins in *C. reinhardtii* should become clearer as additional constructs are made to define the specific sequences important in *C. reinhardtii* genes.

ACKNOWLEDGMENTS

I wish to thank Karen Freriks for excellent technical assistance in every aspect of this project. This work was supported by grant GM41353 from the National Institutes of Health.

REFERENCES

1. Harris, E. H. *The Chlamydomonas Sourcebook,* Academic Press, New York, 1989.
2. Rochaix, J.-D., and van Dillewjin, J. Transformation of the green alga *Chlamydomonas* with yeast DNA, *Nature, 296*: 70–73, 1982.
3. Rochaix, J.-D., van Dillewjin, J., and Rahire, M. Construction and characterization of autonomously replicating plasmids in the green unicellular alga *Chlamydomonas reinhardtii, Cell, 36*: 925–931, 1984.
4. Hasnain, S. E., Manavathu, E., and Leung, W. DNA mediated transformation of *Chlamydomonas reinhardtii* cells, *Molec. Cell. Biol., 5*: 3647–3650, 1985.
5. Mayfield, S. P., and Kindle, K. Stable nuclear transformation of *Chlamydomonas reinhardtii* using a *C. reinhardtii* gene as the selectable marker, *Proc. Natl. Acad. Sci. USA, 87*: 2087–2091, 1990.
6. Erickson, J. M., Rahire, M., and Rochaix, J. D. *Chlamydomonas reinhardtii* gene for the 32,000 mol. wt. protein of photosystem II contains four large introns and is located entirely within the chloroplast inverted repeat, *EMBO J., 3*: 2753–2762, 1984.

7. Kindle, K. L., Schell, R. A., Fernandez, E., and Lefebvre, P. A. Stable nuclear transformation of *Chlamydomonas* using the *Chlamydomonas* gene for nitrate-reductase, *J. Cell. Biol., 109*: 2589–2601, 1989.
8. Kindle, K. L. High-frequency nuclear transformation of *Chlamydomonas reinhardtii, Proc. Natl. Acad. Sci. USA, 87*: 1228–1232, 1990.
9. Debuchy, R., Purton, S., and Rochaix, J.-D. The argininosuccinate lyase gene of *Chlamydomonas reinhardtii*: an important tool for nuclear transformation and for correlating the genetic and molecular maps of arg7 locus, *EMBO J., 8*: 2803–2809, 1989.
10. Mayfield, S. P., Bennoun, P., and Rochaix, J.-D. Expression of the nuclear encoded OEE1 protein is required for oxygen evolution and stability of photosystem II particles in *Chlamydomonas reinhardtii, EMBO J., 6*: 313–318, 1987.
11. Klein, T. M., Wolf, E. D., Wu, R., and Sanford, J. C. High-velocity microprojectiles for delivering nucleic acids into living plant cells, *Nature, 327*: 70–73, 1987.
12. Boynton, J. E., Gillham, N., Harris, E., Hosler, J., Johnson, A., Jones, A., Randolph, B., Robertson, D., Klein, T., Shark, K., and Sanford, J. Chloroplast transformation in *Chlamydomonas* with high velocity microprojectiles, *Science, 240*: 1534–1538, 1988.
13. Blowers, A. D., Bogorad, L., Shark, K. B., and Sanford, J. C. Studies on *Chlamydomonas* chloroplast transformation: foreign DNA can be stably maintained in the genome, *Plant Cell, 1*: 123–132, 1989.
14. Mayfield, S. P. Over-expression of the oxygen-evolving enhancer 1 protein and its consequences on photosystem II accumulation, *Planta, 185*: 105–110, 1991.
15. Brown, L. E., Specher, S. L., and Keller, L. R. Introduction of exogenous DNA into *Chlamydomonas reinhardtii* by electroporation, *Mol. Cell Biol., 11*: 2328–2332, 1991.
16. Diener, D. R., Curry, A. M., Johnson, K. A., Williams, B. D., Lefebvre, P. A., Kindle, K. L., and Rosebbaum, J. L. Rescue of a paralyzed-flagella mutant of *Chlamydomonas reinhardtii* by transformation, *Proc. Natl. Acad. Sci. USA, 87*: 5739–5743, 1990.
17. Mayfield, S. P., Rahire-Schirmer, M., Frank, G., Zuber, H., and Rochaix, J.-D. Analysis of the genes of the OEE1 and OEE3 proteins of the Photosystem II complex from *Chlamydomonas reinhardtii, Plant Mol. Biol., 12*: 683–693, 1989.
18. Goldschmidt, M., and Rahire, M. Sequence, evolution and differential expression of the two genes for encoding variant small subunits of RuBPCase in *Chlamydomonas reinhardtii, J. Mol. Biol., 191*: 421–432, 1986.
19. Bingham, S. E., Cox, J. C., and Strem, M. D. Expression of foreign DNA in *Chlamydomonas reinhardtii, FEMS Micro. Lett., 65*: 77–82, 1989.
20. Dean, C., Favreau, M., Tamaki, S., Bond-Nutter, D., Dunsmuir, P., and Bedbrook, J. Expression of tandem gene fusions in transgenic tobacco plants, *Nucleic Acids Res., 16*: 7601–7617, 1988.
21. Sueoka, N. Mitotic replication of deoxyribonucleic acid in *Chlamydomonas reinhardtii, Proc. Natl. Acad. Sci. USA, 46*: 83–91, 1960.

22. Rochaix, J.-D., Mayfield, S., Goldschmidt-Clermont, M., and Erickson, J. Molecular biology of *Chlamydomonas*, in *Plant Molecular Biology—A Practical Approach* (C. H. Shaw, ed.), IRL Press, London, 1987.
23. Day, A., Schirmer-Rahire, M., Kuchka, M. R., Mayfield, S. P., and Rochaix, J.-D. A transposon with an unusual long terminal repeat arrangement in the green alga *Chlamydomonas reinhardtii, EMBO J., 7*: 1917–1927, 1988.
24. Shillito, R. D., Saul, M. W., Paszkowski, J., Muller, M., and Potrykus, I. High efficient direct gene transfer to plants, *Biotechnology, 3*: 1099–1103, 1985.
25. Khandjian, E. W. U.V. crosslinking of RNA to nylon membrane enhances signal, *Mol. Biol. Rep., 11*: 107–115, 1986.
26. Dower, W. J., Miller, J. E., and Ragsdale, C. W. High efficient transformation of *E. coli* by high voltage electroporation, *Nucleic Acids Res., 16*: 6127–6145, 1988.
27. Fromm, M., Taylor, L., and Walbot, V. Expression of genes transferred into monocot and dicot plant cells by electroporation, *Proc. Natl. Acad. Sci. USA, 82*: 5824–5828, 1985.

8

Microprojectile Bombardment: A Method for the Production of Transgenic Cereal Crop Plants and the Functional Analysis of Genes

Fionnuala Morrish, David D. Songstad,*
Charles L. Armstrong, and Michael Fromm

Monsanto, St. Louis, Missouri

I. INTRODUCTION

Traditional plant breeding has provided us with many improvements in crop plants but is limited by the traits available in sexually compatible varieties and species. The development of genetic engineering has significantly increased the gene pool accessible for crop improvement. In the absence of species and interorganism barriers, a vast array of genes can be made available to improve crop quality, provide resistance to disease, insects, and herbicides, or increase tolerance to stress. Gene transfer techniques for many dicot crops, including cotton, flax, soybean, and potato (1), have been successfully developed by using *Agrobacterium tumefaciens*–mediated gene transfer (2). This technology has allowed the transfer of genes for resistance to virus (3–7), herbicides (8–13), and insects (14–17) and has also allowed the transfer of genes that improve seed protein quality (18,19).

The graminaceous monocots, which include such major food crops as wheat, maize, and rice, are not routinely amenable to gene transfer using *Agrobacterium*. Because there is no natural vector for gene transfer, much effort has been made to deliver foreign genes into protoplasts of these crop plants by free DNA delivery with polyethylene glycol (PEG) (20–23) and by electroporation

**Present affiliation*: Pioneer Hi-Bred International, Inc., Johnston, Iowa

(24–26). Many graminaceous crop species can be readily regenerated from intact cells and remain fertile (27,28). Plant regeneration from protoplasts of these crop species has proved more difficult. Much progress has been made in protoplast regeneration in the last 5 years. The regeneration of maize plants from protoplasts (29) paved the way for the first transgenic maize plants (31). However, the fertility of these regenerants was affected and only recently have fertile plants been recovered from maize protoplasts (30,32,33). Fertile plants have also been recovered from barley (34). Wheat plants have been successfully regenerated from protoplasts, but the fertility of the plants was affected and no sexual progeny were obtained (35,36). Currently, rice is one of the few graminaceous monocots for which a reproducible system for the transformation and regeneration of fertile plants from protoplasts exists (37–39).

Because of the difficulties involved in transformation and regeneration of monocots using protoplasts, a variety of approaches have been tested to introduce foreign DNA into intact plant cells. These include both macroinjection (40) and microinjection (41,42), pollen and embryo DNA imbibition (43–45), electroporation of intact tissues (46), laser microbeam injection (47), pollen-tube pathway (48), agroinfection (49,50), and microprojectile bombardment (51,52) (for critical reviews of these methods see Refs. 53 and 54). Of all these approaches, the most reproducible and successful for the recovery of transformed plants from a wide range of species has proved to be microprojectile bombardment (referred to as the biolistics process). The biolistics process has allowed the recovery of the first fertile transgenic maize plants (55,56). Microprojectile bombardment has also been tested in attempts to improve transformation in systems that are amenable to *Agrobacterium* transformation and has proved to be effective in the transformation of *Populus* (57), cotton (58), and papaya (59). The use of the biolistics gun has also improved the transformation of several soybean genotypes (60,61). Reports of transient gene expression in wheat (62–65), barley (66–68), and rice (63,64) and the recent recovery of stable transformed callus in wheat (69) suggest other crops should follow.

The goal of genetic engineering through plant transformation is the introduction of functional genes into plant cells. Both transient and stable transformation assays are used to assess gene function. Transient assays may be performed within 24 hours after transfer of DNA and allow an analysis of gene expression in the absence of DNA integration. These gene expression assays should be followed by stable analysis and plant regeneration to confirm the expression of introduced genes when integrated into the genome. Transient assays are suitable, however, for determining the efficiency of gene expression from different gene cassettes, that is, different combinations of promoters, enhancers, coding sequences, and other elements that may be used to enhance gene expression.

An array of environmental, developmental, and genetic factors interact to control the expression of genes in plants. The elucidation of the mechanisms by which gene expression is controlled is of fundamental interest to those involved in plant molecular and cell biology. Knowledge of the factors involved in the control of gene expression is also essential for the development of plant vectors for use in plant transformation. Eukaryotic genes consist of an arrangement of functional domains. The critical functional domains involved in the regulation of gene expression can be identified by the selective dissection of genes and a subsequent analysis of their functional activity when they are reintroduced into plant cells. This analysis is most productive when genes are reintroduced into their tissue of origin and are thereby subject to the same physiological, biochemical, and developmental stimuli under which they normally function. Until recently the functional activity of monocot genes was limited to analysis in heterologous systems, such as tobacco, or in monocot protoplasts, which are unlikely to totally reflect the physiological or biochemical state of the cell from which they were derived. Microprojectile bombardment now allows functional transient assays on genes within organized cellular environments in homologous species.

The following review is an assessment of the developments in plant transformation and functional analysis of genes that have been made possible by microprojectile bombardment. The research presented includes studies that have allowed the recovery of transformed plants by the optimization of gun, DNA, target tissue, and selection and regeneration conditions. The important advances, from both transient and stable analysis, in our understanding of the factors that control gene expression in plant cells are also addressed. Although this review is specifically focused on monocot transformation, examples of dicot transformation using microprojectile bombardment are included where appropriate.

II. DNA DELIVERY BY MICROPROJECTILE BOMBARDMENT: INTRODUCTION

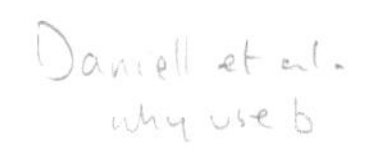

The advantage of microprojectile bombardment over most other methods of gene transfer is that physical penetration of the plant cell wall allows species-independent transfer of DNA into a wide array of target tissues. The microprojectile gun (70) was designed to bombard plant cells with DNA-coated microprojectiles of various high-density metals (gold, tungsten, platinum) (see Fig. 1 and Refs. 50, 70, and 71). These microscopic particles (0.4–2 μm in diameter) are coated with DNA and bombarded at high velocity into intact cells. Macroprojectiles are used as a means of acceleration. The placement of a stopping plate with a small aperture (0.8 mm) in the path of the macroprojectile results in the scattering of particles on impact. A stainless steel mesh screen (100 μm) below the stopping plate further scatters the particles. These particles penetrate the cell wall, a major barrier to DNA transfer, and deliver DNA to the plant cell. The commercially available device (Biolistic PDS-1000, Du Pont) uses a gunpowder discharge to propel the particles. Several other guns using the same principle have been developed; the major modifications are the use of air (64), nitrogen (72,73), helium (74,75), or electrical arc discharge (61,76) instead of gunpowder to propel the projectile. Target tissues that have been bombarded to date include pollen (77), pollen microspore callus (66), meristems (60,78), embryos (59,67,79,80), leaves (64,65), embryogenic callus, and suspension cultures (55,56,58,59,63, 67,68,81) (see Table 1). Transient gene expression has been demonstrated in all these tissues, indicating that DNA is being delivered to cells, and in some cases organelles (64,94,95), in the target tissue. The integration of this DNA into the plant genome and subsequent transfer into sexual progeny has, however, been reported only for soybean (61), tobacco (96), and maize (55,56,87). These encouraging results have prompted efforts to optimize this method by analysis

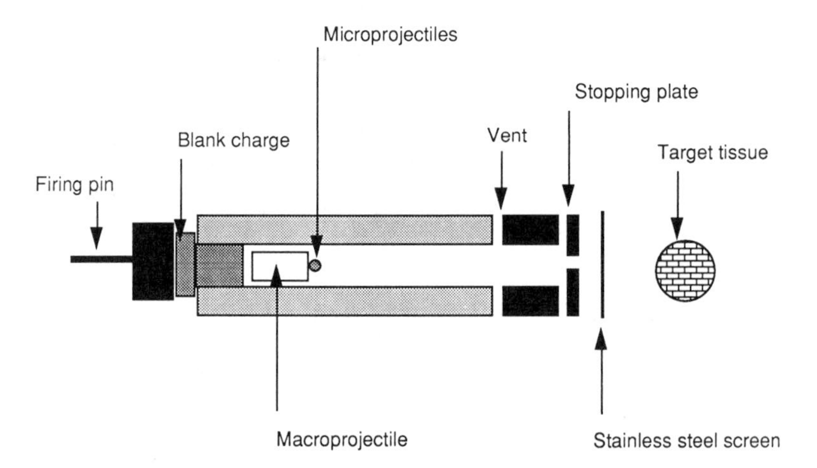

Figure 1 Diagrammatic representation of the microprojectile gun. The vacuum chamber of the gun has not been included.

of the basic parameters involved in successful gene transfer. These parameters are outlined in more detail in the following sections.

A. Vector Parameters Influencing Gene Expression in Plant Cells

1. Vector Construction

The recovery of transformed plant cells requires vectors that ensure efficient production of a functional gene product in the targeted plant cell. Plant vectors used for free DNA delivery typically consist of an expression cassette that contains a promoter region, transcription initiation site, and a portion of the 5′ nontranslated leader of the promoter of interest joined to a synthetic multilinker, followed by a polyadenylation signal. The synthetic multilinker with unique restriction sites allows the insertion of coding sequences derived from prokaryotic or eukaryotic genes, including genomic sequences with introns or cDNAs. This structure allows the construction of a variety of vectors with single or multiple combinations of reporter genes, which allow rapid detection of DNA transfer, and selectable marker genes, which can be used to recover transformed cells under selection. The selection of the individual components that make up a plant vector are critical to the successful expression of introduced genes in plant cells. The individual elements chosen depend on the processing abilities of the plant cell into which a vector is to be introduced. Not all promoters are equally active in all plant species, and certain sequences present in the gene cassette may be efficiently processed in one species but not in another. A detailed discussion of the vector elements important in successful gene expression may be found in Refs. 98–100. A brief review of more recent developments in the design of vectors for use in monocot plants is included here. The reporter genes that are essential in the functional analysis of vector components in plant cells and the selectable markers used in the isolation of stable transformants are also described.

2. Reporter Genes

The presence of DNA in plant cells can be demonstrated within 24 hours after gene transfer by means of transient assays. These assays are based on the expression of genes, called "reporter genes," that are expressed in plant cells without integration into the genome. Hence, they can be used to demonstrate the successful transfer of DNA into plant cells and can also be used in preliminary screening for the effect of vector modification on gene expression. The reporter genes that are most often used include genes encoding enzymes with

Table 1 Tissue Targets, Reporter and Marker Genes, and Species Used for Microprojectile Bombardment

Species	Target Tissue	Reporter gene	Marker gene	Reference
Monocots				
Barley	Suspensions	GUS/*NPT II*		68
	Suspensions	CAT		67
	Embryos	CAT		67
	Microspore callus	GUS		66
Corn	BMS suspensions	GUS		79
	BMS suspensions	CAT		82
	BMS suspensions	GUS	*NPT II*	83,84
	BMS suspensions	GUS	*bar*	85
	BMS suspensions	GUS		64
	Suspensions	CAT		82
	Suspensions	LUC	*ALS*	55
	Suspensions	GUS	*bar*	56
	Callus	GUS	*bar*	55
	Embryos	GUS		79
	Embryos	*R*		78,88
	Embryos/callus	*Bz1*		89,90
	Embryos/callus	*B-I, B-peru*		91
	Embryogenic callus	*C1*		92
Pearl millet	Embryos	GUS		80
Rice	Suspension	CAT/GUS		63
	Callus	GUS		64
Wheat	Suspensions	CAT/GUS		63
	Leaf bases	GUS		64
	Suspensions	GUS	*NPT II*	69
	Leaf	GUS		65
	Callus	GUS		65
	Embryos	GUS		62
Sugarcane	Suspension	GUS	*bar*	81

Table 1 Continued

Species	Target Tissue	Reporter gene	Marker gene	Reference
Dicots				
Papaya	Suspension	GUS	*NPT II*	59
Soybean	Meristems	GUS	*NPT II*	60,61,76,93
Tobacco	Pollen	GUS		77
	Suspensions	GUS		72,73
	Leaf/chloroplast	GUS	*SPC2*	94
	Leaf/chloroplast	GUS		95
	Leaf	GUS	*NPT II*	96,97
Cotton	Suspension	GUS	*APH IV*	58
Populus	Nodules/stems	GUS	*NPT II*	57

distinct substrate specificities that can be monitored by radiochemical (chloramphenicol acetyltransferase, CAT) (101), histochemical or fluorometric (β-glucuronidase, GUS) (102), or luminescence (luciferase, LUC) (103) assays. Reporter genes have also been developed by using the anthocyanin genes of maize, which, when bombarded into cells, result in the production of a purple pigmentation in the cells receiving DNA (88,89). The latter assay has the advantages of being nondestructive and allowing the analysis of the fate of introduced DNA over time. Reporter genes are useful tools in the analysis of gene expression and have been used to identify regulatory sequences in genes (104–109), to characterize the sequences involved in gene responses to environmental and physiological stimuli (110–115), to determine the activity of heterologous promoters in protoplasts from different origins (116), and to evaluate parameters that influence the transfer and expression of DNA (62–64,79,80).

3. Promoters

Promoter strength is critical in allowing high-level transcription of selected coding sequences in plant cells. The use of promoters that drive gene expression in monocot cells ensures the expression of genes introduced into these cells by microprojectile bombardment. Hence, the development of good monocot promoters is one of the components necessary for improving the recovery of stable transformants in cereal crops. Many different promoters, including those derived from viruses, bacteria, and plants, have been used in the construction of plant DNA vectors. These promoters include the 5′ regulatory regions derived from the T-DNA genes for opine synthase (117) and nopaline synthase (118), dual promoter fragments derived from T_R-DNA of octopine Ti plasmids (119), the 35S promoter from cauliflower mosaic virus (CaMV) (103,120,121), and the 35S promoter from figworth mosaic virus (122). Comparative studies with many of these promoters have demonstrated that in most cases the 35S promoter provides constitutive and high levels of expression of an array of genes in plant cells. Hence, the CaMV 35S promoter has become the most widely used promoter in plant transformation studies. The CaMV 35S promoter has been used to drive transcription in monocots and has allowed the successful recovery of transformed callus in wheat (69), maize (24), and barley (23) and transgenic plants in maize (31,55,56) and rice (37,38). Although the 35S promoter is effective in monocot transformation, it is not regarded as the optimum promoter for use with all monocots. In transient assays the 35S promoter is only one-fourth as efficient as the 2′ promoter of the octopine T-DNA in rice protoplasts (123). Furthermore, the alcohol dehydrogenase 1 (ADH1) promoter has allowed the recovery of 10–20 times as many stable lines of rice from rice protoplasts (124). This promoter is, however, anaerobically induced and not present in all tissue. Recently a rice actin promoter was isolated and shown to provide a 6-fold increase in gene expression over the maize ADH1 promoter when introduced into rice protoplasts (125).

Eucaryotic promoters are usually composed of multiple discrete functional modules that each contain one or more recognition sites for proteins that trigger defined frequencies of transcription starts. These units are called *cis*-acting elements, and temporal and spatial regulation of gene expression is possible because of the distinct interactions between these elements (126). Individual promoters may be dissected to determine the functional regions or the presence of specific enhancers or elements that determine tissue specificity. These studies are important in the context of the development of strong and tissue-specific promoters, because they demonstrate the potential for tailoring promoters for specific functional roles in plant cells. The level of expression of a gene can often be increased by adding *cis*-acting sequences, also called enhancer elements, which increase the level of transcription from a promoter. Adding an enhancer element already associated with a gene can also increase expression of the gene, as is shown, for example, by the addition of multiple copies of the anaerobic response element (ARE) to the maize ADH gene (112). Alternatively, enhancers can be obtained from other sources, such as viral or bacterial genomes, and combined with the gene of interest. These combinations can give high gene expression, as found with the octopine synthase (OCS) enhancer from *Agrobacterium tumefaciens*, which gives high expression from the ADH promoter in maize whether or not the protoplasts are anaerobically induced (127). With this strategy promoter constructs have recently been developed by using combinations of the ARE and OCS elements (109). The activity of a series of these promoters was compared with the CaMV 35S promoter by using GUS as a reporter gene in protoplasts of five different monocot species [wheat, maize, rice, einkorn (*Triticum monoccum*), and *Lolium multiflorum*]. From this analysis a 10- to 50-fold increase in expression over the CaMV 35S promoter was found for a promoter that consisted of six ARE and four OCS elements placed upstream of a truncated ADH promoter (206 bp) with an ADH1 intron 1 downstream of the ADH promoter in the untranslated leader (106 bases) (109). When tested in sugarcane, this promoter gave a 50- to 100-fold increase in expression (128). These studies emphasize the potential for mix-and-match promoter constructs that combine enhancers and core promoter sequences for increased gene expression.

The characterization and functional analysis of tissue-specific promoters is also important for developing gene cassettes that will provide tissue-specific expression of introduced genes. Such promoters are important because they allow targeting of gene expression to sites where genes are required, such as the embryo for expression of seed storage protein genes or leaves and stems for protection against insects or viruses. Many monocot promoters have been evaluated in model dicot systems and have been shown to maintain their tissue specificity. These include the tandem promoter of a maize zein gene (129), a high-molecular-weight wheat glutenin promoter (108), the chloroplast FBPase promoter of wheat (130), a pollen-specific promoter from maize (131), the abscisic acid–responsive rice *rab* gene promoter (132), and a series of photosynthetic gene promoters from maize (133). The strength of these promoters relative to those presently used in monocots remains to be tested. There is also evidence that promoters retain their response to regulatory signals, such as hormones or light, when introduced into monocot protoplasts; this is demonstrated in the regulation of the α-amylase promoter by gibberellic acid and abscisic acid (ABA) (111,113), the induction of the EM promoter by ABA (114), and the induction of a rice *rab* gene by ABA (134). It is unclear how these promoters will function when introduced into monocot tissues. The transfer of these and other promoters with appropriate reporter gene sequences into plant cells with the microprojectile gun should allow a rapid functional analysis of these promoters in an array of tissue types and under different environmental stimuli.

4. Introns

Introns are noncoding regions present in many eukaryotic genes, and their inclusion in the transcription units of a number of gene constructs has been shown to increase gene expression in monocot transient expression assays. Introns that have been shown to increase gene expression include the maize *Adh1-S* intron 1 (104,107), the single maize intron of *Bronze* (*Bz1*) (104,107), the first maize intron of *Shrunken1* (106,135), the maize *Adh1* introns 2 and 6 (105), the maize *Adh1-S* intron 6 (136), the intron from the 82-kDa heat-shock gene of maize (137), and the rice introns from *Act1* (125) and *Act3* (107). In all cases tested the inclusion of an intron resulted in increased levels of steady-state RNA. Although the mechanisms underlying intron enhancement of gene expression have not been fully elucidated, it has been proposed that the increased steady-state RNA observed is a result of intron-stimulated transcriptional initiation or increased mRNA stability in the nucleus or increased mRNA export to the cytoplasm. The degree of enhancement has also been shown to depend on the exon sequences flanking the intron; flanking exons influence the efficiency of intron splicing (106,107,125).

In maize, the *Adh1* intron in *Adh1-S* and *Adh1* promoter constructs increased CAT activity 50- to 100-fold. Both the *Adh1* intron and *Bz1* intron, also from maize, stimulated gene expression from other promoters, including the 35S and nopaline synthase (NOS) promoters, and increased expression from regions coding for the reporter genes CAT, LUC, and *NPTII* (104). Enhanced gene expression mediated by the *Adh1* intron has also been demonstrated in heterologous species (135). The stimulation of expression by introns has been shown to be position dependent, and the optimal location for introns is near the 5′ end of the mRNA (104,105).

5. Selectable Markers

Plant vectors suitable for gene expression in plants have been used to introduce selectable markers into plant cells for the recovery of transformed cell lines.

Because of the low stable transformation frequency currently obtainable with microprojectile bombardment [1×10^{-5} to 1×10^{-3} (87,89)], metabolic or visual selection strategies are required for the recovery of transformants. The criteria important for a good selectable marker include clean selection (that is, few escapes) and no adverse effects on plant regeneration and fertility. An array of selectable markers have been used for plant cells. These include the gentomycin acetyltransferase gene (138), bacterial (139) and plant EPSP synthase genes (*EPSP*) (8), the bacterial phosphinothricin acetyltransferase gene (*BAR*) (9), the *Arabidopsis*, maize, and tobacco acetolactate synthase genes (*ALS*) (55,140,141), the bacterial neomycin phosphotransferase II gene (*NPT II*), (24,31,87,123), the bacterial hygromycin B phosphotransferase gene (*HPH*) (38,142), and the mouse dihydrofolate reductase gene (*DHFR*) (143). Selectable markers used to date in the recovery of transformants from microprojectile bombardment includes the *BAR, NPT II, EPSP*, and *ALS* genes (see Table 1). Kanamycin selection using the *NPT II* gene has widely been used in the recovery of transformed dicotyledenous plants. Monocotyledenous cells have, however, been quite resistant. Although transformed calli can be recovered (69,87), there are many escapes, which limit the efficiency of this marker. Basta, bialaphos, and chlorsulfuron selection have both been used successfully to recover transformants in maize (55,56,87) and are most effective when selection is continued through all stages of culture and regeneration. The importance of maintaining selection throughout regeneration was highlighted in a study of the regeneration of maize plants from lines resistant to Basta. All the plants produced without selection were shown to lack the introduced gene, but plants regenerated on selection contained the *bar* gene (55).

B. Gun Parameters Influencing DNA Delivery

The success of microprojectile bombardment depends on the penetration of plant tissues and transfer of DNA with minimum damage to the target cell. Several gun parameters can be modified to optimize DNA transfer and minimize tissue damage. These include particle size and density, DNA quantity, tissue distance from the stopping plate, stopping plate aperture, number of bombardments, method of DNA precipitation, and vacuum.

The efficiency of DNA delivery in different tissues with different gun parameters has been evaluated (62–64,79,80). Transient gene expression has been analyzed in maize, rice, and wheat suspension cells and in wheat and pearl millet embryos. The bacterial GUS gene (102) has generally been the reporter gene of choice for monitoring the transfer of DNA by staining for GUS with the substrate 5-bromo-4-choloro-3-indolyl glucuronide (X-Gluc). This simple transient assay system gives visible evidence of the transfer and expression of the foreign gene, and differences in expression are readily monitored by eye or quantified by a fluorometric assay (102). By using these techniques an array of physical factors have been shown to influence DNA delivery to the target issue.

1. Particle Size

Bombardments with tungsten particles of diameters of 0.6, 1.2, and 2.4 μm indicate that the diameter of microprojectiles influences DNA delivery. Microprojectiles of 0.6 or 2.4 μm were less efficient for gene delivery to BMS cells than the 1.2-μm microprojectiles (79). By using the same range of sizes, gene transfer in wheat was also shown to be most efficient with microprojectiles 1.2 μm in diameter (63). Small differences in diameter evidently influence the ability of the microprojectile to penetrate cell walls and membranes.

2. Particle Density

The efficiency of transformation is affected by the concentration of particles in the DNA–particle mix. Studies with wheat indicate that concentrations of particles of 35.7 and 71.14 mg/ml were 20 times as effective as the lower concentration of 17.8 mg/ml tested (63).

3. DNA Concentration

The amount of DNA precipitated onto microprojectiles can significantly affect the level of particle aggregation. Aggregated particles do not, however, appear to be effective for DNA delivery; increasing the concentration of DNA above 2 μg of DNA per milligram of tungsten had a detrimental effect on gene expression (79).

4. Precipitation Procedure

The absorption of DNA directly to the surface of the microprojectile is essential for efficient DNA delivery. Both $CaCl_2$ and spermidine are necessary for good DNA precipitation onto particles. The concentrations of $CaCl_2$ and spermidine are critical, and optimum concentrations have been shown to be 0.24–1.9 *M* for $CaCl_2$ and 100 m*M* for spermidine (79). The precipitation of DNA onto particles has its drawbacks. Not all particles are coated evenly with DNA, and it is not yet known exactly how much DNA remains on the particle when it finally penetrates the cell. In pearl millet embryos, for example, gold particles have been seen at a distance of 12 cell layers from the surface; however, no expression of the GUS reporter gene was found in these cells (80). Although this may be related to penetrance of the GUS substrate, it is also possible that no DNA remained on the particles delivered to these cells.

5. Particle Volume

The volume of microprojectiles loaded onto the leading face of the polyethylene macroprojectile influences "blow away" of cells. The GUS expression in BMS cells was shown to be highest when small volumes (1.2 or 2.5 μl) were used, whereas volumes of 5 or 10 μl led to a 3-fold reduction in the number of expression units observed. This was related to the blow away of cells because the impact of larger volumes clearly caused severe loss from the central zone of target tissue (79).

6. Vacuum and Distance

Parameters that influence the velocity of the microprojectiles affect their ability to deliver DNA. The degree of vacuum in the sample chamber during bombardment has been shown to influence particle velocity. Studies have indicated that increasing the vacuum from 10 to 28 in. Hg can increase DNA delivery by 350-fold (79). The increase in DNA delivery with increasing vacuum is probably due to the reduction of air resistance that both the macro- and microprojectiles encounter, resulting in more efficient penetration of cell walls and membranes.

The microprojectiles' final velocity is related to air resistance. Hence, the distance traveled will affect the velocity and thereby the penetration of particles. When a sample was placed at various distances from the stopping plate, the highest gene expression in BMS cells was observed when the sample was at a distance of 6 cm. Increasing the distance gave a correlated decrease in DNA delivery with a 3-fold drop in expression at 17 cm (79). Similar results have been demonstrated in pearl millet embryos (80) and in BMS with the air gun apparatus (64).

7. Stopping Plate Aperture

Stopping plates 4.8 mm thick with apertures of 0.8, 1.6, and 2.4 mm were tested with wheat cells. Assays with the GUS gene as a reporter indicated that an aperture of 0.8 mm gave the best results, with a 3- to 4-fold increase over the other two apertures, respectively (63).

8. Number of Bombardments

Bombarding the sample two or three times increases the number of cells giving transient gene expression. In rice and wheat cell suspensions the increase in gene expression is almost proportional to the number of bombardments, with two and three bombardments increasing gene expression by 2- and 3-fold, respectively (63). Wheat embryos (62) and BMS suspensions (82) treated with multiple bombardments also displayed an increase in multiple expression events. Increasing the number of bombardments also increases the tissue damage (80). Hence, it is important to determine the balance between excessive tissue damage and increased DNA transfer for a given tissue target.

9. Method of Particle Propulsion

Many different methods for propelling particles have been invented. Different propulsion devices are intended to improve transformation efficiencies through a reduction in cell injury, selective targeting of cell layers, reduction of gas blasts (which cause blow away of cells), and greater control over acceleration of particles. To date, the only published comparative experiment involves the use of the helium gun. In this report the number of transformants recovered in tobacco by using the helium gun was five times that when the gunpowder gun was used (75). Further comparative studies using monocot cells are needed to determine whether transformation efficiency can be increased by using different methods of particle propulsion.

The studies just outlined have emphasized the importance of optimizing bombardment conditions to give good transient gene expression, which demonstrates the transfer of DNA. The optimization of these parameters is essential before undertaking functional gene expression analysis or selection for the recovery of stable transformants.

C. Target Tissues

A major advantage of particle bombardment transformation is the high degree of flexibility in choice of target tissues. Thus, the most regenerable and selectable tissue systems can be used as targets. Successful transformation involves DNA transfer through a sexual generation and requires the integration of DNA into cells that will give rise to the germline. The tissues that are accessible to bombardment and that appear to be the best targets for germline transfer of integrated DNA are microspores, pollen, and shoot meristems. However, although pollen bombardment has been used to demonstrate the expression of pollen-specific promoters (77), attempts to use bombarded pollen to transfer genes into tobacco have so far been unsuccessful (144). These attempts focused on the use of mature pollen in which the DNA of the sperm (trinucleate pollen) or generative nuclei (binucleate protein) has already completed the S phase of the two cell cycles of pollen development. Given the generally accepted view that DNA replication is required for DNA integration, microspores or early binucleate pollen appear to be better targets because the foreign DNA may become integrated into the genome of the generative nucleus during DNA replication in the S phase. Microspore calli, however, have a tendency to produce albino plants from tissue cultures. The recent improvements in the recovery of green plants from these cultures in both wheat (145) and barley (146) suggest that suitable lines are now available for microprojectile bombardment.

Shoot meristems of soybean are amenable to transformation by the electrical arc discharge gun (60,61). The successful recovery of transformed soybean plants, however, depends on the induction of adventitious shoots from the bombarded meristem. These shoots are subsequently rooted or grafted to produce plants that are chimeric. Transformed R_1 progeny can be produced from these chimeric plants (60,61). Hence, soybean germline cells cannot be specifically targeted, but chimeric shoots can be recovered and used to produce germline events. This procedure depends on the ability to induce adventitious shoots and is not applicable to all species. Attempts to transform the shoot meristem of maize have resulted in the production of chimeric plants in which transformed cells were not germline cells and could not be recovered (78). This demonstrates the difficulty associated with the bombardment of a multicellular shoot apex when we do not yet have the technology to target the distinct germline cells that lie in the LII layer of the maize shoot apex.

In the absence of a suitable target tissue for direct plant production, regenerable tissue culture cells provide the best target tissue currently available for the recovery of transformed plants. Unfortunately, not all commercial varieties can be induced to produce tissue cultures that are amenable to bombardment. In maize, only a narrow range of genotypes produce the type II embryogenic callus, which has been used successfully for transformation by the biolistics process. Similar problems exist in other crop species. Increased efforts have been made to extend the range of genotypes and tissues that can produce these type II cultures, and it has been found that the control of ethylene action by ethylene antagonists promotes the establishment of friable callus cultures from A188 (147,148), the agronomically important B73, and a proprietary B73/BC6 genotype (149). Induction rates as high as 70% have been observed when B73 callus was exposed to 10 μM $AgNO_3$. Ordinarily, less than 0.1% of the B73 explants produce friable embryogenic cultures. The positive effect of $AgNO_3$ has also been observed with cultured maize immature tassel tissues (150). Type II callus production increased by 40% when $AgNO_3$ was included in the medium. Ethylene antagonists have also been shown to promote plant regeneration by 2- to 13-fold in type I (compact) callus cultures of maize (151). Control of ethylene action to promote plant regeneration may be especially important in transgenic cultures that typically have a poor rate of plant regeneration. These studies may also be useful for developing suitable cultures of wheat, barley, and other crops that have not yet been transformed.

D. Target Tissue Treatments for Expression and Integration of DNA

The recovery of transformed plants depends on many target cell factors. The most critical of these are the division of transformed cells, their continued growth on selected media, and the maintenance of regenerative capacity.

Particle bombardment places extensive environmental stress on target tissues. These tissues are subjected to partial vacuum and bombardment with metal particles, both of which may cause extensive cell injury. Because the viability of cells is essential to the recovery of transformed cells, it is important to limit the impact of bombardment on cell viability as much as possible. This can be achieved by limiting the disruption of basic biochemical and physiological processes essential for cell viability. The effect of bombardment on the biochemical and physiological state of cells has not been widely studied, and only a few studies have addressed this important issue.

Cell walls are temporarily damaged by particle penetration, and turgor forces on these weakened cell walls may influence cell viability. Adding mannitol to increase osmolarity has been shown to increase GUS expression in BMS. The optimum concentration of mannitol, from a range of 2% to 8% (w/v), proved to be 6% for transient expression and 2% for the recovery of stable transformants in BMS (152). Mannitol pretreatment has also been effective in increasing the number of stably transformed chloroplasts recovered from tobacco cell bombardment (95).

Bombardment of cells has been shown to induce a dramatic increase in the amount of ethylene released. Studies on ethylene production in B73 × A188 embryogenic maize calli before and after bombardment indicate a 3- to 20-fold increase in ethylene compared with nonbombarded controls (153). An inverse relationship between transient gene expression and ethylene emanation has been observed in these callus cultures 4 days after bombardment. By using the maize *C1* and *B-peru* genes (see Section III), which code for anthocyanins as reporters of gene expression, it was found that anthocyanin spot number per gram fresh weight of callus decreased as the ethylene release increased (Fig. 2). The effect of ethylene on transient gene expression has also been studied in protoplasts of *Solanum tuberosum*, in which the expression of the CAT gene was increased by the addition of silver thiosulfate, an ethylene antagonist (154). Because ethylene production may be partially responsible for cell death

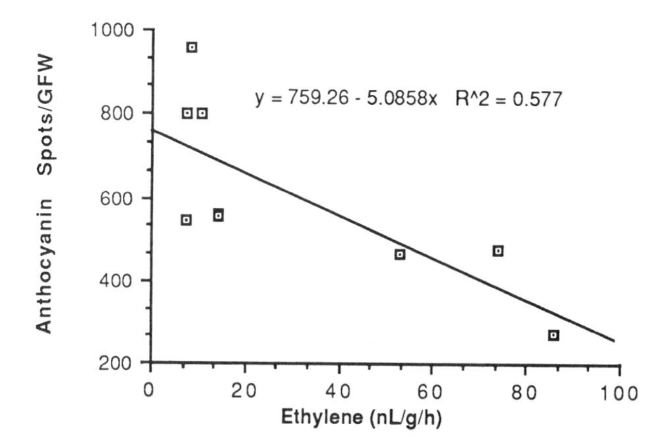

Figure 2 Effect of ethylene emanation from bombarded maize embryogenic B73 × A188 callus cultures on transient anthocyanin expression. Transient activity is quantified as anthocyanin spots per gram fresh weight of callus.

and reduced viability after bombardment, including an ethylene antagonist in the culture medium may limit its detrimental influences.

It is generally believed that the integration of DNA into plant cells requires the presence of actively replicating DNA. The best target tissues, therefore, contain a large population of actively dividing cells. Because of their rapid rate of cell division and regenerative capacity, embryogenic cultures are good targets for transformation (55,56,58,69). These cultures contain a large percentage of regenerable cells and grow in small clusters; this enhances the recovery of transformation events by metabolic selection.

Target tissues may also be pretreated to increase the population of actively dividing cells. Transformation has improved in *Populus* cultures after inducing increased meristematic cell division in target tissues by pretreatments with cytokinins (57). *Populus* transformation was also improved by the use of cell wall inhibitors, which partially synchronize the cells and provide a more uniform cell population (57).

III. USE OF MICROPROJECTILE BOMBARDMENT IN THE FUNCTIONAL ANALYSIS OF GENES

Gene expression in plants is controlled by a wide array of developmental, physiological, and environmental factors that alone or in combination significantly affect transcription and translation in plant cells (99,155–157). Our understanding of differential gene expression in plants has developed dramatically with the use of molecular techniques to characterize genes and evaluate their expression. The functional analysis of these genes can be achieved by creating gene constructs that, upon reintroduction into plant cells, can be used to monitor expression under different developmental, environmental, and physiological stimuli. It is assumed that controls that limit or enhance the expression of endogenous genes will also affect the expression of introduced genes.

Because of the small number of plants that can be transformed, the study of many interesting promoters and coding sequences has been limited to an analysis of their expression in model and sometimes heterologous systems (100,158). The development of microprojectile bombardment has made possible the evaluation of gene expression in a wide array of plant tissues. Since there is much variability between bombardments, great care must be taken when using microprojectile bombardment for evaluating functional gene expression. A definitive comparison of gene expression from a test vector is valid only if an appropriate internal control vector is used to compensate for the variability between bombardments (see Refs. 89–91).

The anthocyanin synthesis pathway of maize provides an interesting system for studying genes that are under environmental, developmental, and genetic control. The interaction of a series of structural and regulatory genes in the biosynthetic pathway results in the production of a purple anthocyanin pigment. The production of anthocyanin pigment in a given tissue depends on the presence of a functional set of structural genes, which include *A1, Bz1, and C2.* These genes require coordinate regulation by the regulatory genes *B, C1, PI*, and *R* (for a review see Ref. 159). For example, the regulatory gene *B* is required for pigmentation in most of the plant body (160), whereas *C1* is required only for pigmentation of the aleurone and embryo (161) and *R* is required for pigmentation in the aleurone, anthers, and coleoptiles (160). An allele of *B, B-peru,* has been shown to substitute for *R* in the aleurone (160), and the *Rscm2* allele of *R* conditions anthocyanin pigmentation in the maize embryo (160). Mutants for the regulatory genes lack anthocyanin pigmentation; when examined at the molecular level, the amounts of mRNA and enzymes encoded by the structural anthocyanin genes were shown to be affected (162). Mutants for the regulatory gene *C1* (*c1*), for example, are colorless and have low levels of mRNA and the enzymes encoded by the *A1* (NADH-dependent reductase) and *Bz1*

(UDP-glucose flavanol 3-*O*-glucosyl transferase) structural genes in aleurones (162,163).

Molecular evidence accumulating from sequencing analysis of the regulatory genes *C1, B, PI,* and *R* supports the proposed regulatory roles of these proteins. Comparisons with sequences of known regulatory proteins, such as the *myc* and *myb* oncogene family, demonstrate the presence of homologous DNA-binding domains. Furthermore, the *C1* gene was shown to contain a carboxy-terminal region rich in acidic amino acid residues, which is analogous to the acidic transcriptional activator domain found in many transcriptional activators (for review see Ref. 164). The elucidation of the function of the regions of *myc*- and *myb*-like homology in these regulatory genes has been undertaken by using microprojectile delivery of genomic clones of both the regulatory and structural genes (89,90,92).

In an initial study, genomic clones of the structural genes *A1* and *Bz1* were introduced into genetic backgrounds that were mutants for these genes (*bz1* and *a1*) and had nonpigmented aleurones. The introduced genes were shown to be functional, on the basis of the production of purple pigmentation by aleurone cells bombarded with these genes (89). This approach was also used to demonstrate the functional transactivation of the *A1* and *Bz1* promoters by genomic clones of the *B-peru* and *B-I* genes (alleles of the regulatory *B* gene) in the aleurone and embryo tissues of genotypes mutant for the *B* regulatory gene (91), and the complementation of the *c1* regulatory mutation in aleurones by bombardment with a *C1* genomic clone (92). These studies demonstrate that genomic clones of both structural and regulatory genes encode functional proteins able to complement mutations in various tissues after being delivered by microprojectile bombardment. In combination with other published results of studies using the *R* (88) anthocyanin regulatory gene, these studies demonstrate that a complex biosynthetic pathway can be activated by microprojectile delivery of the appropriate regulatory gene. Thus, they provide an assay system that could be used to evaluate the interactions between the regulatory and structural genes, which control the coordinate regulation of this complex biosynthetic pathway.

As was previously outlined, the tissue-specific expression of the structural gene *Bz1* is controlled by the products of the regulatory genes *C1* or *PI* and *R* or *B*, which share homology to the *myb* proto-oncogenes and *myc*-like genes, respectively. To determine regions in the *Bz1* promoter critical for *Bz1* expression, a series of *Bz1* promoter deletions were fused to a LUC reporter gene and introduced into embryos with the *C1* and *Rsc* regulatory alleles essential for *Bz1* expression in embryos. This analysis identified two regions essential for

Bz1 expression, both of which were shown to be homologous to the consensus binding sites of the *myb*- and *myc*-like proteins. Site-specific mutations of these *myb*- and *myc*-like sequences reduced *Bz1* expression to 10% and 1% of the wild type, respectively. The presence of critical *myb* and *myc* consensus binding sequences in the structural gene *Bz1* and the *myc* and *myb* homology shared by the regulatory proteins from *C1* and *R* suggest that *C1* and *R* proteins interact with the *Bz1* promoter at these sites.

The function of the domains of the *myb*-homologous *C1* in transcriptional activation of the anthocyanin structural genes has recently been elucidated (92). Codelivery of *C1* constructs with anthocyanin structural gene reporters was used to quantitate the functional activity of various *C1* constructs encoding deletions, frame shifts, and fusion proteins. In summary, the data from these experiments demonstrate the following: (a) The amino terminal and carboxy-terminal acidic regions of the *C1* gene can function as independent domains; (b) the *C1* acidic region functions as a transcriptional activator domain (this was demonstrated by the transcriptional activation of a minimal CaMV 35S promoter, containing the yeast Gal4 DNA recognition sites, by a fusion containing the DNA binding domain of the yeast Gal4 and the acidic carboxyl-terminal domain of *C1*); and (c) the *myc*-homologous basic domain acts with the transcriptional activator to stimulate transcription for the structural gene *Bz1* (this evidence was gained from the results of an experiment in which fusion proteins of the *myc*-homologous basic domain of *C1* and the yeast Gal4 transcriptional activation domain were found to stimulate transcription from a *Bz1* promoter LUC gene construct). These data, together with the demonstrated requirement for an intact *myc* consensus binding site in the structural *Bz1* promoter, strongly suggest that *C1*-regulated expression of the *Bz1* promoter is mediated through a *myb* consensus DNA binding site.

One of the critical environment signals for normal growth and development in plants is light. Light not only acts as a source of energy but also as a stimulus that regulates numerous developmental and metabolic processes. The light regulation of plant gene expression occurs through a set of regulatory photoreceptors. The best characterized of these photoreceptors, phytochrome, is responsible for mediating a large number of the light-induced morphogenic changes in plants (for reviews see Refs. 99, 157, 165, and 166). Phytochrome occurs in two forms, an inactive Pr form that is converted to an active Pfr form by red (R) light. This response is reversible, and the photoreceptor is converted back to its inactive form by far-red (F) light. Phytochrome has been shown to modulate the expression of a number of nuclear genes in either a positive or negative manner (167); however, the transduction mechanism by which phytochrome regulates gene expression is largely unknown. Phytochrome is known to negatively regulate (down-regulate) the transcription of its own *phyA* genes: High levels of Pfr repress *phyA* transcription, and low Pfr levels result in derepression. In monocots the transcription of the *phyA* gene has been shown to be repressed within 5 min of Pfr formation (168). This repression occurs in the absence of protein synthesis and suggests that all the components necessary for transduction exist in the cell before light perception and that *phyA* genes are therefore regulated directly by the primary phytochrome signal. To investigate whether a *phyA* gene isolated from oat would maintain its photoregulatory activity when reintroduced into monocot cells, the activity of the oat *phyA* was assayed by microprojectile-mediated gene transfer into etiolated rice seedlings (169). When bombarded into rice cells, a 1-kb *phyA3* oat promoter fused to CAT responded reversibly to R and F light. Low Pfr concentrations induced high *phyA3*/CAT expression, whereas high Pfr repressed activity to near-basal levels. These results indicate that this cloned promoter fragment contains the sequence necessary for autoregulation of *phyA* genes in rice cells.

To identify regions of the oat *phyA3* gene involved in autoregulation, chimeric constructs containing various deletions and sequence substitution mutants of the oat *phyA3* gene fused to a CAT reporter gene were introduced into etiolated rice seedlings by particle bombardment. Removing *phyA3* sequences 3′ to the transcription start site reduced expression about 5-fold, suggesting that intron 1 of the *phyA3* gene may be required for high activity. The degree of high-Pfr-imposed repression is unaffected by any of a series of deletions of sequence substitutions in the *phyA3* promoter, thus providing no evidence of any Pfr-activated negative elements. In contrast, 5′ and internal deletions identify a minimum of three major positive promoter elements, designated PE1 (-381 to -348 base pairs), PE2 (-635 to -489 bp), and PE3 (-110 to -76 bp), that are necessary for high-level expression in low-Pfr cells (170).

The mRNA levels of genes that are positively controlled by phytochrome show a different expression pattern from that of phytochrome. In these genes the transcript levels increase, rather than decrease, in the presence of R illumination. These responses demonstrate the positive and negative effects that light can have on gene transcription.

The genes encoding the small subunit of ribulose-1,5-bisphosphate carboxylase-oxygenase (*rbcS*) are among the most extensively studied light-responsive genes (for a review see Ref. 99). The control of individual *rbcS* genes by phytochrome has been demonstrated by in vitro transcription experiments. Gene transcripts are present in small amounts in etiolated tissue and increase in abundance after 1 min of R illumination; this effect can be reversed if R illumination is immediately followed by 10 min of FR (171). Recent work on *rbcS* has focused on identifying regulatory DNA sequences within the gene that could mediate transcriptional responses through sequence-specific interactions with DNA-binding proteins (157). The identification of DNA–protein interactions involved in light regulation is an important step toward understanding the pathways of signal transduction and their interconnections.

In the aquatic monocot *Lemna gibba*, the genomic clone *SSU5B* is the most abundantly expressed of the available genomic clones in response to R illumination of dark-treated plants (171). Within the *SSU5B* promoter there are several conserved regions that may function in conferring sensitivity to regulation by phytochrome. To test which regions are sensitive to phytochrome regulation in vivo, a transient assay system using microprojectile bombardment was developed for *Lemna*. First, control experiments were carried out to demonstrate that the SSU5B promoter was light regulated. A control plasmid with a maize ubiquitin promoter fused to the CAT gene with a NOS 3′ untranslated region (ubi-CAT) gave high levels of CAT expression in both light- and dark-treated samples. In contrast, CAT expression from a construct driven by the SSU5B promoter (p5BCATNOS) was very low in fronds incubated in the dark and substantially higher in those incubated in white light (171).

To demonstrate regulation of the p5BCATNOS construct by phytochrome, etiolated *Lemna* fronds were bombarded and then exposed to a variety of light treatments. To compensate for the production of Pfr due to the flash of light produced by the particle gun, all samples received 2-min FR after bombardment. Dark-treated plants showed a low level of expression, which was stimulated 12-fold by a single 1-min R illumination. If FR immediately followed R, little stimulation was observed, thereby demonstrating that expression of the introduced gene was regulated by phytochrome. By using this transient assay and constructs from a 5′ deletion series of the *SSU5B* promoter, a region located between -205 and -83 was shown to be sufficient for phytochrome-regulated expression (171). This result correlates well with studies on the DNA binding factor LRF-1 present in *Lemna* nuclear extracts. The LRF-1 interacts with the SSU5B promoter within this region in a light-regulated manner (172). Thus, this study showed close correlations between DNA elements identified by in vitro binding experiments and phytochrome-regulated transcription in vivo.

In tomato, two anther-specific genes have been characterized and isolated as cDNA clones (*LAT55, LAT58*; see Ref. 173). To characterize *cis*-acting DNA sequences, which are required for the pollen-specific gene expression of these clones, both stable and transient assays have been used (77,173). In transient assays using particle bombardment, the tissue-specific expression of the GUS reporter gene was compared when driven by the 5′ flanking DNA of the pollen-specific gene *LAT52* (pLATR52-7) or with the CaMV 35S promoter (pB1221). In these experiments approximately 3×10^5 mature pollen grains, covering an area of 4 cm^2, were bombarded with pLATR52-7 DNA or pB1221. After a 16-hr incubation period, pollen was stained with X-Gluc. Intense blue staining of pollen grains was observed in samples bombarded with pLATR52-7; in contrast, only pale blue stained pollen grains were observed after bombardment with pB1221. Quantitative measurements of this GUS activity indicate it was approximately 1000-fold lower than that detected in pollen bombarded with pLATR52-7. When intact tomato flowers were bombarded and incubated with X-Gluc, blue-stained cells were consistently detected in anthers and petals with pB1221 but were never observed in flowers bombarded with pLATR52-7. Leaves bombarded with pLATR52-7 also showed no GUS expression. This expression pattern is similar to that found for this construct when stably introduced into tomato (173). These data confirm the correct regulation of the *LAT52* promoter in bombarded tissue and demonstrate the use of this system for analyzing tissue-specific gene regulation.

The *cis*-acting regulatory regions in the promoters of three pollen-specific genes, *LAT52*, *LAT59*, and *LAT56*, were characterized by using transient assays of 5′ promoter deletion constructs driving *GUS* as a reporter gene (174). Only minimal promoter regions of the *LAT52* (-71 to +110 bp) and *LAT59* (-115 to +91 bp) genes are required for their correct, regulated expression during pollen development. Several upstream regions that appear to modulate the activity of the minimal promoters in pollen were also identified. Two upstream activator regions were identified in the *LAT52* promoter, and a 19-bp sequence in one of these regions was shown to act as an enhancer of the heterologous CaMV 35S promoter in pollen (174). This is the first analysis of the organization of *cis*-acting sequences that regulate gene expression during gametophyte development in plants.

IV. INHERITANCE AND EXPRESSION OF INTRODUCED GENES IN MONOCOT PLANTS

Mendelian inheritance and expression of foreign genes is the ultimate evidence needed to confirm the integration of functional genes into plants produced from transformation experiments. The analysis of these events in monocots is especially interesting because in nearly all cases the DNA has been delivered without the aid of *Agrobacterium* Ti borders. Much of the evidence available for gene expression in monocot cells after the integration of DNA (stable gene expression) comes from studies on callus tissues. Owing to the recent developments in monocot transformation, there is also a growing body of research on the expression of genes in transgenic plants and their progeny.

Southern analysis, the polymerase chain reaction (PCR), and dot blots have been used to demonstrate the presence of foreign DNA in monocot genomes. Dot blots and PCR allow the rapid analysis of a large number of samples; however, they do not demonstrate the integration of the DNA into the plant genome and cannot be used to distinguish between different transformation events.

Southern analysis provides the best method presently available to determine the integration, copy number, and physical state of the introduced DNA. Cutting DNA with restriction enzymes, which are not sensitive to methylation and which cut at a single site within the plasmid DNA, allows the presence of bordering genomic DNA fragments to be detected. The presence of an intact gene can be demonstrated by cutting out the gene cassette and probing with all or a portion of the gene coding sequence. If the gene is intact, it should give a distinct hybridizing band the same size as that produced by plasmid DNA cut with the same restriction enzyme.

When introduced into monocot callus, the selectable marker genes *bar* (55,56,85), *NPT II* (24,26,31,87), *ALS* (55,87), *EPSP* (69,87), *APH II* (20,175), and *HPH* (37,38,176) have all proved to be functional and integrated into the genome. Where tested, independent events were shown to have different patterns of integration (26,56,85). In most instances the introduced gene was shown to be associated with characteristic border fragments from the genomic DNA, demonstrating that for each event introduced genes have been integrated into different regions of the plant genome. The number of transgene copies has been shown to vary from 1 to 20 copies. Copy number, however, does not appear to influence gene expression because studies on both the GUS gene and *bar* indicate no correlation between copy number and the level of expression of these genes (26,56). This lack of correlation may be related to position effects due to the site of integration of the introduced genes.

In animal and insect cells the expression of the transgene introduced by free DNA delivery is known to be influenced by position effects that can result in silencing or enhancement of gene expression (177,178). Studies specifically addressing this problem have not yet been conducted in cereal tissue cultures, but the range of variation found in expression of transgenes in different transformed lines (56,85,97) suggests that in plant cells the site of integration also affects gene expression. The genome of plants is known to be highly methylated (179), and it is therefore not surprising that there have been reports of methylation of genes introduced into maize cells, which reduced the expression of the transgene (84). In BMS, methylation of the GUS gene appeared to occur over time; GUS expression was found to decrease with duration in culture (84). An analysis of the introduced gene using a pair of isoschizomers with differential sensitivities to methylation of cytosine has conclusively demonstrated the methylation of the transgene. The significant increase in GUS activity after treatment of cells with 5-azacytidine (an antagonist of DNA methylation) further confirms that methylation did influence the expression of the transgene (84).

Transformation studies in animals (180) and dicot plants (181) using free DNA delivery have demonstrated the concatemerization of plasmid DNA and its introduction into the genome as tandem arrays. The pattern of integration of DNA found in some transformed lines of rice (38) and maize (26) also suggest a similar phenomenon occurs in plant cells. Although the introduced genes remain functional, there is an influence on the level of gene expression. This is evident from the lack of correlation found between copy number and expression of genes. This emphasizes the continuing need, in the development of commercially important crops through genetic engineering, to recover a number of independent transformed lines in order to produce a line with commercial levels of gene expression. It is evident that further developments in vector construction are required to target genes to unmethylated regions of the genome. Transgenes also require protection from the influence of adjacent genomic DNA, which may be achieved by including appropriate border fragments, such as those demonstrated to reduce position effects in *Drosophila* (177).

Progeny analysis of the inheritance of transgenes in cereal crops has been possible only for rice (39,176) and maize (55,56). In maize the transgenic plants produced were male sterile and progeny were derived from outcrosses with pollen from nontransformed control plants. The absence of pollen shed in these transformants is believed to be due to the age of the cultures from which they were derived, and recent evidence indicates that plants derived from younger embryogenic cultures will shed viable pollen (86,87). Enzyme activity, spray data, and Southern analysis of R_1 progeny from transgenic maize plants bearing both the ALS and LUC genes demonstrate the inheritance of these genes in a 1:1 ratio (55). The presence of the *bar* gene and the expression of phosphinothricin acetyltransferase activity was demonstrated in 29 out of 53 R_1 maize plants tested (56). In rice, germinating seedling progeny from selfed transformed plants were shown to have resistance to hygromycin, and the presence of the hygromycin gene was demonstrated by Southern analysis. Inheritance data indicated that in most cases resistance to hygromycin closely conformed to the mendelian inheritance of the trait expected for a gene integrated as a single copy (39).

These studies provide evidence for the inheritance and expression of genes introduced into monocot cells. Further inheritance data and an analysis of the maintenance of expression of introduced genes in subsequent generations are needed to demonstrate the continued expression and genetic stability of these genes.

V. CONCLUSION

Microprojectile bombardment has dramatically improved our ability to transform plant cells. Increases in transformation efficiency will require further developments in gun technology, where there is a need to fine-tune the targeting of microprojectiles. Currently there is poor control over size, aggregation, coating, quantity, dispersal, and velocity of particles, and this lack of uniformity leads to major random fluctuations in results. There is also a need to further develop tissue targets and conditions for growth, selection, and regeneration, which facilitate the rapid recovery of transformed cells that maintain their regenerative capacity and fertility. Improvements in the recovery of green plants from pollen microspores (145,146) and the regeneration of completely fertile transgenic plants from callus cultures (86,87) should encourage further studies to extend the range of suitable target tissues available. Improvements in vector construction, which will ensure optimum gene expression in monocot cells, will only come with a continued analysis of various combinations of genetic elements targeted to different species and tissues. It is unlikely that there will be a universal combination of elements to serve all functions, and only extensive and detailed testing will allow the development of vectors that can be tailored for a specific function. The understanding that has been gained through the functional analysis of genes will greatly facilitate the development of these vectors.

Microprojectile bombardment has been successfully used to demonstrate that cloned plant genes, when reintroduced into plant tissues, are regulated in the same manner as endogenous genes. Microprojectile bombardment, therefore, provides a powerful tool for the rapid study of gene regulation and interaction in plant tissues.

The influence of the position of DNA integration and the interaction of vector and genome elements on the expression of introduced genes needs to be fully evaluated in monocots. With greater understanding of the processes by which plant genomes control gene expression and respond to the presence of an intruding foreign gene it should be possible to design strategies for the more efficient introduction and inheritance of introduced traits in monocot crops.

REFERENCES

1. Hinchee, M. A., Newell, C. A., Connor-Ward, D. V., Armstrong, T. A., Deaton, R., Sato, S. S., and Rozman, R. J. Transformation and regeneration of non-solanaceous crop plants, *Gene Manipulation in Plant Improvement* (J. P. Gustafson, ed.), Plenum Press, New York, 1990.
2. Klee, H., Horsch, R., and Rogers, S. *Agrobacterium* mediated plant transformation and its future applications to plant biology, *Annu. Rev. Plant Physiol. 38*: 467–486, 1987.
3. Nelson, R. S., McCormick, S. M., Delannay, X., Dube, P., Layton, J., Anderson, E. J., Kaniewska, M., Proksch, R. K., Horsch, R., Rogers, S. G., Fraley, R. T., and Beachy, R. N. Virus tolerance, plant growth, and field performance of transgenic tomato plants expressing coat protein from tobacco mosaic virus, *Biotechnology, 6*: 403–409, 1988.
4. Harrison, B. D., Mayo, M. A., and Baulcombe, D. C. Virus resistance in transgenic plants that express cucumber mosaic virus satellite RNA, *Nature, 328*: 799–802, 1987.
5. Cuozzo, M., O'Connel, K. M., Kaniewski, W., Fang, R., Chua, N., and Tumer, N. E. Viral protection in transgenic tobacco plants expressing the cucumber mosaic virus coat protein or its antisense RNA, *Biotechnology, 6*: 549–555, 1988.
6. Abel, P. P., Nelson, R. S., De, B., Hoffmann, N., Rogers, S. G., Fraley, R. T., and Beachy, R. N. Delay of disease development in transgenic plants that express the tobacco mosaic virus coat protein gene, *Science, 232*: 738–743, 1986.
7. Lawson, C., Kaniewski, W., Haley, L., Rozman, R., Newell, C., Sanders, P., and Tumer, N. Engineering resistance to mixed virus infection in a commercial potato cultivar: resistance to potato virus X and potato virus Y in transgenic Russet Burbank, *Biotechnology, 8*: 127–134, 1990.
8. Shah, D., Horsch, R. B., Klee, H. J., Kishore, G. M., Winter, J. A., Tumer, N. E., Hironaka, C. M., Sanders, P. R., Gasser, C. S., Aulent, S., Siegel, N. R., Rogers, S. G., and Fraley, R. T. Engineering herbicide tolerance in transgenic plants, *Science, 233*: 478–481, 1986.
9. De Block, M., Botterman, J., Vandewiele, M., Dockx, J., Thoen, C., Gossele, V. M., Movva, N. R., Thompson, C., Van Montagu, M., and Leemans, J. Engineering herbicide resistance in plants by expression of a detoxifying enzyme, *EMBO J., 6*: 2513–2518, 1987.

10. Fillatti, J. J., Kiser, J., Rose, R., and Comai, L. Efficient transfer of a glyphosate tolerance gene into tomato using a binary *Agrobacterium tumefaciens* vector, *Biotechnology, 5*: 726–730, 1987.
11. Mazur, B. J., and Falco, S. C. The development of herbicide resistant crops, *Annu. Rev. Plant Physiol., 40*: 441–470, 1989.
12. De Block, M., De Brouwer, D., and Tenning, P. Transformation of *Brassica napus* and *Brassica oleracea* using *Agrobacterium tumefaciens* and the expression of the *bar* and *npt* genes in the transgenic plants, *Plant Physiol., 91*: 694–701, 1989.
13. Schulz, A., Wengenmayer, F., and Goodman, H. M. Genetic engineering of herbicide resistance in higher plants, *Crit. Rev. Plant Sci., 9*: 1–15, 1990.
14. Fischhoff, D. A., Bowdish, K. S., Perlak, F. J., Marrone, P. G., McCormick, S. M., Niedermeyer, J. G., Dean, D. A., Kusano-Kretzmer, K., Mayer, E. J., Rochester, D. E., Rogers, S. G., and Fraley, R. T. Insect tolerant transgenic tomato plants, *Biotechnology, 5*: 807–813, 1987.
15. Hidler, V. A., Gatehouse, M. R., Sheermen, S. E., Barker, R. F., and Boulter, D. A novel mechanism of insect resistance engineered into tobacco, *Nature, 330*: 160–163, 1987.
16. Vaeck, M., Reynaerts, A., Hofte, H., Jansen, S., De Beuckeller, M., Dean, C., Zabeau, M., Van Montagu, M., and Leemans, J. Transgenic plants protected from insect attack, *Nature, 328*: 33–37, 1987.
17. Perlak, F. J., Deaton, R. W., Armstrong, T. A., Fuchs, R. L., Sims, S. R., Greenplate, J. T., and Fischoff, D. A. Insect resistant cotton plants, *Biotechnology, 8*: 939–943, 1990.
18. Altenbach, S. B., Pearson, K. W., Meeker, G., Staraci, L. C., and Samuel, S. M. Enhancement of the methionine content of seed proteins by the expression of a chimeric gene encoding a methionine-rich protein in transgenic plants, *Plant Mol. Biol., 13*: 513–522, 1989.
19. De Clercq, A., Vandewiele, M., Van Damme, J., Guerche, P., Van Montagu, M., Vandekerckhove, J., and Krebbers, E. Stable accumulation of modified 2S albumin seed storage proteins with higher methionine content in transgenic plants, *Plant Physiol., 94*: 970–979, 1990.
20. Uchimiya, H., Fushimi, T., Hashimoto, H., Harada, H., Syono, K., and Sugawara, Y. Expression of a foreign gene in callus derived from DNA-treated protoplasts of rice (*Oryza sativa* L.), *Mol. Gen. Genet., 204*: 204–207, 1986.
21. Werr, W., and Lorz, H. Transient gene expression in the Gramineae cell line: a rapid procedure for studying plant promoters, *Mol. Gen. Genet., 202*: 471–475, 1986.
22. Armstrong, C. L., Peterson, W. L., Buchholz, W. G., Bowen, B. A., and Sulc, S. L. Factors affecting PEG-mediated stable transformation of maize protoplasts, *Plant Cell Rep., 9*: 335–339, 1990.
23. Lazzeri, P. A., Brettschneider, R., Luhrs, R., and Lorz, H. Stable transformation of barley via PEG-induced direct DNA uptake into protoplasts, *Theor. Appl. Genet., 18*: 437–444, 1991.

24. Fromm, M. E., Taylor, L. P., and Walbot, V. Stable transformation of maize after gene transfer by electroporation, *Nature, 319*: 791–793, 1986.
25. Ou-Lee, T. M., Turgeon, R., and Wu, R. Expression of a foreign gene linked to either a plant virus or a *Drosophila* promoter after electroporation of protoplasts of rice, wheat and sorghum, *Proc. Nat. Acad. Sci. USA, 83*: 6815–6819, 1986.
26. Lyznik, L. A., Ryan, R. D., Ritchie, S. W., and Hodges, T. K. Stable co-transformation of maize protoplasts with *gus* and *neo* genes, *Plant Mol. Biol., 13*: 151–161, 1989.
27. Morrish, F. M., Vasil, V., and Vasil, I. K. Developmental morphogenesis and genetic manipulation in tissue and cell cultures of the gramineae, *Adv. Genet., 24*: 431–499, 1987.
28. Vasil, I. K. Progress in the regeneration and genetic manipulation of cereal crops, *Biotechnology, 6*: 397–402, 1988.
29. Rhodes, C. A., Lowe, K. S., and Ruby, K. L. Plant regeneration from protoplasts isolated from embryogenic maize cell cultures, *Biotechnology, 6*: 56–60, 1988.
30. Prioli, L. M., and Sondahl, M. R. Plant regeneration and recovery of fertile plants from protoplasts of maize (*Zea mays* L.), *Biotechnology, 7*: 56–60, 1989.
31. Rhodes, C. A., Pierce, D. A., Mettler, I. J., Mascarenhas, D., and Detmer, J. J. Genetically transformed maize plants from protoplasts, *Science, 240*: 204–207, 1988.
32. Shillito, R. D., Carswell, G. K., Johnson, C. M., DiMaio, J. J., and Harms, C. T. Regeneration of fertile plants from protoplasts of elite inbred maize, *Biotechnology, 7*: 581–587, 1989.
33. Morocz, S., Donn, G., Nemeth, J., and Dudits, D. An improved system to obtain fertile regenerants via maize protoplasts isolated from a highly embryogenic suspension culture, *Theor. Appl. Genet., 80*: 721–726, 1990.
34. Jahne, A., Lazzere, P. A., and Horz, H. Regeneration of fertile plants from protoplasts derived from embryogenic cell suspensions of barley (*Hordeum vulgare* L.), *Plant Cell Rep., 10*: 1–6, 1991.
35. Vasil, V., Redway, F. A., and Vasil, I. K. Regeneration of plants from embryogenic suspension cultures protoplasts of wheat (*Triticum aestivum* L.), *Biotechnology, 8*: 429–433, 1990.
36. Chang, Y., Wang, W. C., Warfield, C. Y., Nguyen, H. T., and Wong, J. R. Plant regeneration from protoplasts isolated from long-term cell cultures of wheat (*Triticum aestivum* L.), *Plant Cell Rep., 9*: 611–614, 1991.
37. Datta, S. K., Peterhans, A., Datta, K., and Potrykus, I. Genetically engineered fertile indica-rice recovered from protoplasts, *Biotechnology, 8*: 736–740, 1990.
38. Hayashimoto, A., Li, Z., and Murai, N. A polyethylene glycol-mediated protoplast transformation system for production of fertile transgenic rice plants, *Plant Physiol., 93*: 857–863, 1990.
39. Shimamoto, K., Terada, R., Izawa, T., and Fujimoto, H. Fertile transgenic rice plants regenerated from transformed protoplasts, *Nature, 338*: 274–276, 1989.

40. de la Pena, A., Lorz, H., and Schell, J. Transgenic rye plants obtained by injecting DNA into young floral tillers, *Nature, 325*: 274–276, 1987.
41. Mathias, R. J. Plant microinjection techniques, *Genetic Engineering* (J. K. Setlow, ed.), Plenum Press, New York, 1987, pp. 199–227.
42. Toyoda, H., Yamaga, T., Matsuda, Y., and Ouchi, S. Transient expression of the β-glucuronidase gene introduced into barley coleoptiles by microinjection, *Plant Cell Rep., 9*: 299–302, 1990.
43. Hess, D. Pollen-based techniques in genetic manipulation, *Int. Rev. Cytol., 107*: 367–395, 1987.
44. Topfer, R., Gronenborn, B., Schafer, S., Schell, J., and Steinbiss, H. H. Expression of engineered wheat dwarf virus in seed derived embryos, *Physiol. Plant, 79*: 158–162, 1990.
45. Topfer, R., Gronenborn, B., Schell, J., and Steinbiss, H. H. Uptake and transient expression of chimeric genes in seed-derived embryos, *Plant Cell, 1*: 133–139, 1989.
46. Dekeyser, R. A., Claes, B., De Rycke, R. M. U., Habets, M. E., Van Montagu, M. C., and Caplan, A. Transient gene expression in intact and organized rice tissues, *Plant Cell, 2*: 591–602, 1990.
47. Weber, G., Monajembashi, S., Wolfrum, J., and Greulich, K.-O. Genetic changes induced in higher plant cells by a laser microbeam, *Physiol. Plant, 79*: 190–193, 1990.
48. Luo, Z., and Wu, R. A simple method for the transformation of rice via the pollen-tube pathway, *Plant Mol. Biol. Rep., 6*: 165–174, 1988.
49. Grimseley, N. H., Ramos, C., Hein, T., and Hohn, B. Meristematic tissues of maize plants are most susceptible to agroinfection with maize streak virus, *Biotechnology, 6*: 185–188, 1988.
50. Grimsley, N., Hohn, T., Davies, J. W., and Hohn, B. *Agrobacterium*-mediated delivery of infectious maize streak virus into maize plants, *Nature, 325*: 177–179, 1987.
51. Sanford, J. C. The biolistic process, *Trends Biotechnology, 6*: 299–302, 1988.
52. Klein, T. M., Roth, B. A., and Fromm, M. E. Advances in direct gene transfer into cereals, *Genetic Engineering: Principles and Methods* (J. K. Setlow, ed.), Plenum Press, New York, 1989, pp. 13–31.
53. Potrykus, I. Gene transfer to cereals: an assessment, *Biotechnology, 7*: 535–542, 1990.
54. Potrykus, I., ed. Proceedings of the EMBO workshop "Gene transfer to plants: a critical assessment," *Physiol. Plant, 79*: 125–220, 1990.
55. Fromm, M. E., Morrish, F. M., Armstrong, C., Williams, R., Thomas, J., and Klein, T. M. Inheritance and expression of chimeric genes in the progeny of transgenic maize plants, *Biotechnology, 8*: 833–839, 1990.
56. Gordon-Kamm, W. W., Spencer, T. M., Mangano, M. L., Adams, T. R., Daines, R. J., Start, W. G., O'Brien, J. V., Chambers, S. A., Adams, W. R., Jr., Willets, N. G., Rice, T. B., Mackey, C. V., Krueger, R. W., Kausch, A. P., and Lemaux, P. G. Transformation of maize cells and regeneration of fertile transgenic plants, *Plant Cell, 2*: 603–618, 1990.

57. McCown, B. H., McCabe, D. E., Russell, D. R., Robinson, D. J., Barton, K. A., and Raffa, K. F. Stable transformation of *Populus* and incorporation of pest resistance by electric discharge particle acceleration, *Plant Cell Rep., 9*: 590–594, 1991.
58. Finer, J. J., and McMullen, M. D. Transformation of cotton (*Gossypium hirsutum* L.) via particle bombardment, *Plant Cell Rep., 8*: 586–589, 1990.
59. Fitch, M. M. M., Manshardt, R. M., Gonsalves, D., Slightom, J. L., and Sanford, J. C. Stable transformation of papaya via microprojectile bombardment, *Plant Cell Rep., 9*: 189–194, 1990.
60. McCabe, D. E., Swain, W. F., Martinell, B. J., and Christou, P. Stable transformation of soybean (*Glycine max*) by particle acceleration, *Biotechnology, 6*: 923–926, 1988.
61. Christou, P., Swain, W. F., Yang, N., and McCabe, D. E. Inheritance and expression of foreign genes in transgenic soybean plants, *Proc. Nat. Acad. Sci. USA, 86*: 7500–7504, 1989.
62. Lonsdale, D., Onde, S., and Cuming, A. Transient expression of exogenous DNA in intact, viable wheat embryos following particle bombardment, *J. Exp. Bot., 14*: 1161–1165, 1990.
63. Wang, Y., Klein, T. M., Fromm, M., Cao, J., Sanford, J. C., and Wu, R. Transient expression of foreign genes in rice, wheat and soybean cells following particle bombardment, *Plant Mol. Biol., 11*: 433–439, 1988.
64. Oard, J. H., Paige, D. F., Simmonds, J. A., and Gradziel, T. M. Transient gene expression in maize, rice and wheat cells using an airgun apparatus, *Plant Physiol., 92*: 334–339, 1990.
65. Daniell, H., Krishnan, M., and McFadden, B. F. Transient expression of β-glucuronidase in different cellular compartments following biolistic delivery of foreign DNA into wheat leaves and calli, *Plant Cell Rep., 9*: 615–619, 1991.
66. Creissen, G., Smith, C., Francis, R., Reynolds, H., and Mullineaux, P. *Agrobacterium*- and microprojectile-mediated viral DNA delivery into barley microspore-derived cultures, *Plant Cell Rep., 8*: 680–683, 1990.
67. Kartha, K. K., Chibbar, R. N., Georges, F., Leung, N., Caswell, K., Kendall, E., and Qureshi, J. Transient expression of chloramphenicol acetyltransferase (CAT) gene in barley cell cultures and immature embryos through microprojectile bombardment, *Plant Cell Rep., 8*: 429–432, 1989.
68. Mendel, R. R., Muller, B., Schluze, J., Kolesnikov, V., and Zelenin, A. Delivery of foreign genes to intact barley cells by high-velocity microprojectiles, *Theor. Appl. Genet., 78*: 31–34, 1989.
69. Vasil, V., Brown, S. M., Re, D., Fromm, M. E., and Vasil, I. K. Stably transformed callus lines from microprojectile bombardment of cell suspension cultures of wheat, *Biotechnology, 9*: 743–747, 1991.
70. Klein, T. M., Wolf, E. D., Wu, R., and Sanford, J. C. High velocity microprojectiles for delivering nucleic acids into living cells, *Nature, 327*: 70–73, 1987.
71. Sanford, J. C., Klein, T. M., Wolf, E. D., and Allen, N. Delivery of substances into cells and tissues using a particle bombardment process, *J. Particle Sci. Technol., 5*: 27–37, 1987.

72. Morikawa, H., Iida, A., and Yamada, Y. Transient expression of foreign genes in plant cells and tissues obtained by a simple biolistic device (particle gun), *Appl. Micro. Biotechnol., 31*: 320–322, 1989.
73. Iida, A., Seiki, M., Kamada, M., Yamada, Y., and Morikawa, H. Gene delivery into cultured plant cells by DNA-coated gold particles accelerated by pneumatic particle gun, *Theor. Appl. Genet., 80*: 813–816, 1990.
74. Williams, R. S., Johnston, S. A., Riedy, M., DeVit, M. J., McElligott, S. G., and Sanford, J. C. Introduction of foreign genes into tissues of living mice by DNA-coated microprojectiles, *Proc. Natl. Acad. Sci. USA, 88*: 2726–2730, 1991.
75. Russell, J. A., Roy, M. K., and Sanford, J. C. Optimization of a helium-driven biolistics device for genetic transformation of tobacco cell suspension cultures, *In Vitro, 27*: 97A, 1991.
76. Christou, P., McCabe, D. E., and Swain, W. F. Stable transformation of soybean callus by DNA-coated gold particles, *Plant Physiol., 87*: 671–674, 1988.
77. Twell, D., Klein, T. M., Fromm, M. E., and McCormick, S. Transient expression of chimeric genes delivered into pollen by microprojectile bombardment, *Plant Physiol., 91*: 1270–1274, 1989.
78. Tomes, D. T. Transformation in maize: nonsexual gene transfer, *Annual Meeting Proceedings of 26th Annual Maize Breeders School* (J. Dudley, ed.) University of Illinois, Urbana, 1990, pp. 7–9.
79. Klein, T. M., Gradziel, T., Fromm, M. E., and Sanford, J. C. Factors influencing gene delivery into *Zea mays* cells by high-velocity microprojectiles, *Biotechnology, 6*: 559–563, 1988.
80. Taylor, M. G., and Vasil, I. K. Effect of physical factors, and the histology of transient GUS expression in pearl millet (*Pennisetum glacum* (L.) R. Br.) embryos following microprojectile bombardment, *Plant Cell Rep., 10*: 120–125, 1991.
81. Maretzke, A., Sun, S. S., Nagai, C., and Bidney, D. Development of a transformation system for sugarcane, VII International Congress on Plant Tissue Culture, International Association for Plant Tissue Culture, Amsterdam, 1990, pp. 68, 1990.
82. Klein, T. M., Fromm, M. E., Weissinger, A., Tomes, D., Schaaf, S., Sletten, M., and Sanford, J. C. Transfer of foreign genes into intact maize cells using high velocity microprojectiles, *Proc. Natl. Acad. Sci. USA, 85*: 4305–4309, 1988.
83. Klein, T. M., Koornstein, L., Sanford, J. C., and Fromm, M. E. Genetic transformation of maize cells by particle bombardment, *Plant Physiol., 91*: 440–444, 1989.
84. Klein, T. M., Kornstein, L., and Fromm, M. E. Genetic transformation of maize cells by particle bombardment and the influence of methylation on foreign-gene expression, *Gene Manipulation in Plant Improvement* (J. P. Gustafson, ed.), Plenum Press, New York, 1990, pp. 265–288.
85. Spencer, T. M., Gordon-Kamm, W. J., Daines, R. J., Start, W. G., and Lemaux, P. G. Bialophos selection of stable transformants from maize cell culture, *Theor. Appl. Genet., 79*: 625–631, 1990.
86. Morrish, F. M., Armstrong, C., and Fromm, M. E. (unpublished results).

87. Armstrong, C., Barnason, A., Brown, S., Dean, D., Deaton, R., Dennehey, B., Elmer, S., Fromm, M., Hairston, B., LaVallee, B., Maher, G., Meek, G., Morrish, F., Pajeau, M., Peterson, W., Reedy, M., Sanders, P., Santino, C., Sims, S., and Songstad, D. 1991 symposium of the European Association for Research on Plant Breeding, Reus, Spain (EUCARPIA, Wageningen, The Netherlands). Abstract I.L3, 1991.
88. Ludwig, S. R., Bowen, B., Beach, L., and Wessler, S. R. A regulatory gene as a novel visible marker for maize transformation, *Science, 247*: 449–450, 1990.
89. Klein, T. M., Roth, B. A., and Fromm, M. E. Regulation of anthocyanin biosynthetic genes introduced into intact maize tissues by microprojectiles, *Proc. Natl. Acad. Sci. USA, 86*: 6681–6685, 1989.
90. Roth, B. A., Goff, S. A., Klein, T. M., and Fromm, M.E. C1- and R-dependent expression of the maize Bz1 gene requires sequences with homology to mammalian *myb* and *myc* binding sites, *Plant Cell, 3*: 317–325, 1991.
91. Goff, S. A., Klein, T. M., Roth, B. A., Fromm, M. E., Cone, K. C., Radicella, J. P., and Chandler, V. L. Transactivation of anthocyanin biosynthetic genes following transfer of B regulatory genes into maize tissues, *EMBO J., 9*: 2517–2522, 1990.
92. Goff, S. A., Cinbe, K. C., and Fromm, M. E. Identification of functional domains in the maize transcriptional activator C1: comparison of wildtype and dominant inhibitors proteins, *Genes Dev., 5*: 298–309, 1991.
93. Christou, P. Soybean transformation by electrical discharge particle acceleration, *Physiol. Plant., 79*: 210–212, 1990.
94. Svab, Z., Hajdukiewicz, P., and Maliga, P. Stable transformation of plastids in higher plants, *Proc. Natl. Acad. Sci. USA, 87*: 8526–8530, 1990.
95. Ye, G., Daniell, H., and Stanford, J. C. Optimization of delivery of foreign DNA into higher-plant chloroplasts, *Plant Mol. Biol., 15*: 809–819, 1990.
96. Tomes, D., Wessinger, A. K., Ross, M., Higgins, R., Drimmond, B. J., Schaff, S., Malone-Sconeberg, J., Staebell, M., Flynn, P., Anderson, J., and Howard, J. Transgenic tobacco plants derived by microprojectile bombardment of tobacco leaves, *Plant Mol. Biol., 14*: 261–268, 1990.
97. Klein, T. M., Harper, E. C., Svab, Z., Sanford, J. C., Fromm, M. E., and Maliga, P. Stable genetic transformation of intact *Nicotiana* cells by the particle bombardment process, *Proc. Natl. Acad. Sci. USA, 85*: 8502–8505, 1988.
98. Benfey, P. N., and Chua, N. Regulated genes in transgenic plants, *Science, 244*: 174–181, 1989.
99. Kuhlemeier, C., Green, P. J., and Chua, N. Regulation of gene expression in higher plants, *Annu. Rev. Plant Physiol., 38*: 221–257, 1987.
100. Schell, J. S. Transgenic plants as tools to study the molecular organization of plant genes, *Science, 237*: 1176–1183, 1987.
101. Fromm, M., Taylor, L. P., and Walbot, V. Expression of genes transferred into monocot and dicot plant cells by electroporation, *Proc. Natl. Acad. Sci. USA, 82*: 5824–5828, 1985.

102. Jefferson, R. A. Plant reporter genes: the GUS gene fusion system, *Genetic Engineering: Principles and Methods* (J. K. Setlow, ed.), Plenum Press, New York, 1988, pp. 247–262.
103. Ow, D. W., Jacobs, J. D., and S. H. Functional regions of the cauliflower mosaic virus 35S RNA promoter determined by use of the firefly luciferase gene as a reporter of promoter activity, *Proc. Natl. Acad. Sci. USA, 84*: 4870–4874, 1987.
104. Callis, J., Fromm, M., and Walbot, V. Introns increase gene expression in cultured maize cells, *Genes Dev., 1*: 1183–1200, 1987.
105. Mascarenhas, D., Mettler, I. J., Pierce, D. A., and Lowe, H. W. Intron-mediated enhancement of heterologous gene expression in maize, *Plant Mol. Biol., 15*: 913–920, 1990.
106. Maas, C., Laufs, J., Grant, S., Korfhage, C., and Werr, W. The combination of a novel stimulatory element in the first exon of the maize Shrunken-1 gene with the following intron 1 enhances reporter gene expression up to 1000-fold, *Plant Mol. Biol., 16*: 199–207, 1991.
107. Luehrsen, K. R., and Walbot, V. Intron enhancement of gene expression and the splicing efficiency of introns in maize cells, *Mol. Gen. Genet., 225*: 81–93, 1991.
108. Thomas, M. S., and Flavell, R. B. Identification of an enhancer element for the endosperm-specific expression of high molecular weight glutenin, *Plant Cell, 2*: 1171–1180, 1990.
109. Last, D. I., Brettell, R. I. S., Chamberlain, D. A., Chaudhury, A. M., Larkin, P. J., Marsh, E. L., Peacock, W. J., and Dennis, E. S. pEmu: an improved promoter for gene expression in cereal cells, *Theor. Appl. Genet., 81*: 581–588, 1991.
110. Bruce, W. B., and Quali, P. H. *cis*-Acting elements involved in photoregulation of an oat phytochrome promoter in rice, *Plant Cell, 2*: 1081–1089, 1990.
111. Jacobsen, J. V., and Close, T. J. Control of transient expression of chimeric genes by gibberellic acid and abscisic acid in protoplasts prepared from mature barley aleurone layers, *Plant Mol. Biol., 16*: 713–724, 1991.
112. Olive, M. R., Walker, J. C., Singh, K., Dennis, E. S., and Peacock, W. J. Functional properties of the anerobic responsive element of the maize *Adh*1 gene, *Plant Mol. Biol., 15*: 593–604, 1990.
113. Salmenkallio, M., Hannus, R., and Kauppinen, V. Regulation of α-amylase promoter by gibberellic acid and abscisic acid in barley protoplasts transformed by electroporation, *Plant Cell Rep., 9*: 352–355, 1990.
114. Marcotte, W. R., Russell, S. H., and Quatrano, R. S. Abscisic acid-responsive sequences from the EM gene of wheat, *Plant Cell, 1*: 969–976, 1989.
115. Marcotte, W. R., Jr., Baley, C. C., and Quatrano, R. S. Regulation of a wheat promoter by abscisic acid in rice protoplasts, *Nature, 335*: 454–467, 1988.
116. Hauptmann, R. M., Atkins, P. O., Vasil, V., Tabaeiuzadeh, Z., Rogers, S. G., Horsch, R. B., Vasil, I. K., and Fraley, R. T. Transient expression of electroporated DNA in monocotyledonous and dicotyledonous species, *Plant Cell Rep., 6*: 265–270, 1987.

117. Fraley, R. T., Rogers, S. G., Horsch, R. B., Eicholtz, D. A., Flick, J. S., Adams, S. P., Bittner, M. L., Brans, L. A., Fink, C. L., Fry, J. S., Galluppi, G. R., Goldberg, S. B., Hoffman, N. L., and Woo, S. C. Expression of bacterial genes in plant cells, *Proc. Natl. Acad. Sci. USA, 80*: 4803–4807, 1983.
118. Depicker, A., Stachel, S., Shase, P., Zambryski, P., and Goodman, H. M. Nopaline synthase: transcript mapping and DNA sequence, *J. Mol. Appl. Genet., 1*: 561–574, 1982.
119. Velten, J., Velten, L., Hain, R., and Schell, J. Isolation of a dual plant promoter fragment from the Ti plasmid of *Agrobacterium tumefaciens, EMBO J., 3*: 2723–2730, 1984.
120. Odell, J. T., Knowlton, S., Lin, W., and Mauvais, C. J. Properties of an isolated transcription stimulating sequence derived from cauliflower mosaic virus 35S promoter, *Plant Mol. Biol., 10*: 263–273, 1988.
121. Kay, R., Chan, A., Daly, M., and McPherson, J. Duplication of the CaMV 35S promoter creates a strong enhancer for plants, *Science, 236*: 1299–1302, 1987.
122. Sanger, M., Daubert, S., and Goodman, R. M. Characteristics of a strong promoter from figworth mosaic virus: comparison with the analogous 35S promoter from cauliflower mosaic virus and the regulated mannopine synthase promoter, *Plant Mol. Biol., 14*: 433–443, 1990.
123. Dekeyser, R., Claes, B., Marichal, M., Van Montagu, M., and Caplan, A. Evaluation of selectable markers for rice transformation, *Plant Physiol., 90*: 217–223, 1989.
124. Zhang, W., and Wu, R. Efficient regeneration of transgenic plants from rice protoplasts and correctly regulated expression of foreign gene in the plants, *Theor. Appl. Genet., 76*: 835–840, 1988.
125. McElroy, D., Zang, W., Cao, J., and Wu, R. Isolation of an efficient actin promoter for use in rice transformation, *Plant Cell, 2*: 163–171, 1990.
126. Benfey, P. N., and Chua, N. The cauliflower mosaic virus 35S promoter: combinatorial regulation of transcription in plants, *Science, 250*: 959–966, 1990.
127. Peacock, W. J., Wolstenholme, D., Walker, J. C., Singh, K., Llewellyn, D. J., Ellis, J. G., and Dennis, E. S. Developmental and environmental regulation of the maize alcohol dehydrogenase 1 (Adh1) gene: promoter–enhancer interactions, *Plant Gene Systems and Their Biology* (A. R. Liss, ed.), John Wiley, New York, 1987, pp. 263–277.
128. Rathus, C. (unpublished results cited in Ref. 109).
129. Quattrocchio, F., Tolk, M. A., Coraggio, I., Mol, J. N. M., Viotti, A., and Koes, R. E. The maize zein gene zE19 contains two distinct promoters which are independently activated in endosperm and anthers of transgenic *Petunia* plants, *Plant Mol. Biol., 15*: 81–93, 1990.
130. Lloyd, J. C., Raines, C. A., John, U. P., and Dyer, T. A. The chloroplast FBPase gene of wheat: structure and expression of the promoter in photosynthetic and meristematic cells of transgenic tobacco plants, *Mol. Gen. Genet., 225*: 209–216, 1991.

131. Guerreo, F. D., Crossland, L., Smutter, G. S., Hamilton, D. A., and Mascarenhas, J. P. Promoter sequences from a maize pollen-specific gene direct tissue-specific transcription in tobacco, *Mol. Gen. Genet., 224*: 161–168, 1990.
132. Yamaguchi-Shinozaki, K., Mino, M., Mundy, J., and Chua, N. Analysis of an ABA-responsive rice gene promoter in transgenic tobacco, *Plant Mol. Biol., 15*: 905–912, 1990.
133. Matsuka, M., and Sanada, Y. Expression of photosynthetic genes from the C_4 plant, maize, in tobacco, *Mol. Gen. Genet., 225*: 411–419, 1991.
134. Mundy, J., Yamaguchi-Shinozaki, K., and Chua, N.-H. Nuclear proteins bind conserved elements in the abscisic acid–responsive promoter of a rice *rab* gene, *Proc. Natl. Acad. Sci. USA, 87*: 1406–1410, 1990.
135. Vasil, V., Clancy, M., Ferl, R. J., and Hannah, L. C. Increased gene expression by the first intron of maize *shrunken*-1 locus in grass species, *Plant Physiol., 91*: 1575–1579, 1989.
136. Oard, J. H., Paige, D., and Dvorak, J. Chimeric gene expression using maize intron in cultured cells of breadwheat, *Plant Cell Rep., 8*: 156–160, 1989.
137. Silva, E. M., Mettler, I. J., Dietrich, P. S., and Sinibaldi, R. M. Enhanced transient expression in maize protoplasts, *Genome, 30*(Suppl. 1): A72, 1988.
138. Hayford, M. B., Medford, J. I., Hoffman, N. L., Rogers, S. G., and Klee, H. J. Development of a plant transformation selection system based on expression of genes encoding gentamicin acetyltransferases, *Plant Physiol., 86*: 1216–1222, 1988.
139. della-Cioppa, G., Bauer, S. C., Taylor, M. L., Rochester, D. E., Klein, B. K., Shah, D. M., Fraley, R. T., and Kishore, G. M. Targetting a herbicide-resistant enzyme from *Escherichia coli* to chloroplasts of higher plants, *Biotechnology, 5*: 579–584, 1987.
140. Haughn, G. W., Smith, J., Hazur, B., and Somerville, C. Transformation with a mutant *Arabidopsis* acetolactate synthase gene renders tobacco resistant to sulfonylurea herbicides, *Mol. Gen. Genet., 211*: 266–271, 1988.
141. Lee, K. Y., Townsend, J., Tepperman, J., Black, M., Chui, C. F., Mazur, B., Dunsmuir, M., and Bedbrook, J. The molecular basis of sulfonylurea herbicide resistant in tobacco, *EMBO J., 7*: 1241–1248, 1988.
142. Gritz, L., and Davies, J. Plasmid-encoded hygromycin B resistance: the sequence of hygromycin B phosphotransferase gene and its expression in *Escherichia coli* and *Sacchromyces cerevisiae, Gene, 25*: 179–188, 1983.
143. Eichholtz, D. A., Rogers, S. G., Horsch, R. B., Klee, H. J., Hayford, M., Hoffmann, N. L., Braford, S. B., Fink, C., Flick, J., O'Connell, K. M., and Fraley, R. T. Expression of mouse dihydrofolate reductase gene confers methotrexate resistance in transgenic petunia plants, *Somatic Cell Mol. Genet., 13*: 67–76, 1987.
144. Klein, T., and Twell, D. (unpublished results).
145. Orshinsky, B. R., McGregor, L. J., Johnson, G. I. E., Hucl, P., and Kartha, K. K. Improved embryoid induction and green shoot regeneration from wheat anthers cultured in medium with maltose, *Plant Cell Rep., 9*: 365–369, 1990.

146. Kao, K. N., Saleem, M., Abrams, S., Pedras, M., Horn, D., and Mallard, C. Culture conditions for induction of green plants from barley microspores by anther culture methods, *Plant Cell Rep., 9*: 595–601, 1991.
147. Vain, P., Flament, P., and Soudain, P. Role of ethylene in embryogenic callus initiation and regeneration in *Zea mays* L., *J. Plant Physiol., 135*: 537–540, 1988.
148. Vain, P., Yean, H., and Flament, P. Enhancement of production and regeneration of embryogenic type II callus in *Zea mays* by $AgNO_3$, *Plant Cell Tissue Org. Cult., 18*: 143–151, 1989.
149. Songstad, D. D., Armstrong, C. L., and Petersen, W. L. $AgNO_3$ increases type II callus production from immature embryos of maize inbred B73 and its derivatives, *Plant Cell Rep., 9*: 699–702, 1991.
150. Petersen, W. L., Armstrong, C. L., and Songstad, D. D. High type II callus production from immature maize tassels, *Agronomy Abstracts, 1991*: 200, 1991.
151. Songstad, D. D., Duncan, D. R., and Wildholm, J. M. Effect of 1-aminocyclopropane-1-carboxylic acid, silver nitrate and norobornadiene on plant regeneration from maize callus cultures, *Plant Cell Rep., 7*: 262–265, 1988.
152. Armstrong, C., and Hairston, B. (unpublished results).
153. Songstad, D. (unpublished results).
154. Perl, A., Aviv, D., and Galun, E. Ethylene and in vitro culture of potato: suppression of ethylene generation vastly improves protoplast yield, plating efficiency and transient expression of an alien gene, *Plant Cell Rep., 7*: 403–406, 1988.
155. Coruzzi, G. M. Molecular approaches to the study of amino acid biosynthesis in plants, *Plant Sci., 74*: 145–155, 1991.
156. Goldberg, R. B. Plants: novel developmental processes, *Science, 240*: 1460–1467, 1988.
157. Gilmartin, P. M., Sarokin, L., Memelink, J., and Chua, N. Molecular light switches for plant genes, *Plant Cell, 2*: 369–378, 1990.
158. Dunsmuir, P., Bedbrook, J., Bond-Nutter, D., Dean, C., Gidoni, D., and Jones, J. The expression of introduced genes in regenerated plants, *Genetic Engineering: Principles and Methods* (J. K. Setlow, ed.), Plenum Press, New York, 1987, pp. 45–59.
159. Coe, E. H., Neuffer, M. G., and Hoisington, D. A. The genetics of corn, *Corn and Corn Improvement* (G. F. Spague and J. Dudley, eds.), American Agronomy Society, Madison, Wis., 1988, pp. 81–268.
160. Styles, E. D., Ceska, O., and Seah, K. T. Developmental differences in action of R and B alleles in maize, *Can. J. Genet. Cytol., 15*: 59–72, 1973.
161. Coe, E. H. J. Spontaneous mutation of the aleurone color inhibitor in maize, *Genetics, 47*: 779–783, 1962.
162. Dooner, H. K., and Nelson, O. E. Interaction among C, R, and Vp in the control of the Bz glucosyltransferase during endosperm development in maize, *Genetics, 19*: 309–315, 1979.
163. Dooner, H. K. Coordinate genetic regulation of flavonoid biosynthetic enzymes in maize, *Mol. Gen. Genet., 189*: 136–141, 1983.

164. Ptashne, M. How eukaryotic transcriptional activators work, *Nature, 335*: 683–689, 1988.
165. Kendrick, R. E., and Kronenberg, G. H. M., eds. *Photomorphogenesis in Plants*, Martinus Niijhoff, Dordrecht, The Netherlands, 1986.
166. Tobin, E., and Silverthorne, J. Light regulation of gene expression in higher plants, *Ann. Rev. Plant Physiol., 36*: 569–593, 1985.
167. Nagy, F., Kay, S. A., and Chua, N.-H. Gene regulation by phytochrome, *Trends Genet., 4*: 37–41, 1988.
168. Lissemore, J., and Quail, P. H. Rapid transcriptional regulation by phytochrome of the genes for phytochrome and chlorophyll a/b-binding protein in *Avena sativa, Mol. Cell. Biol., 8*: 4840–4850, 1988.
169. Bruce, W. B., Christensen, A. H., Klein, T., Fromm, M. E., and Quail, P. H. Photoregulation of a phytochrome gene promoter from oat transferred into rice by particle bombardment, *Proc. Natl. Acad. Sci. USA, 86*: 9692–9296, 1989.
170. Bruce, W. B., and Quail, P. *cis*-Acting elements involved in the photoregulation of an oat phytochrome promoter in rice, *Plant Cell, 2*: 1089–1091, 1991.
171. Silverthorne, J., Wimpee, C. F., Yamada, T., Rolfe, S., and Tobin, E. M. Differential expression of individual genes encoding the small subunit of ribulose-1,5-biphosphate carboxylase in *Lemna gibba, Plant Mol. Biol., 15*: 49–59, 1990.
172. Buzby, J. S., Yamada, T., and Tobin, E. M. A light-regulated DNA-binding activity interacts with a conserved region of a *Lemna gibba rbcs* promoter, *Plant Cell, 2*: 805–814, 1990.
173. Twell, D., Yamaguchi, J., and McCormick, S. Pollen-specific gene expression in transgenic plants: coordinate regulation of two different tomato gene promoters during microsporogenesis, *Development, 109*: 705–713, 1990.
174. Twell, D., Yamaguchi, J., Wing, R. A., Ushiba, J., and McCormick, S. Promoter analysis of genes that are coordinately expressed during pollen development reveals pollen-specific enhancer sequences and shared regulatory elements, *Genes Dev., 5*: 496–507, 1991.
175. Toriyama, K., Arimoto, Y., Uchimiya, H., and Hinata, K. Transgenic rice plants after direct gene transfer into protoplasts, *Biotechnology, 6*: 1072–1074, 1988.
176. Murai, N., Li, Z., Kawagoe, Y., and Hayashimoto, A. Transposition of the maize activator element in transgenic rice plants, *Nucleic Acids Res., 19*: 617–622, 1991.
177. Kellum, R., and Schedl, P. A position-effect assay for boundaries of higher order chromosomal domains, *Cell, 64*: 941–950, 1991.
178. Bonnerot, C., Grimber, G., Briand, P., and Nicolas, J.-F. Patterns of expression of position-dependent integrated transgenes in mouse embryos, *Proc. Natl. Acad. Sci. USA, 87*: 6331–6335, 1990.
179. Hepburn, A. G., Belanger, F. C., and Mattheis, J. R. DNA methylation in plants, *Dev. Genet., 8*: 475–493, 1987.

180. Folger, K. R., Wong, E. A., Wahl, G., and Capecchi, M. R. Patterns of integration of DNA microinjected into cultured mammalian cells: evidence for homologous recombination between injected plasmid DNA molecules, *Mol. Cell Biol., 2*: 1372–1387, 1982.
181. Czernilofsky, A. P., Hain, R., Herrera-Estrella, L., Lorz, H., Goyvaerts, E., Baker, B. J., and Schell, J. Fate of selectable marker DNA integrated into the genome of *Nicotiana tabacum, DNA, 5*: 101–113, 1986.

9

Transgenic Rice Plants: Tools for Studies in Gene Regulation and Crop Improvement

Junko Kyozuka and Ko Shimamoto

Plantech Research Institute, Yokohama, Japan

I. INTRODUCTION

Genetic transformation techniques provide new approaches to many fundamental problems in plant biology. In particular, transgenic plants have become essential tools in understanding the in vivo functions of plant genes and the molecular mechanisms of their regulation in dicotyledonous (dicot) species (1). In addition, genetic transformation techniques have been used for crop improvement, and agronomically useful genes, such as herbicide (2,3) viral (4), and insect resistance genes (5,6), have been introduced into a number of dicot species.

In monocotyledonous (monocot) species, production of transgenic plants has been difficult despite extensive efforts over many years. Accordingly, expression of genes derived from monocot species has often been examined in transgenic dicot species (7–13). However, differences in regulation of gene expression between monocot and dicot species became clear in a number of transgenic experiments; that is, monocot genes are not always expressed correctly or at all in dicot plants (8,10). Moreover, obvious anatomical differences between monocot and dicot species often make it difficult to accurately evaluate tissue- and cell-specific expression in transgenic dicot plants (11,12).

Therefore, transgenic monocot plants have been needed for the elucidation of monocot gene expression and the differences between the gene expression systems of monocot and dicot plants. In other words, a monocot species amenable to genetic engineering has been required as a model plant for studies in the regulation of monocot genes. In addition, genetic transformation techniques for monocot species are urgently needed for applied researches, because monocots include major crop species, such as rice, maize, wheat, and barley.

Recently, routine generation of fertile transgenic rice plants and by direct DNA transfer to protoplasts has been achieved (14,15). In addition, rice has several favorable features as a model monocot plant that are shared with *Arabidopsis thaliana* a model dicot plant (16). The linkage map for rice is well developed (17), and a restriction fragment length polymorphism (RFLP) map is being constructed (18,19). Rice possesses a relatively small genome of 200 to 300 Mb per haploid (20), which is three to five times that of *A. thaliana* and contains less repetitive DNA than other plant species (21).

In this chapter, the present status of cereal transformation, methods for regeneration of transgenic rice plants, and the results of recent research making use of transgenic rice are described. Also, the potential of rice as a model monocot species for studies in regulation of gene expression unique to monocot plants is discussed.

II. TRANSGENIC CEREALS

Agrobacterium-mediated transformation is routinely used for producing transgenic dicot plants. However, it cannot be applied to most monocot species, especially cereals, because they are not generally susceptible to *Agrobacterium tumefaciens*. The search for alternative methods for cereal transformation led to examination of various direct transformation methods in many laboratories.

Among the methods studied, protoplast-mediated transformation (14,15) and microprojectile bombardment (22,23) were successfully used to produce fertile transgenic rice and maize, respectively (Table 1). In addition, transgenic plants of maize (28) and orchardgrass (29) were produced by protoplast-mediated transformation. However, their fertility was not described in either report. Protoplast transformation is more frequently used in wide variety of species than is microprojectile bombardment, and stably transformed calli have been obtained by protoplast transformation in other major cereals, including wheat (I. K. Vasil, personal communication) and barley (31). For most cereals, including maize, wheat, and barley, the establishment of efficient regeneration from protoplasts is an important factor limiting the successful production of fertile transgenic plants.

Table 1 Transgenic Cereal Plants[a]

Plant	Method	Selection marker gene	Transmission to progeny	Reference
Rice	Protoplast (EP)	*hph*	+	14,15
Japonica	Protoplast (EP)	*nptII*	?	24
	Protoplast (EP)	*nptII*	?	25
Indica	Protoplast (PEG)	*hph*	+	26
	Protoplast (EP)	*nptII*	?	27
Maize	Protoplast (EP)	*npt II*	–	28
	Microprojectile bombardment	*bar*	+	22
	Microprojectile bombardment	*bar*	+	23
Orchardgrass	Protoplast (EP, PEG)	*hph*	?	29
Rye	Injection to inflorescence	*nptII*	?	30

[a]EP: electroporation; PEG: polyethylene glycol treatment; *hph*: hygromycin phosphotransferase gene; *nptII*: neomycin phosphotransferase gene; *bar*: phosphinothricin acetyltransferase gene.

Regeneration of fertile transgenic maize plants by microprojectile bombardment was reported by two groups (22,23). Both groups adopted embryogenic suspension cells as target materials and used the *bar* gene coding for phosphinotricin acetyltransferase as a selection marker. At present, the success is limited to a specific genotype, and the relatively low frequency of transformation and subsequent plant regeneration from transformed cells seems to be the major problem for the routine production of transgenic plants.

III. GENERATION OF FERTILE TRANSGENIC RICE PLANTS

A. Protoplast-Mediated Transformation

Rice is an exceptional cereal in that plant regeneration from protoplasts has been well established (32–35) and genetically transformed plants can be routinely produced by direct transformation using protoplasts (14,15, 24,25,36).

The "quality" of embryogenic suspension cultures from which protoplasts are isolated is one of the most important factors for successful regeneration of transgenic plants. Normally we use embryogenic suspension cultures derived from callus of mature seeds that show high plating efficiency (~10%). Suspension cultures are renewed every 6 months because they tend to lose their morphogenic capacity during prolonged culture.

Electroporation and polyethylene glycol treatment are used to introduce foreign DNA into rice protoplasts. Under optimum conditions, the difference in efficiency of DNA uptake by protoplasts between these two techniques does not seem to be significant.

The efficiency of generating transgenic rice plants depends on the variety. Differences among varieties in ease of transformation are due to their relative adaptability to in vitro culture and their relative competence for accepting foreign DNA. Although protoplast culture of indica varieties is generally more difficult than for japonica varieties (37), transgenic plants have been obtained in some indica rice (26,27). For the recalcitrant varieties, further improvement of culture conditions is necessary to reproducibly obtain transgenic plants.

B. Selectable Markers

The effectiveness of the selection of transformed cells depends on the selectable markers and selection procedures used. The markers for Hm^r (*hph*) and Km^r (*nptII*) have been used as selectable markers for rice transformation (Table 1). However, effective selection of transformed cells is not always easy with the Km^r marker because rice cells generally have background resistance to kanamycin (38) and, furthermore, albino or sterile plants have often been obtained from rice callus after selection with G418.

In our procedure, selection by hygromycin B is started after 10 to 14 days of culture and is repeated twice for 7 to 10 days each. The Hm^r calli thus selected are transferred to the regeneration medium, and shoots arise from these transformed calli within 2 to 6 weeks. With our procedure transformed rice plantlets can be obtained within 8 to 10 weeks after electroporation. The frequency of plant regeneration from Hm^r calli is approximately 60% to 80% (for a detailed protocol see Ref. 39).

C. Cotransformation

Cotransformation has been efficiently used to introduce nonselectable genes into rice. In cotransformation, a plasmid carrying the nonselectable gene and another carrying the selectable marker gene are mixed and introduced into protoplasts. In our experiments, when *gusA* plasmids were electroporated with Hm^r plasmids in a ratio of 1:1 (50 μg/ml each), the efficiency of cotransformation was 30–50% at the expression level. One advantage of this method is that construction of a plasmid carrying both the selectable marker and the nonselectable gene is not required; thus, it is convenient for generating plants carrying multiple genes.

D. Integration Pattern of Foreign DNA

Integration patterns of foreign DNA in the genome of transgenic rice were closely analyzed by Southern blot analysis using various fragments of Hm^r plasmids as hybridization probes (40). The results indicated that five out of six Hm^r transformants contained one or two functional *hph* sequences, and no other fragmented pieces of the *hph* sequence were detected. Results with respect to the number of copies of the integrated Hm^r gene generally corresponded to those obtained with other transgenic cereals by other groups; most of the transgenic plants contained one or two copies of the transgene. Our study also revealed that there are three different patterns of integration of the Hm^r plasmid into the rice genome. First, almost the entire unit of the plasmid DNA is integrated in a single site (Fig. 1A). Second, a few fragmented pieces of the plasmid DNA are integrated in multiple sites of the rice genome (Fig. 1B). Third, a tandem (4× or 5×) repeat of the entire plasmid is integrated in one site (Fig. 1C). Further analysis of integrated foreign DNA in transgenic rice is required for understanding the mechanisms of DNA integration and for effectively using transgenic plants to introduce economically important genes.

IV. TRANSGENIC RICE PLANTS AS TOOLS FOR ANALYSIS OF MONOCOT GENE EXPRESSION

A. Promoter Analysis

Analysis of regulation of gene expression in a homologous transgenic system has many advantages because introduced genes exhibit their natural properties in transgenic plants. However, such a system is not available for most cereals. Therefore, at the moment rice can be used as a host species to analyze in vivo

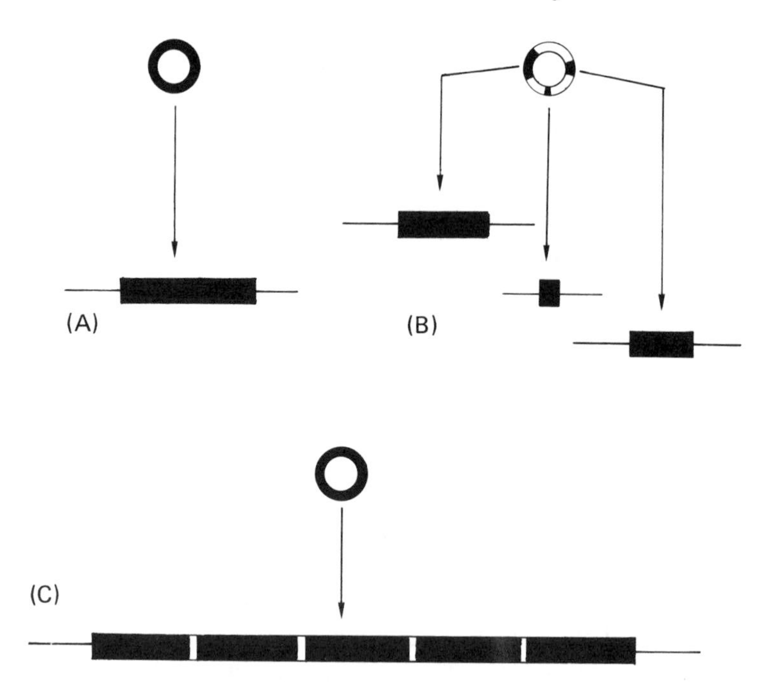

Figure 1 Three different integration patterns of foreign DNA in the genome of transgenic rice. (Top, left) Almost the entire unit of the plasmid DNA is integrated in a single site. (Top, right) A few fragmented pieces of the plasmid DNA are integrated in multiple sites. (Bottom) A tandem (4× or 5×) repeat of the entire plasmid is integrated in one site.

functions of cereal promoters because of its ease of transformation among cereals.

The first promoter whose expression was examined in transgenic rice plants was the 35S promoter of cauliflower mosaic virus (CaMV) (41) because it has been extensively used for expression of agronomically useful genes in dicot species. Quantitative and histochemical analysis of 35S promoter expression in transgenic rice plants and their progeny demonstrated that it is expressed in many tissues, including root, leaf, flower, and seed and that the level of its expression in leaf and root is comparable to that in transgenic tobacco plants. With respect to the cell type specificity in its expression, the 35S promoter tends to be better expressed in vascular tissues than in ground tissues in several of the organs examined. However, a later study indicated that the 35S promoter can be expressed in most cell types in leaves and roots of highly expressed transgenic plants (42). A general conclusion concerning the expression pattern of the 35S promoter is that it is a "constitutive" promoter in rice and that there are no major differences in its expression pattern between rice and tobacco. Thus, the 35S promoter should be useful for introducing agriculturally important genes, such as coat protein genes of viruses (see Section VI).

Some studies demonstrated that the 35S promoter was less effective in cereal cells than in dicot cells (43). Unoptimized transformation conditions or a lack of competence of cells in DNA uptake may have caused insufficient expression of the 35S promoter.

We examined the regulated expression of the maize alcohol dehydrogenase 1 gene (*Adh1*) promoter in transgenic rice plants (44). The maize *Adh1* gene is one of the best studied genes in higher plants. The enzyme alcohol dehydrogenase (ADH) is present in the pollen, embryo, endosperm, and seedling roots of maize plants. Anaerobic induction of ADH proteins has been well documented in several plant species. In maize the synthesis of ADH1 is induced in various parts of roots and other tissues by anaerobiosis, and the induction is primarily at the level of transcription. By studying expression of the promoter in maize protoplasts (45), and transformed tobacco plants (10), the DNA sequences required for anaerobic induction of maize *Adh1* expression have been identified. However, in tobacco the maize *Adh1* promoter did not confer sufficient expression of the reporter gene unless an enhancer sequence was added to the promoter. Furthermore, the degree of induction detected in these studies is much smaller than that observed in maize.

In transgenic rice plants the maize *Adh1* promoter is constitutively expressed in seedling shoots, root caps of seedling roots and mature roots, anthers, anther filaments, pollen, scutellum, and endosperm. The spatial expression pattern of the maize *Adh1* promoter in rice is similar to the distribution of ADH1 protein in maize plants except for the expression in seedlings and mature leaves of transgenic rice.

Anaerobic induction was carefully examined by using 5- to 7-day-old seedlings derived from selfed progenies of primary transgenic plants. The expression was strongly (up to 81-fold) induced in roots of seedlings in response to anaerobic treatment for 24 hr, concomitant with an increase in the level of *gusA* mRNA. Our results indicate that induction in the expression by maize *Adh1* promoter takes place in specific regions of the root in transgenic rice plants and the spatial pattern of expression changes distinctly after induction. The induction started first at the meristem, and after 3 hr there was strong induction in the elongation zone, which is located 1–2 mm above the meristem; the induction then progressed upward from this region. These results generally agree with the finding by in situ staining of ADH1 enzyme activity in the primary root of maize during anaerobic treatment.

These results indicate that the maize *Adh1* promoter is expressed in a regulated manner that reflects the natural property of the promoter. Considering the fact that the maize *Adh1* promoter did not confer sufficient expression in transgenic tobacco plants, it is evident that transgenic rice plants are useful for studying the expression of monocot (cereal) genes.

A slight difference in the expression pattern of the maize *Adh1* promoter between maize plants and transgenic rice plants may provide a clue to the mechanism underlying species differentiation in expression of homologous genes. For instance, low-level expression of the maize *Adh1* promoter is observed in mature leaves of rice, whereas expression is observed in the phloem and epidermis of younger leaves in seedling shoots and induced by anaerobic treatment (Fig. 2). This expression pattern is similar to that of the ADH1 protein in rice, which is abundantly present and anaerobically induced in young leaves. In contrast, neither the ADH protein nor *Adh1* mRNA is

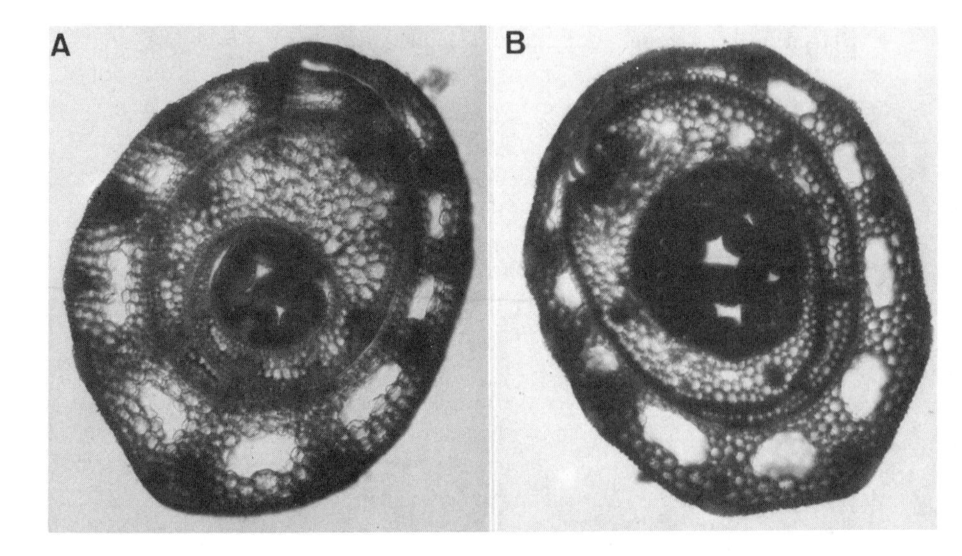

Figure 2 Expression of maize *Adh1* promoter in a seedling shoot of transgenic rice. (A) Constitutive expression; a horizontal section of a 5-day-old shoot stained with X-gluc. (B) Expression after anaerobic treatment; a horizontal section of a 5-day-old shoot stained with X-gluc after 24-hr anaerobic treatment.

detected in either mature or seedling leaves of maize. Consequently, rice may have evolved the mechanism to express the *Adh1* promoter in shoots to survive in water or under anaerobic conditions. The most likely and attractive explanation for this difference is that expression of the *Adh1* promoter in leaves (especially seedling shoots) of transgenic rice may be regulated by cellular factors present or active in rice leaf cells but not in maize leaves.

In addition to the 35S and maize *Adh1* promoters, expression of the *rolC* promoter of the Ri plasmid (46) and the rice *cab* gene promoter (47) was examined in transgenic rice plants. In leaves and roots of transgenic rice plants, the *rolC* promoter exhibited β-glucuronidase (GUS) activity only in vascular tissues. The expression of rice *cab* gene promoter was detected in leaves, stems, and floral organs but not in roots, and its expression was induced by light.

B. Intron Enhancement of Gene Expression

In cereal species, it has been shown that steady-state transcript levels and gene expression are increased by the presence of introns (48). The increasing effect has been described not only for introns from cereal species, such as maize *Adh1* introns 1, 2, 3, 6, 8, and 9 (48–52), maize *Bz1* intron 1 (48,52), maize *Sh1* intron 1 (53), maize actin gene (*Act*) intron 3 (52), and rice actin gene (*Act1*) intron 1 (54), but also for an intron derived from a dicot species, the castor bean catalase gene (*Cat1*) intron 1 (55). The degree of enhancement depends on the intron, the sequences flanking the intron, the reporter gene, and the host cells used in the experiments. Although the intron-mediated enhancement of gene expression in cereals seems to be a general phenomenon, the mechanisms underlying the intron enhancement are not well understood.

The stimulating effect of intron has also been observed in rice. The first intron of maize *Adh1* (50), rice *Act1* intron 1 (54), and castor bean *Cat1* intron 1 increased the expression of a reporter *gusA* gene in transient assays with rice protoplasts. Furthermore, stimulation of the expression of *gusA* by maize *Adh1* intron 1 has been detected in stably transformed rice cells.

The first intron of maize *Adh1* increased the GUS activity in rice protoplasts 4- to 6-fold when the CaMV 35S promoter was used, and an 11- to 18-fold increase was observed with the maize *Adh1* promoter (50). These results are in agreement with the finding reported by Callis et al. in that the first intron of maize *Adh1* enhanced the chloramphenicol acetyltransferase (CAT) activity 16- to 112-fold in maize protoplasts when driven by the *Adh1* promoter;

however, a 5- to 22-fold increase was observed when the CaMV 35S promoter was used (48).

The strong enhancement of gene expression by insertion of the first intron of castor bean *cat1* in the N-terminal region of the coding sequence of *gusA* was demonstrated in transiently as well as stably transformed rice cells. Northern blot analysis showed that the increase of GUS activity was accomplished by the increased level of mature mRNA and efficient splicing. When the same constructs were introduced into tobacco cells, little increase of gene expression was observed and the intron was not spliced efficiently (56). Therefore, the authors suggested that the stimulating effect of the intron is correlated with an efficient splicing of pre mRNA and an increased level of mature mRNA.

In contrast to the reports just described, Peterhans et al. reported that although the intron 3 of the phaseolin gene was efficiently and accurately processed in rice cells, the amount of steady-state mRNA was not increased by the presence of the intron (57). Similarly, the first intron of maize *Adh1* inserted in the untranslated leader sequence of maize *Adh1* promoter enhanced gene expression more than 10-fold; however, it was not efficiently spliced in stably transformed rice cells (J. Kyozuka, unpublished results).

Further analysis of monocot and dicot introns in transformed cereal tissues will be necessary for understanding the mechanism of the intron-mediated enhancement of foreign gene expression in cereal cells.

V. INTRODUCTION AND EXPRESSION OF TRANSPOSABLE ELEMENTS IN TRANSGENIC RICE PLANTS

Transposon tagging is a powerful tool to isolate new genes, and several genes have been isolated in maize (58) and *Antirrhinum majus* (59) by this method. However, because endogenous transposable elements have not been identified in rice so far, the maize autonomous transposable element *Activator* (*Ac*) and nonautonomous element *Dissociation* (*Ds*) were introduced into rice to develop a transposon tagging system.

A phenotypic assay for excision of the *Ac* element was used to introduce *Ac* (60,61). In this assay, excision of the *Ac* element is recognized as the Hm^r (hygromycin-resistant) phenotype because excision of the *Ac* element from the untranslated leader sequence of the *hph* gene reconstitutes a functional Hm^r gene. Excision and reintegration of the *Ac* element in the rice genome was examined by Southern blot analysis, and it was demonstrated that the introduced *Ac* element was active in transformed calli and transgenic plants.

Sequence analysis of its excision sites indicated that the *Ac* element was excised in rice in a manner similar to that in maize.

As an alternative to the use of the autonomous *Ac* element, transposition of a nonautonomous *Ds* element was examined by cotransformation with the *Ac* transposase gene fused with the 35S promoter (35S-*Ac*) (K. Shimamoto et al., unpublished results). This 35S-*Ac* gene is not able to transpose because the ends of *Ac* that contain the 11-bp terminal inverted repeats necessary for transposition were removed from the intact *Ac* element. The *Ds* element was inserted in the chimeric Hm^r gene as was *Ac*, and it was found that the *Ds* element could be excised and reintegrate in the rice chromosome by the action of the transposase produced by the 35S-*Ac* gene. Excision of the *Ds* element from the *hph* gene was monitored, as was the appearance of Hm^r cells. Transposition of the *Ds* element was examined by Southern blot analysis and by sequencing its excision sites and the rice DNA flanking integrated *Ds* elements. The analysis indicated that the *Ds* element actively transposes in rice in the presence of the 35S-*Ac*. Sequences of *Ds* excision sites are similar to those of *Ac* excision sites in transgenic rice, and the duplication of the 8-bp target sequence observed in maize was also found in rice. These results suggest that a two-element (*Ac*/*Ds*) system can also be used for tagging new genes in rice. Interestingly, most of the Hm^r calli did not contain the stably integrated 35S-*Ac* gene, indicating that transiently expressed *Ac* transposase acted on the *Ds* element and caused its transposition into the rice genome. This raises the interesting possibility of generating transgenic plants carrying only the non-autonomous *Ds* element. These *Ds* plants could be used for screening possible mutations caused by *Ds* insertion in the following generation.

VI. INTRODUCTION AND EXPRESSION OF AGRONOMICALLY USEFUL GENES

Genetic transformation is a powerful tool for crop improvement. In some dicot species, such agronomically useful traits as herbicide resistance (2,3), insect resistance (5,6), virus resistance (4), and male sterility (62) have been introduced by the transformation method. Recently there have been a variety of approaches to altering agronomically important traits of crops. However, these approaches have not been applied to cereals until recently because of the lack of a reliable transformation technique.

Establishment of an efficient transformation system in rice has made it possible to apply gene transfer to improve rice, one of the most important crops in the world.

A. Coat Protein–Mediated Viral Resistance

Coat protein–mediated (CP-mediated) protection against virus diseases has been applied to many dicot species since the first report showing that the CP of tobacco mosaic virus (TMV) expressed in transgenic plants conferred resistance to TMV (4). Transgenic rice plants expressing the CP of rice stripe virus (RSV) were generated in our laboratory (63). Rice stripe virus is a member of the *Tenui* virus group and is transmitted by small brown plant hoppers. In Japan, Korea, China, Taiwan, and the former USSR, RSV causes serious damage to rice. The CP expression vector used in the experiment consisted of the 35S promoter, the first intron of *Cat1*, the coding sequence of the CP gene, and the polyadenylation site from the nopaline synthase gene. Western blot analysis using primary transgenic plants revealed that out of 33 independent clones, each of which gave rise to several plants, 19 expressed detectable levels of the CP. The amount of the CP produced in the rice leaves was estimated to be up to 0.5% of the total soluble protein. In the assay for viral resistance, transgenic plants expressing CP did not exhibit disease symptoms, whereas the nontransformed control plants and the transformed plants not expressing the CP showed clear disease symptoms, indicating that the resistance to RSV depended on expression of the introduced CP gene (Fig. 3). The CP gene was stably transmitted to the progeny of primary transgenic plants, and CP expression and the viral resistance were observed in the progeny plants.

This study indicated that introduction of CP genes is a promising approach for introducing viral resistance in cereals. This strategy is applicable to other viruses, such as *Tungro* virus, which is causing severe damage to indica rice in several Asian countries.

B. Introduction of *Bacillus thuringiensis* Toxin Genes

Genes encoding insect control proteins from *Bacillus thuringiensis* (B.t.) have been introduced into several crops for protection against insects. Insect damage in rice is a serious problem in many Asian countries. Two Lepidoptera insects, rice stem borer and rice leaf folder, are largely responsible for the insect damage to rice. Introduction of B.t. toxin genes should be a useful approach for protection against these pests.

Preliminary attempts to express a B.t. toxin gene in rice have been described (64,65). In these studies the B.t. coding sequence was translationally fused with *gusA* and introduced into rice by protoplast transformation. Southern blot analysis of resultant transgenic plants showed integration of the B.t.-*gusA* gene in the rice genome. Furthermore, GUS expression has been detected in roots of

Figure 3 Resistance of the transgenic rice plant to virus infection. A resistant plant with CP expression showing healthy growth and fertility (left) and a susceptible plant with severe disease symptoms (right) at 4 months after virus inoculation.

the primary transgenic plants. Whether the transgenic plants exhibit resistance to any insects that cause yield loss in rice has not been reported yet.

To effectively use the B.t. gene against the major rice pests, many factors influencing their expression need to be considered. For instance it is known that the expression level of B.t. toxin genes in plants is very low and is generally not sufficient to control insects under field conditions (6). One of the reasons for low expression is its codon use, which is substantially different from that of plant genes. Extensive modification of the sequence dramatically improved the expression level of the B.t. gene in plants. Thus, a similar modification may also be needed for applying the B.t. toxin approach to rice. Another consideration is the choice of promoters. Stem borers, for instance, get into the stems of plants and grow inside. This should be taken into account when a B.t. expression vector is constructed. In the future, however, efficient expression of the B.t. gene in rice will be achieved, and this will improve the insect resistance of rice.

C. Other Useful Genes

Starch composition in the endosperm is an economically important trait in rice. Starch of wild-type grain consists of 15–30% amylose and 70–85% amylopectin, whereas the starch in the endosperm of *waxy* mutants completely lacks amylose. Amylose content varies greatly among rice cultivars, and amylose affects grain quality, taste, and cooking properties. Although several other factors contribute to the amylose content of rice seeds, it is primarily determined by the expression level of the *waxy* gene.

A genomic DNA of the rice *waxy* gene has been cloned and sequenced (66). Therefore, opportunities exist for manipulating the amylose content of rice grain by genetic engineering techniques. For example, expression of an introduced *waxy* gene may increase amylose content. On the other hand, if antisense DNA of the rice *waxy* gene is expressed in rice seed, the amylose content will decrease.

The kinds and amounts of proteins present in rice seeds are also important characteristics of seed quality. The seed storage proteins of rice consist of digestible glutelin and undigestible prolamins. Glutelins, which account for 70% of the total proteins in the seed, are targeted to protein body I, and prolamin is found in protein body II. Both proteins are produced from multiple genes, and some of the corresponding genes have been cloned (67,68). By manipulating the expression of these genes, the amounts and the ratio of these two types of proteins may be modified. Also, manipulation of the signal peptide involved in targeting the proteins to protein bodies should be possible.

Two different structural genes encoding phytochrome have been identified in rice (69,70). The *phyA* gene encodes type I phytochrome, which is abundant in etiolated tissues and is down-regulated by light. The second gene, *phyB*, codes for type II phytochrome, which is constitutively expressed at low levels in green tissues. Striking morphological changes observed in transgenic tomato (71) and tobacco (72) expressing high levels of *phyA* suggest that these two phytochrome genes of rice could be exploited to manipulate phytochrome-mediated developmental processes that might lead to beneficial changes in agronomically important traits.

These genes are only a fraction of the isolated genes that are potentially useful for genetically engineering rice. Progress in isolating agronomically useful genes is rapid because of the development of new techniques in molecular biology. This progress, combined with our understanding of various promoters with defined functions, increases the probability of improving rice by genetic engineering in the near future.

VII. PROBLEMS AND FUTURE PROSPECTS

Generation of fertile transgenic plants has been routinely performed by our group and others for the last 3 years. The topics discussed in this chapter clearly indicate that transgenic rice plants are valuable tools for understanding regulation of genes isolated from the graminaceous monocots and for attempting to improve economically important traits by introducing foreign or modified genes. That maize transformation is far from routine and that transformation of wheat and barley has not been achieved suggest that rice will continue to be important to progress in these two aspects of plant biology.

Despite the progress in rice transformation based on protoplast culture, further improvement will be required to fully realize the potential of transgenic approaches in rice research. First, the establishment of a transformation protocol for some of the major indica varieties of rice is urgently needed because indica varieties are grown in the vast majority of rice-growing countries. Recent developments in indica rice transformation and plant regeneration from protoplasts indicates that careful identification of the critical steps in generation of transgenic plants will eventually solve most of the problems, and it will not be long before protoplast-mediated transformation will be possible for most of the major indica varieties of rice.

Second, controlled integration of foreign DNA will be required to maximize the use of transgenic rice plants for several areas of research. At present, our understanding of the factors influencing patterns and efficiency of integration of transgenes is limited, and often deleted or rearranged copies of the plasmid DNA are detected in the chromosome of transgenic rice plants. The use of transposable elements to define integrated DNA might lead to development of transposon-based vectors in higher plants. Viral replications may be useful for expressing certain genes in rice cells, although how long the expression of inserted genes is sustained after their introduction into cells remains to be examined (73,74). In the future, targeting transgenes to a defined locus (75) may become necessary in order not to disrupt the functions of other genes. To this end, however, the transformation frequency should be considerably improved over the present situation.

Rapid progress in the understanding of the rice genome will permit us to use genetic engineering of rice to increase productivity. A near-saturated RFLP map and cloning large DNA (76) will undoubtedly lead to isolation of economically important genes in the future. This technology and transposon tagging will increase the list of useful genes that can be introduced into rice in the future.

Regulated expression of promoters derived from other cereals in transgenic rice plants clearly shows that rice is suitable for the analysis of monocot promoters. Description of expression patterns, determination of *cis*-elements involved in regulated expression with transgenic rice plants, and subsequent isolation of *trans*-acting genes will provide valuable information on monocot gene expression. Rice will also be important in elucidating the mechanisms underlying monocot-specific gene expression.

REFERENCES

1. Schell, J. Transgenic plants as tools to study the molecular organization of plant genes, *Science, 237*: 1176, 1987.
2. Comai, L., Gacciotti, D., Hiatt, W. R., Thompson, G., Rose, R. E., and Stalker, D. Expression in plants of a mutant *aroA* gene from *Salmonella typhimurium* confers tolerance to glyphosate, *Nature, 317*: 741, 1985.
3. De Block, M., Botterman, J., Vandewiele, M., Dockx, J., Thoen, C., Gossele, V., Movva, N. R., Thompson, C., Van Montagu, M., and Leemens, J. Engineering herbicide resistance in plants by expression of a detoxifying enzyme, *EMBO J., 6*: 2513, 1987.
4. Abel, P. P., Nelson, R. S., De, B., Hoffmann, N., Rogers, S. G., Fraley, R. T., and Beachy, R. N. Delay of disease development in transgenic plants that express the tobacco mosaic virus coat protein gene, *Science, 232*: 738, 1986.
5. Fischhoff, D. A., Bowdish, K. S., Perlak, F. J., Marrone, P. G., McCormick, S. M., Niedermeyer, J. G., Dean, D. A., Kusano-Kretzmer, K., Mayer, E. J., Rochester, D. E., Rogers, S. G., and Fraley, R. T. Insect tolerance transgenic tomato plants, *Biotechnology, 5*: 807, 1987.
6. Perlak, F. J., Deaton, R. W., Armstrong, T. A., Fuchs, R. L., Sims, S., Greenplate, J. T., and Fischhoff, D. A. Insect resistant cotton plants, *Biotechnology, 8*: 939, 1990.

7. Lamppa, G., Nagy, F., and Chua, N.-H. Light-regulated and organ-specific expression of wheat Cab gene in transgenic tobacco, *Nature, 316*: 750, 1985.
8. Keith, B., and Chua, N.-H. Monocot and dicot pre-mRNAs are processed with different efficiencies in transgenic tobacco, *EMBO J., 5*: 2419, 1986.
9. Colot, V., Robert, L. S., Kavanagh, T. A., Bevan, M. W., and Thompson, R. D. Localization of sequences in wheat endosperm protein genes which confer tissue-specific expression in tobacco, *EMBO J., 6*: 3559, 1987.
10. Ellis, J. G., Llewellyn, D. J., Dennis, E. S., and Peacock, W. J. Maize *Adh-1* promoter sequences control anaerobic regulation: addition of upstream promoter elements from constitutive genes is necessary for expression in tobacco, *EMBO J., 6*: 11, 1987.
11. Schernthaner, J. P., Matzke, M. A., and Matzke, A. J. M. Endosperm-specific activity of a zein gene promoter in transgenic tobacco plants, *EMBO J., 7*: 1249, 1988.
12. Matsuoka, M., and Sanada, Y. Expression of photosynthetic genes from the C_4 plant, maize, in tobacco, *Mol. Gen. Genet., 225*: 411, 1991.
13. Lloyd, J. C., Raines, C. A., John, U. P., and Dyer, T. A. The chloroplast FBPase gene of wheat: structure and expression of the promoter in photosynthetic and meristematic cells of transgenic tobacco plants, *Mol. Gen. Genet., 225*: 209, 1991.
14. Shimamoto, K., Terada, R., Izawa, T., and Fujimoto, H. Fertile transgenic rice plants regenerated from transformed protoplasts, *Nature, 338*: 274, 1989.
15. Shimamoto, K. Transgenic rice plants, *Molecular Approaches to Crop Improvement* (E. S. Dennis and D. J. Llewellyn, eds.), Springer, Vienna, 1991, p. 1, 1991.
16. Meyerowitz, E. M. *Arabidopsis*, a useful week, *Cell, 56*: 263, 1989.
17. Kinoshita, T. Gene analysis and linkage map, *Biology of Rice* (S. Tsunoda and M. Takahashi, eds.), Japan Scientific Society Press, Tokyo, 1984, p. 187.
18. McCouch, R., Kochert, G., Yu, Z. H., Wang, Z. Y., Khush, G. S., Coffman, W. R., and Tanksley, S. D. Molecular mapping of rice chromosomes, *Theor. Appl. Genet., 76*: 815, 1988.
19. Wang, Z. Y., and Tanksley, S. D. Restriction fragment length polymorphism in *Oryza sativa* L., *Genome, 32*: 1113, 1989.
20. Nishibayashi, S. Is genome of rice small?, *Rice Genet. Newslett., 8*: 152, 1992.
21. Zhao, X., Wu, T., Xie, Y., and Wu, Y. Genome-specific repetitive sequences in the genus *Oryza, Theor. Appl. Genet., 78*: 201, 1989.
22. Fromm, M., Morrish, F., Armstrong, C., Williams, R., Thomas, J., and Klein, T. M. Inheritance and expression of chimeric genes in the progeny of transgenic maize plants, *Biotechnology, 8*: 833, 1990.
23. Gordon-Kamm, W. J., Spencer, T. M., Mangano, M. L., Adams, T. R., Daines, R. J., Start, W. G., O'Brien, J. V., Chambers, S. A., Adams, W. R., Willetts, N. G., Rice, T. B., Mackey, C. J., Krueger, R. W., Kausch, A. P., and Lemaux, P. G. Transformation of maize cells and regeneration of fertile transgenic plants, *Plant Cell, 2*: 603, 1990.
24. Toriyama, K., Arimoto, Y., Uchimiya, H., and Hinata, K. Transgenic rice plants after direct gene transfer into protoplasts, *Biotechnology, 6*: 1072, 1988.

25. Zhang, H. M., Yang, H., Reach, E. L., Golds, T. J., Davis, A. S., Mulligan, B. J., Cocking, E. C., and Davey, M. R. Transgenic rice plants produced by electroporation-mediated plasmid uptake into protoplasts, *Plant Cell Rep., 7*: 379, 1988.
26. Datta, S. K., Peterhans, A., Datta, K., and Potrykus, I. Genetically engineered fertile indica-rice recovered from protoplasts, *Biotechnology, 8*: 738, 1990.
27. Peng, J., Lyznik, L. A., Lee, L., and Hodges, T. Transformation of indica rice protoplasts with *gusA* and *neo* genes, *Plant Cell Rep., 9*: 168, 1990.
28. Rhodes, C. A., Pierce, D. A., Mettler, I. J., Mascarenhas, D., and Detmer, J. J. Genetically transformed maize plants from protoplasts, *Science, 240*: 204, 1988.
29. Horn, M. E., Shillito, R. D., Conger, B. V., and Harms, C. T. Transgenic plants of orchardgrass (*Dactylis glomerata* L.) from protoplasts, *Plant Cell Rep., 7*: 469, 1988.
30. de la Pena, A., Lörz, H., and Schell, J. Transgenic rye plants obtained by injecting DNA into young floral tillers, *Nature, 325*: 274, 1987.
31. Lazzeri, P. A., Brettschneider, R., Luhrs, R., and Lörz, H. Stable transformation of barley via PEG-induced direct DNA uptake into protoplasts, *Theor. Appl. Genet., 81*: 437, 1991.
32. Abdullah, R., Cocking, E. C., and Thompson, J. A. Efficient plant regeneration from rice protoplasts through somatic embryogenesis, *Biotechnology, 4*: 1087, 1986.
33. Toriyama, K., Hinata, K., and Sasaki, T. Haploid and diploid plant regeneration from protoplasts of anther callus in rice, *Theor. Appl. Genet., 73*: 16, 1986.
34. Yamada, Y., Yang, Z. Q., and Tang, D. T. Plant regeneration from protoplast-derived callus of rice (*Oryza sativa* L.), *Plant Cell Rep., 5*: 85, 1986.
35. Kyozuka, J., Hayashi, Y., and Shimamoto, K. High frequency plant regeneration from rice protoplasts by novel nurse culture methods, *Mol. Gen. Genet., 206*: 408, 1987.
36. Tada, Y., Sakamoto, M., and Fujimura, T. Efficient gene introduction into rice by electroporation and analysis of transgenic plants: use of electroporation buffer lacking chloride ions, *Theor. Appl. Genet., 80*: 475, 1990.
37. Kyozuka, J., Otoo, E., and Shimamoto, K. Plant regeneration from protoplasts of indica rice: genotypic differences in culture response, *Theor. Appl. Genet., 76*: 887, 1988.
38. Dekeyser, R., Claes, B., Marichal, M., Montague, M. C., and Caplan, A. Evaluation of selectable markers for rice transformation, *Plant Physiol., 90*, 217, 1989.
39. Kyozuka, J., and Shimamoto, K. Transformation and regeneration of rice protoplasts, *Plant Tissue Culture Manual* (K. Lindsey, ed.), Kluwer, Boston, B2: 1, 1992.
40. Fujimoto, H., Izawa, T., Terada, R., Yu, R., Suzuki, M., and Shimamoto, K. (unpublished results).
41. Terada, R., and Shimamoto, K. Expression of CaMV35S-GUS gene in transgenic rice plants, *Mol. Gen. Genet., 220*: 389, 1990.
42. Battraw, M. J., and Hall, T. C. Histochemical analysis of CaMV 35S promoter-β-glucuronidase gene expression in transgenic rice plants, *Plant Mol. Biol., 15*: 527, 1990.

43. Hauptmann, R. M., Ozias-Akins, P., Vasil, V., Tabaeizadeh, Z., Rogers, S. G., Horsch, R. B., Vasil, I. K., and Fraley, R. T. Transient expression of electroporated DNA in monocotyledonous and dicotyledonous species, *Plant Cell Rep., 6*: 265, 1987.
44. Kyozuka, J., Fujimoto, H., Izawa, T., and Shimamoto, K. Anaerobic induction and tissue-specific expression of maize Adh1 promoter in transgenic rice plants and their progeny, *Mol. Gen. Genet., 228*: 40, 1991.
45. Walker, J. C., Howard, E. A., Dennis, E. S., and Peacock, W. J. DNA sequences required for anaerobic expression of the maize alcohol dehydrogenase 1 gene, *Proc. Natl. Acad. Sci. USA, 84*: 6624, 1987.
46. Matsuki, R., Onodera, H., Yamauchi, T., and Uchimiya, H. Tissue-specific expression of the *rolC* promoter of the Ri plasmid in transgenic rice plants, *Mol. Gen. Genet., 220*: 12, 1989.
47. Tada, Y., Sakamoto, M., Matsuoka, M., and Fujimura, T. Expression of a monocot LHCP promoter in transgenic rice, *EMBO J., 10*: 1803, 1991.
48. Callis, J., Fromm, M., and Walbot, V. Introns increase gene expression in cultured maize cells, *Genes Dev., 1*: 1183, 1987.
49. Oard, J. H., Paige, D., and Dvorak, J. Chimeric gene expression using maize intron in cultured cells of breadwheat, *Plant Cell Rep., 8*: 156, 1989.
50. Kyozuka, J., Izawa, T., Nakajima, M., and Shimamoto, K. Effect of the promoter and the first intron of maize *Adh1* on foreign gene expression in rice, *Maydica, 35*: 353, 1990.
51. Mascarenhas, D., Mettler, I. J., Pierce, D. A., and Lowe, H. W. Intron-mediated enhancement of heterologous gene expression in maize, *Plant Mol. Biol., 15*: 913, 1990.
52. Luehrsen, K. R., and Walbot, V. Intron enhancement of gene expression and the splicing efficiency of introns in maize cells, *Mol. Gen. Genet., 225*: 81, 1991.
53. Vasil, V., Clancy, M., Ferl, R. J., Vasil, I. K., and Hannah, L. C. Increased gene expression by the first intron of maize *Shrunken-1* locus in grass species, *Plant Physiol., 91*: 1575, 1989.
54. McElroy, D., Zhang, W., Cao, J., and Wu, R. Isolation of an efficient actin promoter for use in rice transformation, *Plant Cell, 2*: 163, 1990.
55. Tanaka, A., Mita, S., Ohta, S., Kyozuka, J., Shimamoto, K., and Nakamura, K. Enhancement of foreign gene expression by a dicot intron in rice but not in tobacco is correlated with an increased level of mRNA and an efficient splicing of the intron, *Nucleic Acids Res., 18*: 6767, 1990.
56. Ohta, S., Mita, S., Hattori, T., and Nakamura, K. Construction and expression in tobacco of a β-glucuronidase (GUS) reporter gene containing the intron within the coding sequence, *Plant Cell Physiol., 31*, 805, 1990.
57. Peterhans, A., Datta, S. K., Datta, K., Goodall, G. J., Potrykus, I., and Paszkowski, J. Recognition efficiency of *Dicotyledoneae*-specific promoter and RNA processing signals in rice, *Mol. Gen. Genet., 222*: 361, 1990.

58. Wienand, U., and Saedler, H. Plant transposable elements: unique structures for gene tagging and gene cloning, *Plant DNA Infectious Agents* (T. Hohn and J. Schell, eds.), Springer, Vienna, 1988, p. 205.
59. Carpenter, R., and Coen, E. S. Floral homeotic mutations produced by transposon-mutagenesis in *Antirrhinum majus, Genes Dev., 4*: 1483, 1990.
60. Izawa, T., Miyazaki, C., Yamamoto, M., Terada, R., Iida, S., and Shimamoto, K. Introduction and transposition of the maize transposable element *Ac* in rice (*Oryza sativa* L.), *Mol. Gen. Genet., 227*: 391, 1991.
61. Murai, N., Li, Z., Kawagoe, Y., and Hayashimoto, A. Transposition of the maize *activator* element in transgenic rice plants, *Nucleic Acids Res., 19*: 617, 1991.
62. Mariani, C., De Beuckeleer, M., Truettner, J., Leemans, J., and Goldberg, R. B. Induction of male sterility in plants by a chimeric ribonuclease gene, *Nature, 347*: 737, 1990.
63. Hayakawa, T., Zhu, Y., Itoh, K., Kimura, Y., Izawa, T., Shimamoto, K., and Toriyama, S. Genetically engineered rice resistant to rice stripe virus, an insect transmitted virus (submitted for publication).
64. Yang, H., Guo, S. D., Li, J. X., Chen, X. J., and Fan, Y. L. Transgenic rice plants produced from protoplasts following direct uptake of *Bacillus thuringiensis*–endotoxin protein gene, *Rice Genet. Newslett., 6*: 159, 1989.
65. Xie, D. X., Fan, Y. L., and Ni, P. C. Transgenic rice plant obtained by transferring the *Bacillus thuringiensis* toxin into a Chinese rice cultivar Zhonghua 11, *Rice Genet. Newslett., 7*: 147, 1990.
66. Wang, Z. Y., Wu, Z. I., Xing, Y. Y., Zheng, F. G., Guo, X. I., Zhang, W. G., and Hong, M. M. Nucleotide sequence of rice waxy gene, *Nucleic Acids Res., 18*: 5898, 1990.
67. Okita, T. W., Hwang, Y. S., Hnilo, J., Kim, W. T., Aryan, A. P., Larson, R., and Krishnan, H. B. Structure and expression of the rice glutelin multigene family, *J. Biol. Chem., 264*: 12573, 1989.
68. Masumura, T., Hibino, T., Kidzu, K., Mitsukawa, N., Tanaka, K., and Fujii, S. Cloning and characterization of a cDNA encoding a rice 13 kDa prolamin, *Mol. Gen. Genet., 221*: 1, 1990.
69. Kay, S. A., Keith, B., Shinozaki, K., Chye, M.-L., and Chua, N.-H. The rice phytochrome gene: structure, autoregulated expression, and binding of GT-1 to a conserved site in the 5′ upstream region, *Plant Cell, 1*: 351, 1989.
70. Dehesh, K., Tepperman, J., Christensen, A. H., and Quail, P. H. *phyB* is evolutionarily conserved and constitutively expressed in rice seedling shoots, *Mol. Gen. Genet., 225*: 305, 1991.
71. Boylan, M. T., and Quail, P. H. Oat phytochrome is biologically active in transgenic tomato, *Plant Cell, 1*: 765, 1989.
72. Keller, J., Shanklin, J., Vierstra, R. D., and Hershey, H. P. Expression of a functional monocotyledonous phytochrome in transgenic tobacco, *EMBO J., 4*: 1005, 1989.

73. Matzeit, V., Schaefer, S., Kammann, M., Schalk, H.-J., Schell, J., and Gronenborn, B. Wheat dwarf virus vectors replicate and express foreign genes in cells of monocotyledonous plants, *Plant Cell, 3*: 247, 1991.
74. Ugaki, M., Ueda, T., Timmermens, C. P ., Vieira, J., Elliston, K. O., and Messing, J. Replication of a geminivirus derived shuttle vector in maize endosperm cells, *Nucleic Acids Res., 19*: 371, 1991.
75. Paszkowski, J., Baur, M., Bogucki, A., and Potrykus, I. Gene targeting in plants, *EMBO J., 7*: 4021, 1988.
76. Sobral, B. W., Honeycutt, R. J., Atherly, A. G., and McClelland, M. Analysis of rice (*Oryza sativa* L.) genome using pulsed-field gel electrophoresis and rare-cutting restriction endonucleases, *Plant Mol. Biol. Rep., 8*: 253, 1991.

10

Amplification, Movement, and Expression of Genes in Plants by Viral-Based Vectors

Thomas H. Turpen

Biosource Genetics Corporation, Vacaville, California

William O. Dawson

University of Florida, Lake Alfred, Florida

I. INTRODUCTION

As a method of gene transfer, the use of viral vectors provides a rapid alternative to stable integrative transformation and regeneration of plants. Several advantages of this approach have been discussed previously (1–5). In some instances plant viruses catalyze the replication and movement of their genomes from a few initially infected cells to virtually every cell of the host over a period of days to weeks. Viruses have adapted their infection strategies to the architecture of the plant. The cells of a mature plant organ do not move relative to each other but are connected through plasmodesmata to the rest of the soma forming one large compartment, the symplast. Plant viruses can alter the molecular exclusion limits of plasmodesmata, allowing movement of their genomes between adjacent cells (6). Long-distance movement occurs through phloem tissues. Thus, a suitable virus might be engineered to transfer genes into individual plants in the time period of a growing season without requiring seed production. In contrast, genetically altered seed is a necessary first product of most plant genetic engineering efforts.

II. BACKGROUND VIROLOGY

To develop a plant virus as a gene vector, one must be able to manipulate molecular clones of viral genomes and retain the ability to generate infectious recombinants. From a technical perspective these cloning manipulations appeared to be easier for the relatively rare plant viruses that encapsidate DNA as the infectious genome—the geminiviruses and caulimoviruses—than for viruses with an RNA genome. The geminiviruses encapsidate single-stranded DNA and are truly DNA-based replicons. The caulimoviruses encapsidate double-stranded DNA and replicate through an RNA intermediate by a viral-encoded reverse transcriptase enzyme. Caulimoviruses are therefore members of the broader family of pareretroviruses.

As both of these groups were discovered, characterized, and molecularly cloned, their performance as gene vectors was assessed. Infectious chimeric viruses can be generated by transfection from mechanically inoculated naked DNA or from cloned viral genomes inserted into *Agrobacterium*. There have been several reports of the successful expression of foreign genes in whole plants by using these viruses in laboratory-scale experiments. The expression potential and problems associated with these viral groups are the subject of recent reviews (5,7–9) and will not be considered in detail here.

The vast majority (94%) of plant viruses have RNA-based genomes (10). Many RNA viruses have levels of gene expression or host ranges that could be useful for development as vectors (3,11). The techniques required to genetically engineer RNA viruses have progressed rapidly. At present the RNA composition of the genomes is no longer a serious limitation in their use as gene vectors.

Ahlquist et al. cloned into a specialized bacterial plasmid (pPM1) full-length cDNA copies of each of the three RNAs composing the genome of brome mosaic virus (BMV) (12). The resulting plasmids were linearized with restriction enzymes at unique sites introduced at the 3′ end of the viral cDNA. Upon transcription of the individual cDNA clones in vitro, RNA molecules nearly identical to those normally encapsidated in virion particles were synthesized. The transcription products were mixed, used to mechanically inoculate plants, and found to be about 10% as infectious (by weight) as the native RNA extracted from virion particles (13).

These first experiments used an *Escherichia coli* RNA polymerase and a promoter derived from phage lambda. Recently additional coliphage RNA polymerase promoters and enzymes have been characterized and made commercially available (SP6, T7, T3). By using these tools, infectious transcripts from an increasing number of RNA viruses have been synthesized (Table 1).

All of the viruses listed in Table 1 are positive-strand RNA viruses. The RNA strand encapsidated in virions functions as messenger RNA. Based on similarities in nonstructural protein sequences, positive-strand viruses have been grouped into "superfamilies" suggesting an evolutionary relationship with viruses that infect vertebrates (48). These viruses have either a cap structure (7-methyl-GpppG) at the 5′ end of the genome, as do cellular mRNAs, or a small peptide covalently linked to the 5′ nucleotide (VPg: virion protein, genome linked). Telomere function is provided at the 3′ end by either a polyadenylation region or a tRNA-like structure. Individual viral genes are expressed by a combination of strategies including segmentation of the genome, proteolysis of large "polyprotein" precursors, translational frame-shifting, readthrough of stop codons, and synthesis of subgenomic mRNAs (49). Theoretically, cDNA clones from any of these viruses (or combinations thereof) might be adapted to express foreign genes in plants.

A reliable transfection efficiency greatly facilitates vector development. The number of infectious molecules synthesized in a transcription reaction is most often estimated either in a dilution end-point assay or by counting the number of chlorotic or necrotic lesions forming on mechanically inoculated leaves of an appropriate host plant. Lesion-forming units (LFU) on leaf surfaces can be used in quantitative comparisons in much the same way that plaque-forming units (PFU) are used in cultured lawns of bacterial or animal cells. The major difference is that LFU assays are inherently inefficient on a molar basis and are more variable than most PFU assays. Yields of LFU are often smaller for some recombinant viruses than for wild-type viruses. Although small numbers of LFU may be adequate for most experiments, many researchers have worked to optimize the yields of infectivity from their cDNA clones. One way to quantitatively estimate transfection efficiencies is to compare the specific infectivity (LFU/RNA weight) of transcripts synthesized in vitro with genomic RNA purified from virions. For the viruses listed in Table 1, transfection efficiencies can vary over three orders of magnitude (10^0–10^{-3}) when described as the ratio [specific infectivity (transcripts)/specific infectivity (genomic RNA)]. A discussion of possible sources of this wide variation follows.

Not all cDNA clones derived from a population of viral RNA have the same biological activity. One of two cDNA clones of BMV RNA 1 produced noninfectious transcripts and was found to contain multiple-point mutations (13). Similar results have been obtained for tobacco mosaic virus (TMV) (26,27,31) and for other viruses (20,22,32,42,44). Whether or not sequence variation observed in cDNA clones reflects sequence variation in the viral

Table 1 Infectious Transcripts from cDNA Clones of RNA Viruses Infecting Plants

Group/Virus[a]	Strain	Promoter[b]	Reference
Superfamily A			
Alfalfa mosaic virus (RNA 3)	S	T7	Dore et al. (14,15)
Bromovirus			
Brome mosaic virus	Russian	Lambda P_M	Ahlquist et al. (13)
	Russian	SP6, T7	Janda et al. (16)
	ATCC66	CaMV 35S, T7	Mori et al. (17)
Cowpea chlorotic mottle virus	Type	T7	Allison et al. (18)
Cucumovirus			
Cucumber mosaic virus	Fny	T7	Rizzo et al. (19)
	Q	T7	Hayes et al. (20)
Furovirus			
Beet necrotic yellow vein virus			
RNA 3, 4		T7	Ziegler-Graff et al. (21)
RNA 1, 2		T7	Quillet et al. (22)
Hordeivirus			
Barley stripe mosaic virus	Type	T7	Petty et al. (23)
	ND 18	T7	
Potexvirus			
Potato virus X		T7	Hemenway et al. (24)
White clover mosaic virus	O	SP6	Beck et al. (25)
Tobamovirus			
Tobacco mosaic virus	U1	Lambda P_M	Dawson et al. (26)
	L, $L_{11}A$	Lambda P_M	Meshi et al. (27)
	L, $L_{11}A$	CaMV 35S	Yamaya et al. (28,29)
	M	T7	Holt et al. (30)
	U1	T7	Holt et al. (31)

Tobravirus			
Tobacco rattle virus			
RNA 1		Lambda P_M	Hamilton et al. (32)
RNA 2		Lambda P_M	Angenent et al. (33)
Pea early browning virus	SP5	T7	MacFarlane et al. (34)
Tymovirus			
Turnip yellow mosaic virus	Type	T7	Weiland et al. (35)
Superfamily B			
Carmovirus			
Turnip crinkle virus		Lambda P_M, T7	Heaton et al. (36)
Dianthovirus			
Red clover necrotic mosaic virus	Australian	T7	Xiong et al. (37)
Luteovirus			
Barley yellow Dwarf virus	PAV	T7	Young et al. (38)
Tombusvirus			
Cucumber necrosis virus		T7	Rochon et al. (39)
Cymbidium ringspot virus		T7	Burgyan et al. (40)
Tomato bushy stunt virus	Cherry	T7	Hearne et al. (41)
Superfamily C			
Comovirus			
Cowpea mosaic virus		T7	Vos et al. (42)
			Eggen et al. (43)
Potyvirus			
Plum pox virus	Rankovic	T7	Riechmann et al. (44)
Tobacco vein mottling virus		T3, T7	Domier et al. (45)

[a]The proposed superfamily classification system of viral groups is based on similarities in amino acid sequence motifs in putative RNA polymerase and nucleic acid helicase proteins (46). Superfamilies have sequence similarities to viruses infecting vertebrates: A = Sindbis-like, C = Picornavirus-like, and B = intermediate to A and C. Adapted from Matthews (47).
[b]Transcription from the eukaryotic RNA pol II promoter (CaMV 35S) occurs in vivo. The in vivo promoter functioning in (14) was not characterized. All other transcripts are synthesized in vitro.

RNA population, such variation might be expected because of the error rates ($\sim 10^{-4}$–10^{-5}) of the enzymes used in cDNA synthesis (reverse transcriptase, Klenow fragment of DNA polymerase I) or subsequent amplification (Taq DNA polymerase) (50–52).

As a further possible complication, sequence variation may be selected in the cloning process. Difficulties associated with cloning viral cDNAs are often attributed to toxicity effects on *E. coli* host cells or structural properties of the sequences or both (28,34). In this regard, it is not necessary to reconstruct full-length cDNA clones if cDNA subclones can be ligated to yield microgram amounts of full-length template. Ligation products can subsequently be used as a source of transcription templates (22).

Not surprisingly, the more a transcription product resembles the native virus, the higher the infectivity. However, some deviations are better tolerated than others. A 5′-cap structure, known to stabilize mRNA and enhance translation, is often required for detectable infectivity or increases LFU by 10–100-fold. An unmethylated dinucleotide derivative will substitute for the authentic cap with a 2-fold or less decrease in infectivity (4,23,34). Interestingly, the members of the superfamily B viruses that have been characterized to date do not show a significant increase in specific infectivity for capped versus uncapped transcripts. For some of the coliphage promoters, full promoter activity includes nucleotides on the 3′ side of the transcription initiation site. Depending on the 5′ end sequence of a given virus, this can create a conflict between the goals of synthesizing transcripts in high yields and synthesizing transcripts identical to the viral 5′ end sequence. Transcripts with 0–2 additional nonviral nucleotides at the 5′ end are relatively highly infectious, whereas longer extension may decrease infectivity by 10–100-fold or to undetectable levels.

Adequate fidelity at the 3′ end has been obtained in most cases by using linearized template ("run-off transcripts") having <7–12 additional nucleotides or by the insertion of self-cleaving sequences (53). For a few viral genomes with polyA tails, long 3′ extensions (>100 bp) do not significantly reduce infectivity (25,54). A few nucleotides removed from 3′ tRNA-like structures or polyA tails can be added to the genome *in planta* (55,56). In all cases examined, terminal nucleotide additions, present in the synthetic genome used as inoculum, are not detected in the virus population after transfection.

Reconstituting virion particles from capsid protein and transcribed RNA is reported to increase infectivity for TMV (27). This process may not be practical or necessary in most cases because detailed reconstitution procedures have not been published for many other viruses and the infectivity of RNA transcribed in vitro is usually sufficient. Perhaps other general nucleic acid binding proteins will prove to be effective as capsid substitutes.

Another approach to infecting plants from cDNA is to design the insertion of full-length viral cDNA behind a plant RNA polymerase II promoter such that transcription initiates at the 5′ nucleotide of viral cDNA. Once it is introduced into a plant cell, capping would be predicted to occur in vivo. Transcription termination might be nonspecific or include normal polyadenylation signals from characterized cellular genes. To generate a precise viral 3′ terminus, such an in vivo transfection cassette could include self-cleaving sequences. Recently such a vector was successfully designed to express BMV RNA from the 35S promoter and termination signals of cauliflower mosaic virus (17). Plasmid DNAs at concentrations of micrograms per microliter yielded hundreds of LFU on leaves of *Chenopodium hybridum*. However, no transfection occurred on similarly inoculated barley plants, a natural systemic host for BMV. It is possible that different hosts vary in their susceptibility to naked-DNA transfection. This may provide a useful alternative to the synthesis of infectious transcripts.

If the in vivo transfection cassettes just described are not infectious as naked DNA, infectivity can be generated in some cases by introducing the construction into *Agrobacterium. Agrobacterium* can be used in the stable transformation of plants or to deliver an RNA virus from inoculated wound sites (28,57). In the case of stable integration of viral cDNA, transfection events occur during development of plants. Alternatively, transfection cassettes may be delivered by other gene transfer techniques, such as electroporation or biolistics.

III. PROTOTYPE VECTORS

As a prerequisite for vector development, one must identify the types of cDNA sequence alterations (duplications, deletions, insertions, gene fusions, etc.) a virus can tolerate while retaining the ability to replicate. The nucleotide sequences of several plant viral genomes reveal a compact coding strategy. Large RNA polymerase open reading frames (ORFs) are often required *in trans* for RNA replication, and small noncoding sequences or structures are required *in cis*. Once these genomic regions have been mapped, other genes may be added to the virus or substituted for ORFs nonessential for replication. By using this approach, several genes have been successfully expressed in plant cells (Table 2). However, the substitution of foreign genes for capsid protein genes and capsid protein fusion experiments in additional vector

Table 2 Plant RNA Viruses Used as Gene Vectors

Group/virus	Gene[a]	Reference
Bromovirus		
Brome mosaic virus	*cat*	French et al. [58] Ahlquist et al. [59]
Furovirus		
Beet necrotic yellow vein virus	*gus*	Jupin et al. [60]
Hordeivirus		
Barley stripe mosaic virus	*luc*	Joshi et al. [61]
Tobamovirus	*cat*	Takamatsu et al. [62]
Tobacco mosaic virus	*cat*	Dawson et al. [63]
	ENK	Takamatsu et al. [64]
	gus	Cassidy et al. [65]

[a]Genes are *cat*: chloramphenicol acetyltransferase; *gus*: β-glucuronidase; *luc*: firefly luciferase; and ENK: coding sequence for the pentapeptide Leu-enkephalin.

designs have led to debilitated systemic movement of the recombinant viruses in whole plants. Gene expression was confirmed only in inoculated leaves or protoplasts.

Although there are important similarities among plant RNA viruses, many details of their molecular biology are unpredictably different and must be experimentally established for each virus. For example, one cannot assume that the full activity of *cis*-acting replication elements will always be located in terminal noncoding portions of the genomes. For BMV an ~200-base internal intergenic sequence is necessary (66). In a related bromovirus, cowpea chlorotic mottle virus (CCMV), only the terminal sequences are necessary for full activity (67). For barley stripe mosaic virus (BSMV) and BMV, *cis*-acting sequences required for replication may extend into adjacent coding regions (61,68). For alfalfa mosaic virus (AlMV), a few molecules of coat protein must be added to initiate an infection with virion or transcript RNA (69).

Packaging constraints are an important consideration in the development of non-rod-shaped viruses as gene vectors. The origin of assembly and capsid protein coding sequences of TMV can be substituted into the genome of BMV and tobacco rattle virus (TRV), yielding TMV-type rods in infected tissues (70,71). In the case of the chimeric BMV/TMV genome, virion structure is transformed from icosahedral to rod shaped. Systemic movement of the chimeric virus does not occur in either case. Perhaps additional functions can be transferred between RNA viruses of different groups.

A. Manipulating the TMV Genome

This section focuses on the use of TMV as a gene vector. The genome of TMV is a 6.4-kb plus-sense RNA molecule. The 5′ end is capped, and the 3′ end contains a tRNA-like structure with an extended pseudoknot region. The virus encodes information necessary for its replication, encapsidation, and movement in plant cells (72,73). Functions associated with five ORFs have been described in detail. *cis*-Acting elements required for replication are thought to be positioned at the termini (74,75). Additional *cis*-acting sequences include the origin of assembly and promoters regulating subgenomic mRNA synthesis for movement and capsid proteins. Several biological activities, such as viral movement and viral resistance, have been demonstrated for TMV proteins in transgenic hosts (31,76–79).

Tobacco leaves can accumulate >10 mg of TMV per gram of fresh-weight tissue in 7–14 days after inoculation (80). Capsid protein synthesis reaches a maximum rate of 70% of the total cellular protein synthesis and can accumulate to 10% of the leaf dry weight (81,82). No specific shutoff of host protein synthesis is associated with infections (83).

Many alterations, including gene and subgenomic mRNA promoter duplications, have been constructed in the TMV genome while retaining infectivity of in vitro transcripts. Capsid protein–related peptides synthesized from mutants deficient in virion assembly can accumulate to levels comparable to those of the wild-type virus (63). Therefore, encapsidation of genomic RNA is not a requirement for high-level peptide accumulation. The entire coding sequence for the capsid protein can be deleted from TMV without loss of replicative function or subgenomic RNA synthesis. The resulting noncoding transcript can accumulate to ~5% of total mRNA (57). Naked-RNA viruses, including those containing reporter genes in this position, are largely confined to the inoculated leaves of the plant (62,63,84).

To allow systemic spread, a virus was designed to express both the capsid gene and *cat* by duplicating the capsid subgenomic mRNA promoter of TMV. This vector was found to retain movement functions, but during systemic invasion of the plant, homologous recombination removed the repeated and intervening reporter gene sequences so that only recombinants without the foreign ORF infected upper leaves (85).

To circumvent the recombination problem, a promoter with homologous function but low sequence identity (45%) to the U1 strain of TMV was cloned from a distantly related tobamovirus, odontoglossum ringspot virus (ORSV). These two capsid protein promoter regions were used in tandem to construct the prototype vector TB2 (Fig. 1) (86). The U1 promoter controlled synthesis of subgenomic mRNA for genes to be expressed from the adjacent insertion site, whereas the ORSV promoter was used for coat protein mRNA synthesis. Viruses containing ORFs for production of the enzymes neomycin phosphotransferase (NPTII) and dihydrofolate reductase (DHFR), as well as other genes, invaded whole plants from inoculated leaves. Plants accumulated virion particles containing chimeric genomes at levels comparable to those of wild-type TMV, indicating no substantial inhibition of replication. A reduced frequency of recombination between the repeated subgenomic mRNA promoters allowed expression of the inserted coding sequences through several plant-to-plant passages over a period of months. A single passage in *Nicotiana benthamiana* can easily result in an amplification of $>10^{12}$ particles over several weeks.

B. Genetic Stability

Point mutations and deletions are potential sequence instabilities occurring in RNA viral vector systems. Deletions are thought to result from recombination as the RNA polymerase switches templates during replication (87,88). Recombination in mammalian RNA viruses is well documented (89), and

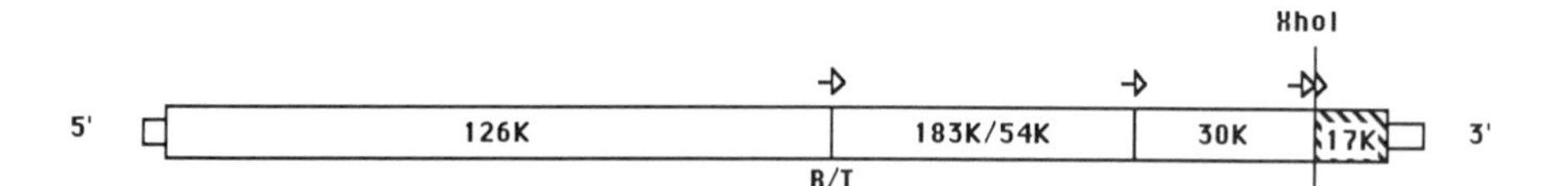

Figure 1 TB2: Autonomous virus and systemic vector. Several replication functions have been assigned to the 126K/183K and 54K reading frames of TMV. The 126K protein is translated directly from genomic RNA. The 183K protein is a readthrough product (R/T). Subgenomic-size mRNAs are synthesized from minus-strand promoters (arrows) for the 54K, 30K (movement protein), and 17K (coat protein) ORFs. TB2 is a chimeric virus between the U1 and O strains of TMV. The coat protein ORF and subgenomic mRNA promoter of the O strain replace the coat protein ORF of the U1 strain. The subgenomic mRNA promoter for the coat protein gene of the U1 strain drives expression of sequences after insertion into the XhoI site of viral cDNA and in vitro synthesis of infectious RNA.

recent work has demonstrated recombination in plant RNA viruses as well (55,90,91).

Point mutations are a particular concern with RNA vectors because viral RNA polymerases are thought to have a relatively low replication fidelity. In DNA synthesis the mutation frequency is 10^{-7}–10^{-11} misincorporations per base pair replicated (51), whereas in viral RNA synthesis the mutation frequency is reported to be within the range 10^{-4}–10^{-6} for a number of bacterial and animal viruses (92,93). It is difficult to evaluate and compare such figures, because the number of viral replication cycles in vivo, the competitive fitness of the mutants relative to the initial virus, and the homogeneity of the starting inoculum are often undetermined. However, poliovirus RNA-dependent RNA polymerase fidelity has been measured in vitro, and very high rates of 10^{-4} were obtained (94). Also, a careful in vivo study of the same virus yielded comparable misincorporations per base for one passage (95).

The potential instability of RNA viral vectors has been regarded as a critical barrier to their utility. It has been proposed that low fidelity is an inherent property of RNA viruses owing to the absence of proofreading functions characteristic of DNA replication systems. Thus, foreign sequences would be maintained only by selection, so RNA viruses would be of little use as expression vectors (96). However, there is some doubt about the implications of these fidelity measurements even in animal cell systems, because vectors based on sindbis or influenza virus replicons can amplify *cat* efficiently, producing large amounts of the enzyme in tissue culture cells even after multiple passages (97,98).

1. Marker Gene Stability–Recombination

Analysis of foreign sequences carried by the prototype vector TB2 provides answers to fundamental questions of genetic stability in RNA genomes amplified in plant cells (86). The virus TBD4 contains the coding sequence for DHFR (230 bp) inserted at the XhoI site in TB2. The TBD4 virus was serially cultured through up to 10 passages (170 days) on the systemic host *N. benthamiana*. The presence of the reporter gene in the virion population was determined by a combination of techniques, including RNase protection assays, RNA sequencing, and the PCR amplification of cDNA, and no detectable deletions were found.

In contrast, deletions of portions (420–650 bp) of the NPTII sequence (800 bp) from the virus TBN62 were more readily detected, occurring within 1–3 serial passages. However, some plants infected for 35 days contained virus with no deletions.

2. Marker Gene Stability—Point Mutations

The available experimental evidence for sequence drift of marker genes in RNA vectors is limited. After several passages of TBD4 and TBN62 in plants, the DHFR and NPTII ORFs were cloned from PCR-amplified cDNA and were sequenced. Very few point mutations were found in this sampling of the "gene population." For example, a line that had been serially passaged in plants 9 times yielded 25 clones, but only 7 of these contained a point mutation (one had two mutations) in the DHFR gene; all the other clones had a DHFR sequence identical to the published sequence. Thus, 72% of the RNA population had a perfectly intact DHFR sequence after 9 passages.

C. Expression

Expression of NPTII was analyzed in systemically infected leaf tissue. Phenotypic expression was apparent by the ability of leaf discs to regenerate plantlets on kanamycin-containing media. Western blot analysis showed comigration of NPTII from infected plants with the pure enzyme from commercial sources at yields similar to those generally observed from chromosomally integrated genes (0.01–0.1% of total leaf extract protein). The yields of immunoprecipitable protein and total enzyme activity indicate that NPTII produced by TMV transfection has approximately the same order-of-magnitude specific activity as NPTII produced from bacterial sources.

It appears that these vectors may be more stable than originally predicted (96). For the plant RNA viruses, the frequency of base mutations per replication cycle needs to be accurately determined. Discrepancies between error rates may reflect differences in replicase fidelity between RNA viruses or uncharacterized RNA proofreading systems in plants. Alternatively, the error rates may be similar, and the paradox of genetic stability explained by founder effects. The products of a transcription reaction used to initiate these infections are essentially homogeneous, providing a well-defined "genetic bottleneck." Unless sequence variants are strongly selected, the numbers of correct molecules produced in the first few rounds of replication may often continue to predominate in derivative populations.

IV. FUTURE DIRECTIONS—CROPPING WITH GENES THAT MOVE IN PLANTS

Tobacco mosaic virus is one of the best understood viruses in all aspects of its biology. Largely because of this historical experience with TMV, the ability to predict effects of engineered, derivative, viruses in the environment has increased. Members of the tobamovirus group have a relatively broad host range, which is useful for genetic engineering purposes. Tobacco mosaic virus is not known to be efficiently transmitted in tobacco culture through seed, by insects, or by other organisms except humans (99). Containment of genetically engineered viruses is likely to be of greater concern in cases in which related strains are serious pathogens with the capability of being vectored by organisms such as insects into or from the target site. However, only through carefully designed and publicly reviewed field testing can the efficacy and safety of such vectors be established.

After extensive public review, prototype expression vectors based on the TMV genome have been approved for field testing by the North Carolina and the U.S. Departments of Agriculture. These experiments, initiated in May 1991, establish the first field test of a genetically engineered plant virus.

In addition to the use of autonomous viruses, other strategies currently being developed may differ in their levels of biological containment and genetic stability. These alternative strategies are based on genetic complementation, whereby viral encoded functions are separated onto different genomic segments, either as replicating RNA or as sequences integrated into the host chromosome.

A. Complementation Between Viral Segments

Satellite virus RNAs (SV RNAs) and defective interfering-like RNAs (DI RNAs) present additional options for vector engineering. The DI RNAs are helper-dependent RNA molecules occurring in high-multiplicity passages of some members of the tombusvirus and carmovirus groups (100–102). The DI RNAs can be used in the design of helper-dependent vector RNAs (103). No DI RNAs for TMV have been observed in nature. However, RNAs with large deletions in the replicase and movement regions can be replicated in the presence of a helper virus (104). Genetic complementation might also be forced by dividing movement, structural, or other functions into component RNAs (105).

The SV RNAs exist in mixed infections with helper viruses in nature and often influence symptoms. They encode information for their own structural proteins and possibly for other proteins, and they have limited sequence similarity with their specific helper virus. Some aspects of their genome structure, such as the presence of a 5′-hydroxyl, suggest that an autocatalytic cleavage step is a component of their replication cycles (106). This may add another level of complexity in genetically engineering molecules to function as gene vectors. Transcripts that are infectious in the presence of helper virus have been synthesized from cloned cDNA of a satellite virus of tobacco necrosis virus (TNV) and TMV (107,108). The use of these systems in gene transfer experiments is being investigated.

B. Complementation in Transgenic Hosts

Viral genes can be transferred to the host by chromosomal integration to provide the genetic complementation functions of a helper virus. As first reported for AlMV, replicase genes can be transferred to "helper plants" (109) and function *in trans* to replicate RNA 3 and deletion mutants of RNA 3 (91,110,111). Foreign genes are currently being tested for expression in this system. The cell-to-cell movement protein and capsid proteins of TMV complement, at least partially, respectively deficient viral mutants in transgenic tobacco. For some viral functions it may be difficult to produce the amounts of complementation activity necessary to restore mutants to wild-type phenotypes (31,77,112–114). For example, few systemic infection sites are observed in transgenic plants expressing wild-type TMV coat protein when inoculated with viral coat protein deletion mutants.

Success in maintaining a complementation system depends on levels of RNA recombination. Although viral RNA recombination *in planta* was not reported to occur between cellular mRNA and viral RNA in the complementation experiments just described, the size of populations lacking detectable recombinants was not reported. Recent reports prove that recombination can occur between movement protein mRNA expressed in a transgenic plant and movement-defective in vitro transcripts such that an autonomous progeny virus is obtained in systemically infected leaves (115). Therefore, if viral genomic RNA and cellular mRNA are not compartmentalized, recombination is a strong possibility. If the progeny of the recombination event is an autonomous virus, the possible biological containment advantages of releasing defective genomes would not be realized.

Finally, if foreign ORFs were inserted into the host chromosome with the necessary *cis*-acting flanking sequences, they should be amplified and expressed in the presence of a helper virus. Some observations support this possibility. Satellite RNAs in the presence of helper virus (116,117) and replication-competent derivatives of TMV (28,29,57) can transfect plants from chromosomally integrated copies. The construction of host–cell transcripts that can be replicated in the presence of a helper virus has been described in a bacteriophage Qβ/*E. coli* system (118).

V. CONCLUSION

To evaluate the potential of RNA-based transient vectors, more information on limiting factors is required. More data are needed on the effects of specific sequences and sequence arrangements on replication rates, gene stability, and gene expression levels. The effects of recombination in different contexts, including complementation schemes, need to be measured. The specific activities of pure enzymes produced by RNA viruses should be compared with chromosomal DNA-encoded sources. The final protein products must be characterized in detail as the ultimate assessment of genetic stability (119).

Plant viral vectors are subject to the constraints of the biology of the individual viruses and plants hosts from which they are derived. When developed appropriately for their expression capabilities, these vectors will become useful tools for molecular biology and industry. In the laboratory, RNA vectors may be used to rapidly produce from a plant cell desired quantities of a gene product for various assays. Studies of structure–activity relationships in biology are often limited by the time required to accumulate mutant gene products for testing. These vectors might also be used to inactivate host gene expression through antisense, ribozyme, or other technologies for studies on specific host gene functions. Perhaps the greatest potential utility is in the production of complex molecules by using a higher eukaryote photoautotroph as a host. This production may be adjusted for yields as required by the product application on a virtually unlimited scale, from individually infected plants to acres of a new crop. There are opportunities to find alternative uses for crops, such as tobacco, through efficient, rapid, and flexible gene transfer with RNA viral vectors.

ACKNOWLEDGMENTS

We are grateful to those who contributed research results from work in progress or in press, to our colleagues at Biosource Genetics Corporation for critical readings of the manuscript, and to Dr. Chris Kearney for his contribution on genetic stability studies.

REFERENCES

1. Grill, L. K. Utilizing RNA viruses for plant improvement, *Plant Mol. Biol. Rep., 1*: 17, 1983.
2. van Vloten-Doting, L. Advantages of multipartite genomes of single-stranded RNA plant viruses, in nature, for research, and for genetic engineering, *Plant Mol. Biol. Rep., 1*: 55, 1983.
3. Siegel, A. Plant-virus-based vectors for gene transfer may be of considerable use despite a presumed high error frequency during RNA synthesis, *Plant Mol. Biol., 4*: 327, 1985.
4. Ahlquist, P., French, R., and Bujarski, J. J. Molecular studies of brome mosaic virus using infectious transcripts from cloned cDNA, *Adv. Virus Res., 32*: 215, 1987.
5. Joshi, R. L., and Joshi, V. Strategies for expression of foreign genes in plants, *FEBS Lett., 281*: 1, 1991.
6. Robards, A. W., and Lucas, W. J. Plasmodesmata, *Annu. Rev. Plant Physiol. Plant Mol. Biol., 41*: 369, 1990.
7. Hull, R. The use and misuse of viruses in cloning and expression in plants, *Recognition and Response in Plant–Virus Interactions* (R. S. S. Fraser, ed.), NATO ASI Series, Vol. H 41, Springer-Verlag, Berlin, 1990, pp. 443–457.
8. Walden, R., and Schell, J. Techniques in plant molecular biology–progress and problems, *Eur. J. Biochem., 192*: 563, 1990.
9. Hohn, T., and Goldbach, R. Viruses as vectors in plant biology, *Encyclopedia of Virology* (R. G. Webster and A. Granoff, eds.), Saunders Scientific Publications, Philadelphia (in press).
10. Zaitlin, M., and Hull, R. Plant virus–host interactions, *Annu. Rev. Plant Physiol., 38*: 291, 1987.
11. Ahlquist, P., and Pacha, R. F. Gene amplification and expression by RNA viruses and potential for further application to plant gene transfer, *Physiol. Plant., 79*: 163, 1990.
12. Ahlquist, P., and Janda, M. cDNA cloning and *in vitro* transcription of the complete brome mosaic virus genome, *Mol. Cell. Biol., 4*: 2876, 1984.
13. Ahlquist, P., French, R., Janda, M., and Loesch-Fries, S. Multicomponent RNA plant virus infection derived from cloned viral cDNA, *Proc. Natl. Acad. Sci. USA, 81*: 7066, 1984.
14. Dore, J.-M., and Pinck, L. Plasmid DNA containing a copy of RNA3 can substitute for RNA3 in alfalfa mosaic virus RNA inocula, *J. Gen. Virol., 69*: 1331, 1988.
15. Dore, J.-M., Erny, C., and Pinck, L. Biologically active transcripts of alfalfa mosaic virus RNA 3, *FEBS Lett., 264*: 183, 1990.
16. Janda, M., French, R., and Ahlquist, P. High efficiency T7 polymerase synthesis of infectious RNA from cloned brome mosaic virus cDNA and effects of 5′ extensions on transcript infectivity, *Virology, 158*: 259, 1987.
17. Mori, M., Mise, K., Kobayashi, K., Okuno, T., and Furusawa, I. Infectivity of plasmids containing brome mosaic virus cDNA linked to the cauliflower mosaic virus 35S RNA promoter, *J. Gen. Virol., 72*: 243, 1991.

18. Allison, R. F., Janda, M., and Ahlquist, P. Infectious *in vitro* transcripts from cowpea chlorotic mottle virus cDNA clones and exchange of individual RNA components with brome mosaic virus, *J. Virol., 62*: 3581, 1988.
19. Rizzo, T. M., and Palukaitis, P. Construction of full-length cDNA clones of cucumber mosaic virus RNAs 1, 2 and 3: generation of infectious transcripts, *Mol. Gen. Genet., 222*: 249, 1990.
20. Hayes, R. J., and Buck, K. W. Infectious cucumber mosaic virus RNA transcribed *in vitro* from clones obtained from cDNA amplified using the polymerase chain reaction, *J. Gen. Virol., 71*: 2503, 1990.
21. Ziegler-Graff, V., Bouzoubaa, S., Jupin, I., Guilley, H., Jonard, G., and Richards, K. Biologically active transcripts of beet necrotic yellow vein virus RNA-3 and RNA-4, *J. Gen. Virol., 69*: 2347, 1988.
22. Quillet, L., Guilley, H., Jonard, G., and Richards, K. *In vitro* synthesis of biologically active beet necrotic yellow vein virus RNA, *Virology, 172*: 293, 1989.
23. Petty, I. T. D., Hunter, B. G., Wei, N., and Jackson, A. O. Infectious barley stripe mosaic virus RNA transcribed *in vitro* from full-length genomic cDNA clones, *Virology, 171*: 342, 1989.
24. Hemenway, C., Weiss, J., O'Connell, K., and Tumer, N. E. Characterization of infectious transcripts from a potato virus X cDNA clone, *Virology, 175*: 365, 1990.
25. Beck, D. L., Forster, R. L. S., Bevan, M. W., Boxen, K. A., and Lowe, S. C. Infectious transcripts and nucleotide sequence of cloned cDNA of the potexvirus white clover mosaic virus, *Virology, 177*: 152, 1990.
26. Dawson, W. O., Beck, D. L., Knorr, D. A., and Grantham, G. L. cDNA cloning of the complete genome of tobacco mosaic virus and production of infectious transcripts, *Proc. Natl. Acad. Sci. USA, 83*: 1832, 1986.
27. Meshi, T., Ishikawa, M., Motoyoshi, F., Semba, K., and Okada, Y. *In vitro* transcription of infectious RNAs from full-length cDNAs of tobacco mosaic virus, *Proc. Natl. Acad. Sci. USA, 83*: 5043, 1986.
28. Yamaya, J., Yoshioka, M., Meshi, T., Okada, Y., and Ohno, T. Expression of tobacco mosaic virus RNA in transgenic plants, *Mol. Gen. Genet., 211*: 520, 1988.
29. Yamaya, J., Yoshioka, M., Meshi, T., Okada, Y., and Ohno, T. Cross protection in transgenic tobacco plants expressing a mild strain of tobacco mosaic virus, *Mol. Gen. Genet., 215*: 173, 1988.
30. Holt, C. A., Hodgson, R. A. J., Coker, F. A., Beachy, R. N., and Nelson, R. S. Characterization of the masked strain of tobacco mosaic virus: identification of the region responsible for symptom attenuation by analysis of an infectious cDNA clone, *Mol. Plant–Microbe Interact., 3*: 417, 1990.
31. Holt, C. A., and Beachy, R. N. *In vivo* complementation of infectious transcripts from mutant tobacco mosaic virus cDNAs in transgenic plants, *Virology, 181*: 109, 1991.
32. Hamilton, W. D. O., and Baulcombe, D. C. Infectious RNA produced by *in vitro* transcription of a full-length tobacco rattle virus RNA-1 cDNA, *J. Gen. Virol., 70*: 963, 1989.

33. Angenent, G. C., Posthumus, E., and Bol, J. F. Biological activity of transcripts synthesized *in vitro* from full-length and mutated DNA copies of tobacco rattle virus RNA2, *Virology, 173*: 68, 1989.
34. MacFarlane, S. A., Wallis, C. V., Taylor, S. C., Goulden, M. G., Wood, K. R., and Davies, J. W. Construction and analysis of infectious transcripts synthesized from full-length cDNA clones of both genomic RNAs of pea early browning virus, *Virology, 182*: 124, 1991.
35. Weiland, J. J., and Dreher, T. W. Infectious TYMV RNA from cloned cDNA: effects *in vitro* and *in vivo* of point substitutions in the initiation codons of two extensively overlapping ORFs, *Nucleic Acids Res., 17*: 4675, 1989.
36. Heaton, L. A., Carrington, J. C., and Morris, T. J. Turnip crinkle virus infection from RNA synthesized *in vitro, Virology, 170*: 214, 1989.
37. Xiong, Z., and Lommel, S. A. Red clover necrotic mosaic virus infectious transcripts synthesized *in vitro, Virology, 182*: 388, 1991.
38. Young, M. J., Kelly, L., Larkin, P. J., Waterhouse, P. M., and Gerlach, W. L. Infectious *in vitro* transcripts from a cloned cDNA of barley yellow dwarf virus, *Virology, 180*: 372, 1991.
39. Rochon, D. M., and Johnston, J. C. Infectious transcripts from cloned cucumber necrosis virus cDNA: evidence for a bifunctional subgenomic mRNA, *Virology, 181*: 656, 1991.
40. Burgyan, J., Nagy, P. D., and Russo, M. Synthesis of infectious RNA from full-length cloned cDNA to RNA of cymbidium ringspot tombusvirus, *J. Gen. Virol., 71*: 1857, 1990.
41. Hearne, P. Q., Knorr, D. A., Hillman, B. I., and Morris, T. J. The complete genome structure and synthesis of infectious RNA from clones of tomato bushy stunt virus, *Virology, 177*: 141, 1990.
42. Vos, P., Jaegle, M., Wellink, J., Verver, J., Eggen, R., van Kammen, A., and Goldbach, R. Infectious RNA transcripts derived from full-length DNA copies of the genomic RNAs of cowpea mosaic virus, *Virology, 165*: 33, 1988.
43. Eggen, R., Verver, J., Wellink, J., de Jong, A., Goldbach, R., and van Kammen, A. Improvements of the infectivity of *in vitro* transcripts from cloned cowpea mosaic virus cDNA: impact of terminal nucleotide sequences, *Virology, 173*: 447, 1989.
44. Riechmann, J. L., Laín, S., and Garciá, J. A. Infectious *in vitro* transcripts from a plum pox potyvirus cDNA clone, *Virology, 177*: 710, 1990.
45. Domier, L. L., Franklin, K. M., Hunt, A. G., Rhoads, R. E., and Shaw, J. G. Infectious *in vitro* transcripts from cloned cDNA of a potyvirus, tobacco vein mottling virus, *Proc. Natl. Acad. Sci. USA, 86*: 3509, 1989.
46. Habili, N., and Symons, R. H. Evolutionary relationship between luteoviruses and other RNA plant viruses based on sequence motifs in their putative RNA polymerases and nucleic acid helicases, *Nucleic Acids Res., 17*: 9543, 1989.
47. Matthews, R. F. *Plant Virology*, 3rd ed., Academic Press, San Diego, 1991, p. 678.

48. Goldbach, R. Genome similarities between positive-strand RNA viruses from plants and animals, *New Aspects of Positive-Strand RNA Viruses* (M. A. Brinton and F. X. Heinz, eds.), American Society for Microbiology, Washington, D.C., 1990, pp. 3–11.
49. Matthews, R. F. *Plant Virology*, 3rd ed., Academic Press, San Diego, 1991, pp. 143–270.
50. Fields, S., and Winter, G. Nucleotide-sequence heterogeneity and sequence rearrangements in influenza virus cDNA, *Gene, 15*: 207, 1981.
51. Loeb, L. A., and Kunkel, T. A. Fidelity of DNA synthesis, *Annu. Rev. Biochem., 51*: 429, 1982.
52. Keohavong, P., and Thilly, W. G. Fidelity of DNA polymerase in DNA amplification, *Proc. Natl. Acad. Sci., USA, 86*: 9253, 1989.
53. Dzianott, A. M., and Bujarski, J. J. Derivation of an infectious viral RNA by autolytic cleavage of *in vitro* transcribed viral cDNAs, *Proc. Natl. Acad. Sci. USA, 86*: 4823, 1989.
54. Wellink, J., Eggen, R., Verver, J., Goldbach, R., van Kammen, A. Use of full-length copies in the study of expression and replication of the bipartite RNA genome of cowpea mosaic virus, *New Aspects of Positive-Strand RNA Viruses* (M. A. Brinton and F. X. Heinz, eds.), American Society for Microbiology, Washington, D.C., 1990, pp. 116–122.
55. Hall, T. C., Rao, A. L. N., Pogue, G. P., Huntley, C. C., and Marsh, L. E. Replication, repair and recombination of brome mosaic virus RNA, *New Aspects of Positive-Strand RNA Viruses* (M. A. Brinton and F. X. Heinz, eds.), American Society for Microbiology, Washington, D.C., 1990, pp. 47–54.
56. Guilford, P. J., Beck, D. L., and Forster, R. L. S. Influence of the poly(A) tail and putative polyadenylation signal on the infectivity of white clover mosaic potexvirus, *Virology, 182*: 61, 1991.
57. Turpen, T. H., Turpen, A. M., della-Cioppa, G., Grill, L. K. Viral vectors for gene expression in plant cells, *Proc. Biotech USA '90*, Conference Management Corporation, Norwalk, 1990, pp. 112–117.
58. French, R., Janda, M., and Ahlquist, P. Bacterial gene inserted in an engineered RNA virus: efficient expression in monocotyledonous plant cells, *Science, 231*: 1294, 1986.
59. Ahlquist, P., French, R., and Sacher, R. Gene expression vectors derived from plant RNA viruses, *Viral Vectors* (Y. Gluzman and S. H. Hughes, eds.), Cold Spring Harbor Laboratory, New York, 1988, pp. 183–189.
60. Jupin, I., Richards, K., Jonard, G., Guilley, H., and Pleij, C. W. A. Mapping sequences required for productive replication of beet necrotic yellow vein virus RNA 3, *Virology, 178*: 273, 1990.
61. Joshi, R. L., Joshi, V., and Ow, D. W. BSMV genome mediated expression of a foreign gene in dicot and monocot plant cells, *EMBO J., 9*: 2663, 1990.

62. Takamatsu, N., Ishikawa, M., Meshi, T., and Okada, Y. Expression of bacterial chloramphenicol acetyltransferase gene in tobacco plants mediated by TMV-RNA, *EMBO J., 6*: 307, 1987.
63. Dawson, W. O., Bubrick, P., and Grantham, G. L. Modifications of the tobacco mosaic virus coat protein gene affecting replication, movement and symptomatology, *Phytopathology, 78*: 783, 1988.
64. Takamatsu, N., Watanabe, Y., Yanagi, H., Meshi, T., Shiba, T., and Okada, Y. Production of enkephalin in tobacco protoplasts using tobacco mosaic virus RNA vector, *FEBS Lett., 269*: 73, 1990.
65. Cassidy, B., and Nelson, R. Construction of a viral vector for rapid analysis of gene expression in whole plants (abstract), *Phytopathology, 80*: 1037, 1990.
66. French, R., and Ahlquist, P. Intercistronic as well as terminal sequences are required for efficient amplification of brome mosaic virus RNA3, *J. Virol., 61*: 1457, 1987.
67. Pacha, R. F., Allison, R. F., and Ahlquist, P. *cis*-Acting sequences required for *in vivo* amplification of genomic RNA3 are organized differently in related bromoviruses, *Virology, 174*: 436, 1990.
68. Traynor, P., Young, B. M., and Ahlquist, P. Deletion analysis of brome mosaic virus 2a protein: effects on RNA replication and systemic spread, *J. Virol., 65*: 2807, 1991.
69. Bol, J. F., van Vloten-Doting, L., and Jaspars, E. M. J. A functional equivalence of top component α RNA and coat protein in the initiation of infection by alfalfa mosaic virus, *Virology, 46*: 73, 1971.
70. Sacher, R., French, R., and Ahlquist, P. Hybrid brome mosaic virus RNAs expression and are packaged in tobacco mosaic virus coat protein *in vivo, Virology, 167*: 15, 1988.
71. Guilford, P. J., Ziegler-Graff, V., and Baulcombe, D. C. Mutation and replacement of the 16-kDa protein gene in RNA-1 of tobacco rattle virus, *Virology, 182*: 607, 1991.
72. Dawson, W. O., and Lehto, K. M. Regulation of tobamovirus gene express, *Adv. Virus Res., 38*: 307, 1990.
73. Okada, Y., Meshi, T., and Watanabe, Y. Structure and function of tobacco mosaic virus RNA, *Viral Genes and Plant Pathogenesis* (T. P. Pirone and J. G. Shaw, eds.), Springer-Verlag, New York, 1990, pp. 23–28.
74. Takamatsu, N., Watanabe, Y., Iwasaki, T., Shiba, T., Meshi, T., and Okada, Y. Deletion analysis of the 5′ untranslated leader sequence of tobacco mosaic virus RNA, *J. Virol., 65*: 1619, 1991.
75. Takamatsu, N., Watanabe, Y., Meshi, T., and Okada, Y. Mutational analysis of the pseudoknot region in the 3′ noncoding region of tobacco mosaic virus RNA, *J. Virol., 64*: 3686, 1990.
76. Abel, P. P., Nelson, R. S., De, B., Hoffmann, N., Rogers, S. G., Fraley, R. T., and Beachy, R. N. Delay of disease development in transgenic plants that express the tobacco mosaic virus coat protein gene, *Science, 232*: 738, 1986.

77. Deom, C. M., Oliver, M. J., and Beachy, R. N. The 30-kilodalton gene product of tobacco mosaic virus potentiates virus movement, *Science, 237*: 389, 1987.
78. Wolf, S., Deom, C. M., Beachy, R. N., and Lucas, W. J. Movement protein of tobacco mosaic virus modifies plasmodesmatal size exclusion limit, *Science, 246*: 377, 1989.
79. Golemboski, D. B., Lomonossoff, G. P., and Zaitlin, M. Plants transformed with a tobacco mosaic virus nonstructural gene sequence are resistant to the virus, *Proc. Natl. Acad. Sci. USA, 87*: 6311, 1990.
80. Fraenkel-Conrat, H. The history of tobacco mosaic virus and the evolution of molecular biology, *The Plant Viruses, Vol. 2: The Rod-Shaped Plant Viruses* (M. H. V. van Regenmortel and H. Fraenkel-Conrat, eds.), Plenum Press, New York, 1986, p. 8.
81. Siegel, A., Hari, V., and Kolacz, K. The effect of tobacco mosaic virus infection and virus-specific protein synthesis in protoplasts, *Virology, 85*: 494, 1978.
82. Fraser, R. S. S. *Biochemistry of Virus-Infected Plants*, Research Studies Press, Letchworth, England, 1987, pp. 1–7.
83. Fraser, R. S. S., and Gerwitz, A. Tobacco mosaic virus infection does not alter the polyadenylated messenger RNA content of tobacco leaves, *J. Gen. Virol., 46*: 139, 1980.
84. Saito, T., Yamanaka, K., and Okada, Y. Long-distance movement and viral assembly of tobacco mosaic virus mutants, *Virology, 176*: 329, 1990.
85. Dawson, W. O., Lewandowski, D. J., Hilf, M. E., Bubrick, P., Raffo, A. J., Shaw, J. J., Grantham, G. L., and Desjardins, P. R. A tobacco mosaic virus-hybrid expresses and loses an added gene, *Virology, 172*: 285, 1989.
86. Donson, J., Kearney, C. M., Hilf, M. E., and Dawson, W. O. Systemic expression of a bacterial gene by a tobacco mosaic virus-based vector, *Proc. Natl. Acad. Sci. USA, 88*: 7204, 1991.
87. Lazzarini, R. A., Keene, J. D., and Schubert, M. The origins of defective interfering particles of the negative-strand RNA viruses, *Cell, 26*: 145, 1981.
88. Kirkegaard, K., and Baltimore, D. The mechanism of RNA recombination in poliovirus, *Cell, 47*: 433, 1986.
89. King, A. M. Q. Genetic recombination in positive strand RNA viruses, *RNA Genetics*, Vol. II (E. Domingo, J. J. Holland, and P. Ahlquist, eds.), CRC Press, Boca Raton, Florida, 1988, pp. 149–165.
90. Allison, R., Thompson, C., and Ahlquist, P. Regeneration of a functional RNA virus genome by recombination between deletion mutants and requirement for cowpea chlorotic mottle virus 3a and coat genes for systemic infection, *Proc. Natl. Acad. Sci. USA, 87*: 1820, 1990.
91. van der Kuyl, A. C., Neeleman, L., and Bol, J. F. Complementation and recombination between alfalfa mosaic virus RNA 3 mutants in tobacco plants, *Virology, 183*: 731, 1991.
92. Domingo, E., and Holland, J. J. High error rates, population equilibrium, and evolution of RNA replication systems, *RNA Genetics*, Vol. III, (E. Domingo, J. J. Holland, and P. Ahlquist, eds.), CRC Press, Boca Raton, Florida, 1988, pp. 3–36.

93. Smith, F. I., and Palese, P. Influenza viruses: high rate of mutation and evolution, *RNA Genetics*, Vol. III (E. Domingo, J. J. Holland, and P. Ahlquist, eds.), CRC Press, Boca Raton, Florida, 1988, pp. 123–135.
94. Ward, C. D., Stokes, A. M., and Flanegan, J. B. Direct measurement of the poliovirus RNA polymerase error frequency *in vitro, J. Virol., 62*: 558, 1988.
95. de la Torre, J. C., Wimmer, E., and Holland, J. J. Very high frequency of reversion to guanidine resistance in clonal pools of guanidine-dependent type 1 poliovirus, *J. Virol., 64*: 664, 1990.
96. van Vloten-Doting, L., Bol, J. F., and Cornelissen, B. Plant-virus-based vectors for gene transfer will be of limited use because of the high error frequency during viral RNA synthesis, *Plant Mol. Biol., 4*: 323, 1985.
97. Xiong, C., Levis, R., Shen, P., Schlesinger, S., Rice, C. M., and Huang, H. V. Sindbis virus: an efficient, broad host range vector for gene expression in animal cells, *Science, 243*: 1188, 1989.
98. Luytjes, W., Krystal, M., Enami, M., Parvin, J. D., and Palese, P. Amplification, expression, and packaging of a foreign gene by influenza virus, *Cell, 59*: 1107, 1989.
99. Gooding, G. V., Jr. Epidemiology and control, *The Plant Viruses, Vol. 2: The Rod-Shaped Plant Viruses* (M. H. V. van Regenmortel and H. Fraenkel-Conrat, eds.), Plenum Press, New York, 1986, pp. 133–152.
100. Li, X. H., Heaton, L. A., Morris, T. J., and Simon, A. E. Turnip crinkle virus defective interfering RNAs intensify viral symptoms and are generated *de novo, Proc. Natl. Acad. Sci. USA, 86*: 9173, 1989.
101. Burgyan, J., Rubino, L., and Russo, M. *De novo* generation of cymbidium ringspot virus defective interfering RNA, *J. Gen. Virol., 72*: 505, 1991.
102. Knorr, D. A., Mullin, R. H., Hearne, P. Q., and Morris, T. J. *De novo* generation of defective interfering RNAs of tomato bushy stunt virus by high multiplicity passage, *Virology, 181*: 193, 1991.
103. Levis, R., Huang, H., and Schlesinger, S. Engineered defective interfering RNAs of sindbis virus express bacterial chloramphenicol acetyltransferase in avian cells, *Proc. Natl. Acad. Sci. USA, 84*: 4811, 1987.
104. Raffo, A. J., and Dawson, W. O. Construction of tobacco mosaic virus subgenomic replicons that are replicated and spread systemically in tobacco plants, *Virology, 184*: 277, 1991.
105. Geigenmüller-Gnirke, U., Weiss, B., Wright, R., and Schlesinger, S. Complementation between sindbis viral RNAs produces infectious particles with a bipartite genome, *Proc. Natl. Acad. Sci. USA, 88*: 3253, 1991.
106. Bruening, G., Passmore, B. K., van Tol, H., Buzayan, J. M., and Feldstein, P. A. Replication of a plant virus satellite RNA: evidence favors transcription of circular templates of both polarities, *Mol. Plant–Microbe Interact., 4*: 219, 1991.
107. van Emmelo, J., Ameloot, P., and Fiers, W. Expression in plants of the cloned satellite tobacco necrosis virus genome and of derived insertion mutants, *Virology, 157*: 480, 1987.

108. Mirkov, T. E., Mathews, D. M., Ellicott, K., Dodds, J. A., and Fitzmaurice, L. Factors affecting efficient infection of tobacco with *in vitro* RNA transcripts from cloned cDNAs of satellite tobacco mosaic virus, *Virology, 179*: 395, 1990.
109. van Dun, C. M. P., van Vloten-Doting, L., and Bol, J. F. Expression of alfalfa mosaic virus cDNA1 and 2 in transgenic tobacco plants, *Virology, 163*: 572, 1988.
110. Taschner, P. E. M., van der Kuyl, A. C., Neeleman, L., and Bol, J. F. Replication of an incomplete alfalfa mosaic virus genome in plants transformed with viral replicase genes, *Virology, 181*: 445, 1991.
111. van der Kuyl, A. C., Neeleman, L., and Bol, J. F. Deletion analysis of *cis*- and *trans*-acting elements involved in replication of alfalfa mosaic virus RNA3 in vivo, *Virology, 183*: 687.
112. Osbourn, J. K., Plaskitt, K. A., Watts, J. W., and Wilson, T. M. A. Tobacco mosaic virus coat protein and reporter gene transcripts containing the TMV origin-of-assembly sequence do not interact in double-transgenic tobacco plants: implications for coat protein-mediated protection, *Mol. Plant–Microbe Interact., 2*: 340, 1989.
113. Osbourn, J. K., Sarkar, S., and Wilson, T. M. A. Complementation of coat protein-defective TMV mutants in transgenic tobacco plants expressing TMV coat protein, *Virology, 179*: 921, 1990.
114. Dore, J.-M., van Dun, C. M. P., Pinck, L., and Bol, J. F. Alfalfa mosaic virus RNA3 mutants do not replicate in transgenic plants expressing RNA3-specific genes, *J. Gen. Virol., 72*: 253, 1991.
115. Lommel, S. A., and Xiong, Z. Reconstitution of a functional red clover necrotic mosaic virus by recombinational rescue of the cell-to-cell movement gene expressed in a transgenic plant (abstract), *J. Cell. Biochem.*, Suppl. 15A: 151, 1991.
116. Baulcombe, D. C., Saunders, G. R., Bevan, M. W., Mayo, M. A., and Harrison, B. D. Expression of biologically active viral satellite RNA from the nuclear genome of transformed plants, *Nature, 321*: 445, 1986.
117. Gerlach, W. L., Llewellyn, D., and Haseloff, J. Construction of a plant disease resistance gene from the satellite RNA of tobacco ringspot virus, *Nature, 328*: 802, 1987.
118. Mills, D. R. Engineered recombinant messenger RNA can be replicated and expressed inside bacterial cells by an RNA bacteriophage replicase, *J. Mol. Biol., 200*: 489, 1988.
119. Burstyn, D. G., Copmann, T., Dinowitz, M., Garnick, R., Losikoff, A., Lubiniecki, A., Rubino, M. S., and Wiebe, M. Assessment of genetic stability for biotechnology products, *Pharmacy Technol., 15*: 34, 1991.

IV

MOLECULAR FARMING

11

Assembly of Multimeric Proteins in Plant Cells: Characteristics and Uses of Plant-Derived Antibodies

Andrew Hiatt

The Scripps Research Institute, La Jolla, California

Keith Mostov

University of California, San Francisco, San Francisco, California

I. PROPERTIES OF ANTIBODIES

Antibodies are a group of blood proteins produced by vertebrate animals in response to foreign materials or infectious agents. Research during the last two decades has revealed the structure of antibody molecules, the complex mechanisms by which these immunoglobulin proteins are synthesized, and the manner in which they bind to foreign material as the body's first line of biochemical defense. Knowledge of these processes has broadened our understanding of the immune system and permitted the exploitation of antibodies as immunochemical reagents for a wide variety of applications.

There are many categories of antibodies that share structural and functional properties but differ in important characteristics. The most extensively studied antibodies, and the antibodies most commonly used as reagents, belong to the immunoglobulin G (IgG) family. The IgGs are the class of antibodies investigated in the following experiments. These proteins are composed of two pairs of subunits of different molecular weights and different amino acid sequences. Each IgG molecule has two identical large-molecular-weight (heavy-chain) subunits linked together by disulfide bonds. Each heavy chain is also bonded by disulfide bonds to one of two identical lower-molecular-weight (light-chain) subunits. In addition, there are extensive intramolecular disulfide bonds, which are necessary for antigen binding. The resulting tetrameric protein is capable of high-affinity binding of a specific molecule.

Some antibodies can distinguish between molecules that differ by substitution of a single atomic residue or even those that differ only by enantiomeric form. In addition to this high degree of specificity, two properties of antibodies are central to their general usefulness as reagents. First, purified antibodies have the ability to bind tightly to the molecules they recognize. Second, animals can produce antibodies that recognize synthetic molecules. Consequently, an almost limitless repertoire of antibodies is available for development and use in specific applications.

II. USE OF ANTIBODIES

Antibodies are used as therapeutic agents, as vaccines, as binding moieties for affinity separations, and as the active agents in commercial assays. Because antibodies can have such a high degree of specificity, they can be used as the "detector" for both quantitative and qualitative assays. The fields of application for such assays is nearly unrestricted because synthetic compounds can be used to immunize animals, resulting in the production of antibodies against many compounds of interest in medicine, research, and forensic science.

Recently, molecular biologists have begun to exploit this ability to generate antibodies against synthetic compounds, which do not normally induce an immune response. They have induced the production of antibodies that are not constrained to the recognition of "natural" compounds or to the "natural" array of biochemical reactions that occur in cells. Therefore, it is now possible to envision enzymelike antibodies ("abzymes") that will catalyze new reactions. Such antibodies are now being developed (1–4). This advancement has further expanded the range of possible uses of antibodies.

Most reagent antibodies, such as those used in diagnostic immunoassays, are produced in animals and isolated from serum or ascites fluid or are produced in culture by specialized immune cells (hybridoma cells). With the advent of recombinant DNA technology it became possible to clone the genes for the antibody proteins and express them in other organisms. This process was most easily achieved in cells of vertebrate organisms because the specialized cellular machinery for processing the antibodies already existed in such cells. Part of the difficulty with the synthesis of antibodies in recombinant organisms is that the production of a normal, functional antibody requires the synthesis of two proteins and therefore the expression of two genes. Both of these genes must be transferred to the recombinant organism to achieve antibody expression in the foreign host. This heterologous expression requires the orderly assembly of the protein tetramer. The fidelity of such assembly is ensured in cells that normally synthesize immunoglobulins because of the recognition of processing and signal sequences in the immunoglobulin genes and because of specialized cellular machinery to accommodate these functions. Expression in heterologous systems that do not possess these characteristics is somewhat more difficult and is still poorly understood.

III. PRODUCTION OF ANTIBODIES IN RECOMBINANT MICROORGANISMS

Expression of immunoglobulin genes in yeast has been reported by a number of different laboratories (5–7). In general, the results have shown that immunoglobulin chains can be produced at varying levels and assembled to form antibodies that are targeted for either secretion or intracellular accumulation. In one report, antibodies directed against an intracellular yeast alcohol dehydrogenase reduced the intracellular dehydrogenase activity (5). In another report, yeast invertase leader sequences were introduced onto immunoglobulin cDNAs (6). Expression of proteins depended on the yeast phosphoglycerate kinase promoter, resulting in the secretion of fully assembled, functional antibodies and Fab fragments. These results have shown that in yeast there are no special requirements for assembly of functional antibodies; however, overall levels of production may be quite low.

Expression of normal, functioning antibodies in bacteria has not been achieved. This may be because of the lack of a specific glycosylation system and the absence of secretory and processing organelles. Fab fragments, however, can be efficiently secreted from microorganisms. In addition, high-level expression of antibodies has been achieved in insect cell cultures and a variety of cultured mammalian cells (8).

To achieve expression of functional antigen-binding proteins within bacteria (i.e., nonsecreted), immunoglobulin chains were engineered to be expressed as a single polypeptide in which the gamma and kappa antigen-binding regions were joined by a short polypeptide linker segment (9,10). Several different linker segments of varying lengths and amino acid sequences have been used with considerable success. Single-chain antigen-binding proteins can be stably expressed in bacteria, where they accumulate in intracellular inclusion bodies. Generally, these proteins maintain a high avidity for the antigen used to elicit the original antibody. In one instance, using an anti-fluorescein antibody and its single-chain derivative, the engineered antigen-binding protein was estimated to have a K_a of 1.1×10^9 M^{-1} compared with a value of 8.0×10^9 M^{-1} for the equivalent antibody derived from a mouse cell culture. These results demonstrate that in the absence of a mechanism for assembly of immunoglobulin chains (for example, antibodies directed to intracellular accumulation in higher organisms) Fv regions can be engineered to express an antigen-binding capability nearly equivalent to that of the whole antibody. This is potentially significant for expression of intracellular antibodies in plants because synthesis into the endoplasmic reticulum secretion appears to be a prerequisite for efficient assembly of separate gamma and kappa chains. The results from bacterial expression offer an alternative method for introducing antibody derivatives in an intracellular environment.

IV. CHARACTERIZATION OF AN ANTIBODY EXPRESSED IN TOBACCO

A catalytic IgG_1 antibody (6D4) was chosen for expression in tobacco (1). This antibody recognizes a synthetic phosphonate ester, P3, and can catalyze the hydrolysis of certain carboxylic esters. Gamma and κ cDNAs derived from the 6D4 hybridoma were cloned into the plant expression vector, pMON530 (11). The strategy used with mouse immunoglobulin genes was to transform tobacco leaf discs and regenerate separate plants expressing either the heavy or light chains. These plants were then sexually crossed to produce progeny expressing functional antibody (12,13). Transformation of tobacco was mediated by cocultivation of leaf segments with *Agrobacterium tumefaciens* containing the γ or κ constructs (14). Tobacco has routinely been used as the recipient plant because of its efficient transformation and regeneration (15). Transformed plant cells were regenerated to produce sexually fertile transgenic plants by manipulating the growth-regulating hormones in the medium (16).

Transgenic plants containing γ or κ chains were crossed to produce progeny expressing both chains (12). The F_1 plants contained assembled functional antibody as determined by the following criteria: (a) Western blots of plant extracts under reducing conditions contained equimolar amounts of γ and κ chains that migrated at 50 kDa and 25 kDa, respectively. Under nonreducing conditions both γ and κ bands migrated at about 160 kDa. (b) Enzyme-linked immunoabsorbent assays (ELISA) in which extracts were added to microtiter plates coated with goat anti-mouse γ and then detected with goat anti-mouse κ-HRPO indicated the presence of equimolar amounts of γ bound to κ with no detectable free γ or κ (Tables 1 and 2). (c) Similar results were obtained from ELISA using a P3–BSA conjugate as antigen: The affinity of γ-κ

Table 1 Accumulation of Immunoglobulin Gamma and Kappa Chains in Transgenic Tobacco Plants

Chain[b]	Accumulation (ng/mg total protein)[a]	
	Mean ± S.D.[c]	Maximum[d]
γNL	30 ± 16	60
γL	1412 ± 270	2400
γL (κL)	3330 ± 2000	12800
γNL (κNL)	32 ± 26	60
κNL	1.4 ± 1.2	3.5
κL	56 ± 5	80
κL (γL)	3700 ± 2300	12800
κNL (γNL)	6.5 ± 5	20

[a]Accumulation of individual gamma and kappa chains was estimated by ELISA. Microtiter wells were coated with goat anti-mouse heavy- or light-chain-specific IgG. Plant leaves were homogenized with a mortar and pestle at 4°C after removal of their midveins. Bound gamma or kappa chains from the homogenates were reacted with goat anti-mouse heavy- or light-chain-specific antibodies conjugated to horseradish peroxidase and were then detected according to the manufacturer's instructions. All values are derived from the quantity of purified 6D4 antibody required to give an equivalent color development in ELISA. Total protein was determined by Bio-Rad Coomassie assay. At least nine plants in each category were assayed.
[b]γNL and κNL refer to cDNAs containing no leader sequences, and γL and κL refer to cDNAs with leader sequences: γ(κ) refers to gamma chains in a plant that also expresses kappa chains, and κ(γ) refers to kappa chains in a plant that also expresses gamma chains.
[c]Values are derived from at least two determinations per plant and do not include transformants producing no detectible gamma or kappa chain.
[d]Values for plants with the highest levels of accumulation.

complexes for P3 was identical to that of the hybridoma-derived antibody. The specificity for P3 was indicated by inhibition of P3–BSA binding by free P3 in which half-maximal inhibition was about 10 μM for plant-derived or hybridoma-derived antibody. Assembled antibody accumulated to greater than 1% of total extractable leaf protein. A key factor leading to this high level of expression was the inclusion of the sequences encoding the mouse leader peptide in the transferred immunoglobulin genes. Leader peptides, in general, are the molecular signals that direct synthesis to occur on the endoplasmic reticulum; the newly synthesized protein is sequestered with the organelle (17,18). In the absence of the mouse leader peptide, no assembly of heavy and light immunoglobulin chains occurred in tobacco and, more important, accumulation of the individual chains was very poor. Clearly, targeting the

Table 2 Distribution and Assembly of Immunoglobulin Gamma and Kappa Chains in Sexual Crosses of Transgenic Tobacco Plants[a]

	Number of plants with progeny expressing γ or κ chains				γκ percent assembly
Cross	γ only	κ only	γκ	Null	(mean ± S.D.)
κNL × γNL	4	6	3	5	0
κL × γL	3	10	11	4	95 ± 16

[a]The ELISA for assembly used anti-kappa-chain-specific antibodies conjugated to horseradish peroxidase to detect antigen bound to microtiter wells coated with unlabeled anti-gamma-chain-specific antibodies, and vice versa. Values derived from these assays were used to calculate the percent assembly by comparison with the purified 6D4 antibody. This was determined at least three times for each γκ plant.

synthesis of the protein to the endoplasmic reticulum is essential in both stabilizing and assembling components of multimeric proteins. In addition, assembly of the complete tetrameric antibody entailed an increase in protein accumulation, indicating that the native conformation of proteins is also an important determinant of their stability.

V. EXPRESSION OF ANTIBODIES IN PLANT CELL CULTURES

The best 6D4-producing plant was also propagated as a callus suspension culture (13). The callus profilerates as de-differentiated cells (hormonally induced from leaf cells) that grow at high density in a defined medium consisting primarily of sucrose, nitrate, and the appropriate plant hormone. We have used these suspension cultures to incorporate ^{35}S-methionine into the antibody to determine whether secretion of intracellular localization is the predominant fate of the antibody molecule. Our results have indicated that most of the antibody pool is being secreted into the medium (13) (Table 3). The question of whether all of the antibody is ultimately secreted or if some fraction is retained within the cell was addressed by pulse chase experiments that demonstrated quantitative secretion of antibody into the growth medium. The culture medium of stationary-phase cells contain high levels of antibody presumably derived from both active secretion and cell death. Functional antibody purified from the spent growth medium is present at concentrations up to 20 mg/liter.

Table 3 Secretion of 6D4 Antibody from Tobacco Protoplasts and Callus Cells[a]

		^{35}S-Methionine incorporation into 6D4 antibody (%)	
Cell type	Method	Cells	Medium
At 2 hr			
Protoplasts	Protein A	75	25
	SDS–PAGE	76	24
Callus suspension	Protein A	72	28
	SDS–PAGE	80	20
After 2-hr chase			
Protoplasts	Protein A	13	87
	SDS–PAGE	14	86
Callus suspension	Protein A	27	73
	SDS–PAGE	32	68

[a]Results are expressed as the percentage of total radiolabel in immunoglobulin associated with cells or medium as determined by binding to protein A–sepharose or SDS–PAGE mobility as described in Section VI. Equal numbers of cells were used in each labeling experiment. Cell viability of protoplasts was measured by using fluorescein diacetate as a marker and was found to be 80–90% throughout the labeling period.

VI. CATALYTIC ACTIVITY OF THE PURIFIED PLANT ANTIBODY

The plant antibody was further analyzed after purification on Sephacryl–FPLC and protein A–sepharose. A crude homogenate of tobacco leaf in Tris buffered saline (pH 8) with 1 mM PMSF was centrifuged to remove insolubles and concentrated by filtration. Sephacryl–FPLC fractions were assayed by ELISA, pooled, and adsorbed to protein A–sepharose by using buffers that enhance binding of IgG_1s. The eluted antibody was evaluated by Coomassie blue staining and Western blotting and was found to be virtually pure. The quantitative retention of the antibody on protein A–sepharose indicates that the interdomain conformation of Fc recognized by protein A (between CH2 and CH3) is intact. The purified antibody eluted from the protein A column was dialyzed against the appropriate buffers and used for analysis of catalysis, glycosylation, and amino acid sequence.

Catalytic activity was measured by incubation of the antibody with substrate in the presence and absence of inhibitor as described in Ref. 1. The results

Table 4 Catalytic Activity of the 6D4 Antibody Produced in Tobacco

Antibody	K_M ($\times10^{-6}$ M)	V_{max} ($\times10^{-6}$ M sec^{-1})	Kr ($\times10^{-6}$ M) (competitive)	K_{cat} (sec^{-1})
Tobacco	1.41	0.057	0.47	0.008
Mouse ascites	9.8	0.31	1.06	0.025

demonstrate that the antibody maintains its catalytic capability; differences between hybridoma and plant antibody may be due to inaccuracy in assessing the degree of purification and antibody concentration (Table 4).

VII. GLYCOSYLATION

Glycosylation of the heavy chain was investigated by lectin binding analysis (19,20). Purified antibody was first Western blotted to nitrocellulose; various biotinylated lectins were then incubated with the blots, after which lectin binding was visualized with strepatvidin–alkaline phosphatase and bromochloride phosphate. Concanavalin A (Con A) binding (specific for mannose, glucose) was equivalent in equal amounts of hybridoma and plant antibody. Once Con A bound to the plant-derived γ chain, whereas the ascites γ chain was recognized by Con A and by the lectins from *Ricinus communis* (specific for terminal galactose and *N*-acetylgalactosamine) and wheat germ agglutinin (*N*-acetylglucosamine dimers, terminal sialic acid) (21). The lectins from *Datura stromonium* (*N*-acetyl glucosamine oligomers, *N*-acetyl lactosamine) and *Phaseolus vulgaris* (galactose-β1,4-*N*-acetyl glucosamine-β1,2-mannose) did not bind to either the plant- or ascites-derived γ chain. Elution of the lectin from the blots with α-methylglucoside was used to compare the relative affinity of Con A binding to the plant and ascites heavy chains (20). The results showed that the two antibodies are indistinguishable by Con A affinity or by the quantity of Con A bound per microgram of γ chain. Blots in which the antibodies were first digested with endoglycosidase H (21,22) displayed no reduction in Con A binding under conditions in which Con A binding to ovalbumin (containing a high-mannose-type carbohydrate) was diminished. Because endoglycosidase H resistance is characteristic of complex carbohydrates processed in the Golgi apparatus, these results indicate that the transgenic antibody is processed in a fashion similar to that of complex mammalian glycoproteins. Future experiments will employ proton nuclear magnetic resonance spectroscopy to determine the exact molecular structure of the glycan (23). The results to date are consistent with a complex carbohydrate containing terminal fucose or xylose attached to the plant antibody.

VIII. NH2-TERMINAL AMINO ACID SEQUENCE OF THE LIGHT-CHAIN AND HEAVY-CHAIN FRAGMENTS

Evidence that the antibody has been cotranslationally secreted into the lumen of the endoplasmic reticulum is provided by the N-terminus amino acid sequence. Purified antibody was blotted onto a polyvinylidene difluoride (PVDF) membrane (24), after which the heavy- or light-chain bands were located on the blot by Coomassie blue staining. Pieces of PVDF membrane with bound immunoglobulin were then subjected to automated sequence analysis. The heavy chains from both the plant antibody and 6D4 were intractable to sequencing, which indicated a blocked N terminus. The light-chain N-terminal sequence was Asp-Val-Val-Leu for both the plant and mouse antibodies. This demonstrates appropriate proteolytic processing of the mouse leader sequence by the plant endoplasmic reticulum.

On Western blots the purified antibody contained four peptide bands with molecular weights of 50, 38, 30, and 24 kDa, respectively. The high-molecular-weight band comigrated with the full-length heavy chain, whereas the 25-kDa band comigrated with the light chain. The two additional bands were recognized by a heavy-chain-specific antibody. These two bands accounted for approximately 10–30% of the total heavy-chain protein and were in equal abundance. The N-terminal sequences of the two bands were analyzed after blotting to PVDF membranes; Ile-Val-Pro constituted the NS2 terminus of each. Inspection of the complete amino acid sequence derived from the nucleotide sequence of the 6D4 heavy chain revealed that the I-V-P triplet is adjacent to a Lys-Lys pair, which would be a target site for a serine protease. Inclusion of PMSF in the homogenization buffer greatly reduces, but does not eliminate, the lower-molecular-weight heavy-chain bands.

IX. USES OF ANTIBODIES IN PLANTS

Of the variety of bioactive compounds expressed in transgenic plants, antibodies may offer the widest range of possible applications. Envisioned applications in plants that depend solely on the binding affinity of the antibody include pathogen resistance (viral, fungal, and insect) and modulation of metabolic pathways to produce new developments or nutritional characteristics (25). Production of antibodies by plant cells offers a variety of new possibilities for basic research in plant biology as well as for large-scale production of antibodies for use as therapeutic, diagnostic, or affinity reagents. The unparalleled capacity and flexibility of agricultural production suggests that antibodies derived from plants may be significantly less expensive than antibodies from any other source. Moreover, antibodies in plants may become useful reagents for isolating and processing environmental contaminants and industrial biproducts.

Antibodies produced in plants possess all the characteristics of and are virtually indistinguishable from antibodies produced by hybridoma cells. Expression of catalytic antibodies may make it possible to introduce new catalytic capabilities into plants and produce catalytic antibodies useful in industrial processes. An intriguing possibility for using functional antibodies in plants would be to direct those antibodies against environmental pollutants. One of the key differences between plant cells and other organisms is the structure and characteristics of the surrounding cell wall. The mechanical strength and contiguous nature of plant cell walls are largely responsible for the rigidity of the entire plant. The diameter of pores in the cell wall imposes a restriction on the size of molecules that can freely permeate it. This exclusion limit is between 35 and 50 Å, which corresponds to a molecular weight of less than 20,000 for a globular protein. Antibodies may be too large to freely permeate plant cell walls (26). Consequently, expression of an antibody in a plant cell is equivalent to producing a binding and retention capacity within a semipermeable membrane. Any antigen with a molecular weight less than 20,000 (e.g., environmental pollutants, industrial byproducts, pesticides, herbicides) might be collected and retained by a plant expressing an antibody that is functional in situ. Current research on applications of antibodies as biofilters is aimed at characterizing the functional properties of the antibody as it resides within the boundaries of the cell wall. Future efforts will be aimed at enhancing the functionality of antibodies in plants to enable catalytic processing of molecules retained within the cell.

With respect to relevant antigens in plants, hormones are an obvious target for in situ binding and possibly subsequent inactivation by an endogenous antibody. Plant growth and development is controlled by a few low-molecular-weight hormones, such as indoleacetic acid, ethylene, benzylaminopurine, and a variety of more complex organic molecules (27). Little is known about the biosynthetic pathways or the mechanism of action of plant hormones. However, by expression within the plant cell of monoclonal antibodies that recognize these hormones, one could potentially evaluate developmental and metabolic events that are controlled by their free titer. Ideally, one would control the expression of the antibody as well as target expression to different organs or subcellular locations. In this way, activities of the hormone at various developmental stages could be unraveled. We do not yet know whether the immunoglobulin components of an antibody can be assembled in organelles other than the endoplasmic reticulum. Experiments are under way to determine conditions for assembly in vacuoles, chloroplasts, and cytosolic components of the cell.

Although many fungal, bacterial, and viral pathogens have been characterized with respect to the genetics of host–pathogen interactions, few have been thoroughly investigated at the biochemical level. In some instances, however, pathogen-related proteins or other organic molecules have been shown to be required for pathogenesis (28,29). Expression of an intracellular or selected antibody that binds antigens essential for pathogenesis might ameliorate the symptoms of the infection by reducing the functional titer. This strategy has two advantages: First, it does not require isolation of genes involved in synthesis of the target antigen; second, pools of antigen that may be localized in subcellular compartments can be the specific target and leave other pools unaffected. Clearly, the success of this approach will depend on a much more detailed understanding of the behavior of antibodies in plants. When we have a clear picture of the assembly, stability, and functionality of targeted immunoglobulins, we can devise appropriate strategies for localized antigen binding.

A key determinant of the success of producing valuable foreign proteins agriculturally is the extent of accumulation of the protein in the transgenic host plant. This is true whether the desired product is needed in vast quantities or in relatively small amounts. The usefulness of any transgene product in a plant will be affected by its specific abundance because this will directly influence the acreage to be harvested as well as subsequent operations, such as purification.

Although often detectable, yields of transgene plant products are also often disappointingly low (12,30). Relatively little is known about the determinants of protein stability and half-life in plants, and for that reason strategies of increasing levels of mRNA transcription are often emphasized. To be successful, this approach would ultimately demand a large energetic contribution to protein synthesis from the cell's resources. An alternative strategy requires determining factors that augment protein stability. This approach has the advantage of not requiring excessive energy for biosynthesis. Encouraging results in experiments aimed at maximizing antibody accumulation in tobacco suggest that assembly of the components of a multimer and secretion are crucial in augmenting accumulation (12). Other results have suggested that targeting accumulation to a specific organ (e.g., seed) can also enhance accumulation (31).

Injection of purified antibodies directed against cell surface antigens of tumors has long been recognized as a possible strategy for combating cancers (32,33). This strategy, however, requires relatively large amounts of antibody per individual. More than a gram could be required to treat an adult, alternatively, administering smaller amounts of antibodies over many years (assuming the antibodies are nonimmunogenic) is also a possible strategy. Total requirements for antibodies could conceivably amount to hundreds of kilograms. The high capacity and flexibility of agricultural production offers many advantages for producing antibodies that are destined for purification. Genetically stable seed stocks of antibody-producing plants can be isolated and stored indefinitely at low cost; the seed stock can be converted into a harvest of any quantity of antibody within one growing season. Although tobacco has been used as the principal research tool to initiate the study of antibodies in plants, there may be more appropriate plants for production. A variety of common crop plants can be used as production hosts. Acreages of perennial forage crops could be generated by clonal propagation or from seed and could be harvested many times in a growing season. The choice of species may depend on the quantity and nature of contaminants encountered during purification. Some candidates are alfalfa, soybean, tomato, and potato. Because large-scale production of antibodies is not yet commonplace, appropriate techniques for purification of hundreds or thousands of grams have not yet been perfected.

Finally, antibodies that can be obtained in large quantities could potentially be used for industrial-scale purification processes. In addition to the huge repertoire of antigens that can be used to elicit antibodies, one also has the option of obtaining antibodies with any desired K_a for any particular antigen. This would be crucial for affinity separations using immobilized antibodies on an industrial scale because one would want to release the purified product without damaging the antibody. An antibody could be obtained with the appropriate K_a for the desired product such that repeated purifications could be performed by using appropriate conditions for binding and release.

X. NEW TYPES OF ANTIBODIES FOR EXPRESSION IN PLANTS

One potential application of antibodies produced in plants is the passive immunization of mucosal surfaces. Most infectious agents enter the mammalian body through exposed mucosal surfaces, such as those of the gastrointestinal, respiratory, urinary, reproductive, and ocular systems. Mammals have a specialized mucosal or secretory immune system (34). The primary antibody components of this system are secretory IgA and, in some cases, secretory IgM. These antibodies are unusually complex and consist of polypeptides normally synthesized by plasma and epithelial cells. When found in secretions, IgA and IgM are usually polymers, and for this reason they are known as polymeric immunoglobulins (pIg) (35). Polymeric IgA is classically a dimer (i.e., four heavy or "a" chains and four light chains), although there are also higher polymers. This IgA usually contains an extra polypeptide, known as the J or joining chain, a peptide with a molecular weight of 15,600. Typically one molecule of dimeric IgA contains one J chain. The H chain has a weak homology to the basic immunoglobulin fold. The immunoglobulin IgM is usually a pentamer that also contains one or more J chains.

In mammals the pIgs containing J chains are synthesized by plasma cells. These plasma cells tend to underlie exposed mucosal surfaces. The epithelial cells that line these surfaces express a pIg receptor (pIgR) that transports the pIg across the epithelial cells and releases it at the mucosal or luminal surface (36). This delivers the pIg to the optimal location, where it can form the first line of defense against invading organisms.

The structure and function of the pIgR have been extensively studied (37). The pIgR is an integral membrane protein that spans the membrane once. The amino-terminal, extracellular, ligand-binding domain contains five immunoglobulinlike domains. The carboxy-terminal, cytoplasmic segment of the pIgR contains 103 amino acids. The intracellular pathway and transport function of the pIgR were examined originally by studying rat liver, in which pIg is transported from blood to bile. The cloned cDNA for the pIgR has also been expressed in cultured epithelial Madin–Darby canine kidney cells, which have permitted detailed molecular studies (38). The receptor follows the normal biosynthetic pathway for plasma membrane proteins, starting in the rough endoplasmic reticulum and traveling to the Golgi and trans-Golgi network. From there it is sent to the basolateral surface, where the pIgR can bind pIg previously released by plasma cells. The receptor–ligand complex is then endocytosed and carried by vesicles to the apical or mucosal surface by a process called transcytosis. When the receptor–ligand complex reaches the apical surface, the extracellular, ligand-binding domain of the pIgR is endoproteolytically cleaved off the membrane-spanning portion of the receptor. This cleaved fragment, called the secretory component (SC), is released with the pIg into the external secretions. The complex of SC and pIg is generically called secretory immunoglobulin; secretory IgA (sIgA) and secretory IgM (sIgM) refer to SC complexed to IgA and IgM, respectively.

Secretory immunoglobulins are exposed to harsh conditions, such as digestive enzymes in the intestinal tract, acid in the stomach, and bacterial proteases. Unlike serum immunoglobulins, such as IgG, these secretory immunoglobulins are ideally suited for harsh environments (35). The SC complexed to IgA protects the IgA against proteolysis and denaturation, thereby increasing the functional half-life of the active antibody.

Secretory immunoglobulins have several mechanisms that prevent infection (34). These include agglutination, direct killing, inhibition of attachment and invasion of epithelial surfaces, opsonization, and inhibition of bacterial toxins and enzymes. The antibodies also inhibit entry of other antigens, thereby modulating the immune response.

An ideal method to produce secretory immunoglobulins for passive immunization would be to express them in plants. One might express the full-length pIgR and rely on a bacterial protease to cleave off SC. Alternatively, a truncated, soluble version of pIgR lacking its membrane-spanning and cytoplasmic portions could be expressed directly, obviating the need for the protease (39). All four secretory immunoglobulin chains (α or m, light, J, and SC) could be coexpressed in the same cell. This might be achieved by transfecting genes for the individual polypeptides into separate plants and then creating a strain expressing all four by sexual crosses and selections.

All four chains should assemble into intact, functional secretory antibodies. In many species, SC becomes disulfide-linked to the a chain, although this is not required to form sIgA. The disulfide-exchanging enzyme needed is apparently present in the secretory pathway in plants.

The assembled secretory immunoglobulin would not need to be purified from the plants. Indeed, it might be preferable and less expensive to simply eat the plant or some processed portion of it. Processing (e.g., homogenization) should not include steps that would inactivate the antibody, e.g., extensive heating.

The plants just described would provide passive oral immunity to various enteric pathogens. They could be used to protect farm animals as well as humans. They would be especially useful in newborn humans and animals, which are particularly susceptible to enteric illnesses, such as viral-caused diarrhea (34). Premature human infants are especially susceptible to overwhelming intestinal infections in a disease called necrotizing enterocolitis. Conventional antibodies cannot protect against viruses.

One might construct plant strains that simultaneously express several secretory antibodies to a spectrum of pathogens or strains of one pathogen. Once the heavy and light chains of appropriate specificity are cloned and expressed, plants producing the desired panel of specific antibodies could be generated by sexual crosses and selection. These antibodies might be also used for protection of other mucosal surfaces, such as those of the respiratory, reproductive, and ocular systems. In that case, purification of the antibodies would probably be necessary.

REFERENCES

1. Tramontano, A., Janda, K., and Lerner, R. A. Catalytic antibodies, *Science, 234*: 1566–1570, 1986.
2. Hilvert, D., Carpenter, S. H., Nared, K. D., and Auditor, M. T. Catalysis of concerted reactions by antibodies: The Claisen rearrangement, *Proc. Natl. Acad. Sci. USA, 85*: 4953–4955, 1988.
3. Shokat, K. M., Leumann, C. J., Sugasawara, R., and Schultz, P. G. A new strategy for the generation of catalytic antibodies, *Nature, 338*: 269–271, 1989.
4. Schultz, P. G., Lerner, R. A., and Benkovic, S. J. Catalytic antibodies, *Chem. Eng. News, 68*: 26–40, 1990.
5. Carlson, J. A new means of inducibly inactivating a cellular protein, *Mol. Cell. Biol., 8*: 2638–2646, 1988.
6. Horwitz, A. H., Chang, C. P., Better, M., Hellstrom, K. E., and Robinson, R. R. Secretion of functional antibody and Fab fragment from yeast cells, *Proc. Natl. Acad. Sci. USA, 85*: 8678–8682, 1988.
7. Wood, C. R., Boss, M. A., Kenten, J. H., Calvert, J. E., Roberts, N. A., and Emtage, J. S. The synthesis and *in vivo* assembly of functional antibodies in yeast, *Nature, 314*: 446–449, 1985.
8. Pulitz, J. Z., Kabasek, W. L., Duchene, M., Marget, M., von Specht, B.-U., and Domdey, H. Antibody production in baculovirus-infected insect cells, *Biotechnology, 8*: 651–654, 1990.
9. Huston, J. S., Levinson, D., Mudgett-Hunter, M., Tai, M. S., Novotny, J., Margolies, M. N., Ridge, R. J., Bruccoleri, R. E., Haber, E., Crea, R., et al. Protein engineering of antibody during sites: recovery of specific antibody in anti-digoxin single-chain Fv analogue produced in *Escherichia coli, Proc. Natl. Acad. Sci. USA, 85*: 5879–5883, 1988.

10. Bird, R. E., Hardman, K. D., Jacobson, J. W., Johnson, S., Kaufman, B. M., Lee, S. M., Lee, T., Pope, S. H., Riordan, G. S., and Whitlow, M. Single-chain antigen-binding proteins, *Science, 242*: 423–426, 1988 (erratum, *Science, 244*: 409, 1989).
11. Rogers, S. G., Klee, H. J., Horsch, R. B., and Fraley, R. T. Improved vectors for plant transformation: expression cassette vectors and new selectable markers, *Meth. Enzymol., 153*: 253–276, 1987.
12. Hiatt, A. C., Cafferkey, R., and Bowdish, K. Production of antibodies in transgenic plants, *Nature, 342*: 76–78, 1989.
13. Hein, M. B., Tang, Y., McLeod, D. A., Janda, K. D., and Hiatt, A. C. Evaluation of immunoglobulins from plant cells, *Biotechnol. Prog., 7*: 455–461, 1991.
14. Horsch, R. B., et al. A simple and general method for transferring genes in plants, *Science, 227*: 1229–1231, 1985.
15. Bhozwani, S. S., and Razdan, M. K. Plant tissue culture: theory and practice, *Developments in Crops Science*, Vol. 5, Elsevier, Amsterdam, 1983.
16. Vasil, V., and Hildebrandt, A. Differentiation of tobacco plants from single, isolated cells in microculture, *Science, 150*: 889–892, 1985.
17. Walter, P., and Lingappa, V. R. Mechanism of protein translocation across the endoplasmic reticulum membrane, *Annu. Rev. Cell Biol., 2*: 499–516, 1986.
18. von Heijne, G. Signal sequences—the limits of variation, *J. Mol. Biol., 184*: 99–105, 1985.
19. Goldstein, I. J., and Hayes, C. E. Lectins: carbohydrate-binding proteins of plants and animals, *Adv. Carbohydr. Chem. Biochem., 35*: 120–340, 1978.
20. Faye, L., and Crispeels, M. J. Characterization of N-linked oligosaccharides by affinoblotting with concanavalin A-peroxidase and treatment of the blots with glycosidases, *Anal. Biochem., 149*: 218–224, 1985.
21. Kijimopto-Ochiai, S., Katagiri, Y. U., Hatae, T., and Okuyama, H. Type analysis of the oligosaccharides chains on microheterogeneous components of bovine pancreatic DNAase by the lectin-nitrocellulose sheet method, *Biochem. J., 257*: 43–49, 1989.
22. Trimble, R. B., and Maley, F. Optimizing hydrolysis of N-linked high-mannose oligosaccharides by endo-b-N-acetyl glucosaminidase H, *Anal. Biochem., 141*: 515–522, 1984.
23. Matsudaira, P. Sequence from picomole quantities of proteins electroblotted onto polyvinylidene difluoride membranes, *J. Biol. Chem., 262*: 10035–10038, 1987.
24. Sturm, A., Kuik, A. V., Vliegenthart, J. F. G., and Crispeels, M. J. Structure, position, and biosynthesis of the high mannose and the complex oligosaccharide side chains of the bean storage protein phaseolin, *J. Biol. Chem., 262*: 13392–13403, 1987.
25. Hiatt, A. C. Antibodies produced in plants, *Nature, 344*: 469–470, 1990.
26. Carpita, N., Sabularse, D., Montezinos, D., and Helmer, D. P. Determination of the pore size of cell walls of living plant cells, *Science, 205*: 1144–1147, 1979.
27. Varner, J. E., and Tuan-Hua Ho, D. Hormones, *Plant Biochemistry* (W. M. Bonner and J. E. Varner, eds.), Academic Press, New York, 1976, pp. 713–770.

28. Deom, C. M., Oliver, M. J., and Beachy, R. N. The 30-kilodalton gene product of tobacco mosaic virus potentiates virus movement, *Science, 237*: 294–389, 1987.
29. Yoder, O. C. Toxins in pathogenesis, *Ann. Rev. Phytopathol., 18*: 103–129, 1990.
30. Hiatt, A. The potential of antibodies in plants, *Agbiotechnol. News Informat., 2*: 653–655, 1990.
31. Vandekerckhove, J., et al. Enkephalins produced in transgenic plants using modified 2S seed storage proteins, *Biotechnology, 7*: 929–932, 1989.
32. Clark, M., Gilliland, L., and Waldmann, H. Hybrid antibodies for therapy, *Prog. Allergy, 45*: 31–49, 1988.
33. Reisfeld, R., and Cheresh, D. Human tumor antigens, *Adv. Immunol., 40*: 323–378, 1987.
34. Childers, N. K., Bruce, M. G., and McGhee, J. R. Molecular mechanisms of immunoglobulin A defense, *Annu. Rev. Microbiol., 43*: 503–536, 1989.
35. Underdown, B. J., and Schiff, J. M. Immunoglobulin A: strategic defense initiative at the mucosal surface, *Annu. Rev. Immunol., 4*: 389–417, 1986.
36. Mostov, K. E., and Simister, N. E. Transcytosis, *Cell, 43*: 389–390, 1985.
37. Mostov, K. E., Friedlander, M., and Blobel, G. The receptor for transepithelial transport of IgA and IgM contains multiple immunoglobulin-like domains, *Nature, 308*: 37–43, 1984.
38. Mostov, K. E., and Deitcher, D. L. Polymeric immunoglobulin receptor expressed in MDCK cells transcytoses IgA, *Cell, 46*: 613–621, 1986.
39. Mostov, K., Breitfeld, P., and Harris, J. M. An anchor-minus form of the polymeric immunoglobulin receptor is secreted predominantly apically in Madin–Darby canine kidney cells, *J. Cell Biol., 105*: 2031–2036, 1987.

12

Protein Production in Transgenic Crops: Analysis of Plant Molecular Farming

Jan Pen and Peter C. Sijmons

Mogen International NV, Leiden, The Netherlands

Albert J. J. van Ooijen

Gist-brocades NV, Delft, The Netherlands

André Hoekema

Mogen International NV, Leiden, The Netherlands

I. INTRODUCTION

Plants have long been used as a natural source of a wide variety of products. Besides being directly used as food and feed, plants are producers of oils and fats, polymers, industrial chemicals, medicines, textiles, dyes, detergents, and cosmetics. Among the materials of plant origin that are used are such diverse compounds as secondary metabolites, alcohols, fatty acids, lipids, carbohydrates, fibers, and proteins.

The chemistry of these natural compounds is very diverse. There has been an enormous increase in our knowledge of the chemical structures and functional properties of plant materials that can be applied in various processes. Also, there has been a vast amount of research on (chemical) modifications that make plant materials especially suitable for specific processes.

The use of wild plant species as raw materials for economic applications is limited by the fact that specific compounds can be obtained only in limited quantities, from specific organs of the plant, or from crops that require special climates. There has been a long-standing worldwide effort to improve crops in this respect by using classical breeding techniques; in many instances, however, these limitations remain.

With the advent of plant genetic engineering, it has become possible to add an extra dimension to the traditional exploitation of plants for economic use, namely, targeted modification of plants. The first examples of this new technology were in the engineering of disease-resistance traits, such as virus resistance (1–5) and insect resistance (6–8). Recently, research has also begun to focus on the improvement of crop qualities for specific purposes, e.g., taste, processing characteristics, solids content, color, and shelf life. However, the understanding of biochemical pathways in plants needed for designed improvements in quality, is still limited.

Gene transfer to plants has become possible for a variety of plant species, including many economically important crops (9). Combined with efficient expression systems for the introduced genes, gene transfer could turn plants into a suitable alternative for the production of desirable products. Plants could be exploited either for the production of "new" compounds or for the overproduction, derivatization, or "bioconversion" of compounds that naturally occur in the host plant.

In terms of cost–effectiveness for producing biomass, the growing of crops in the field can generally compete with any other system: It is inexpensive, it can be done in bulk quantities, and it requires limited infrastructure. This observation suggests that the exploitation of arable crops for the production of e.g., food, feed, or processing materials, would be very attractive.

As a first step in exploring the potential of field crops in this respect, we have started to compare a direct gene product, i.e., a recombinant protein expressed in transgenic plants, with data on existing microbial production systems. To allow comparisons of aspects of expression and purification, application, etc., two examples are presented: a pharmaceutical protein and an industrial enzyme.

II. HUMAN SERUM ALBUMIN

A. Expression in Transgenic Plants

Human serum albumin (HSA) was chosen as a first target because it is a protein with a relatively high value, broad clinical use, and a worldwide demand estimated to be in ton quantities. This protein is a large, complex polypeptide of 66.5 kDa that requires posttranslational processing. In humans, HSA is initially synthesized as pre-pro-albumin and is processed by removal of the amino-terminal pre-peptide of 18 amino acids and subsequent cleavage of the six residues of the pro-peptide by a serine proteinase. This results in the secretion of the mature polypeptide of 585 amino acids. The expression of the human pre-pro-HSA gene has been demonstrated in various microbial expression systems, such as *Saccharomyces cerevisiae* (10), *Bacillus subtilis* (11), and *Escherichia coli* (12). These data allow a direct comparison of plants and microbial production systems.

Potato was chosen as a target crop because the wet milling processes used in the potato starch industry are compatible with the separation of proteins present in the liquid fraction of the process. This fraction is essentially a waste product. As such, transgenic starch tubers would constitute an inexpensive source of biomass for protein production. At laboratory scale, reconstitution experiments with HSA added to ground, untransformed tubers have not shown any HSA degradation, which indicates that HSA is stable during the industrial process.

We have modified the gene encoding HSA to allow expression and processing of the pre-pro-polypeptide in the plant cell. A schematic diagram of the chimeric gene constructs is shown in Fig. 1a.

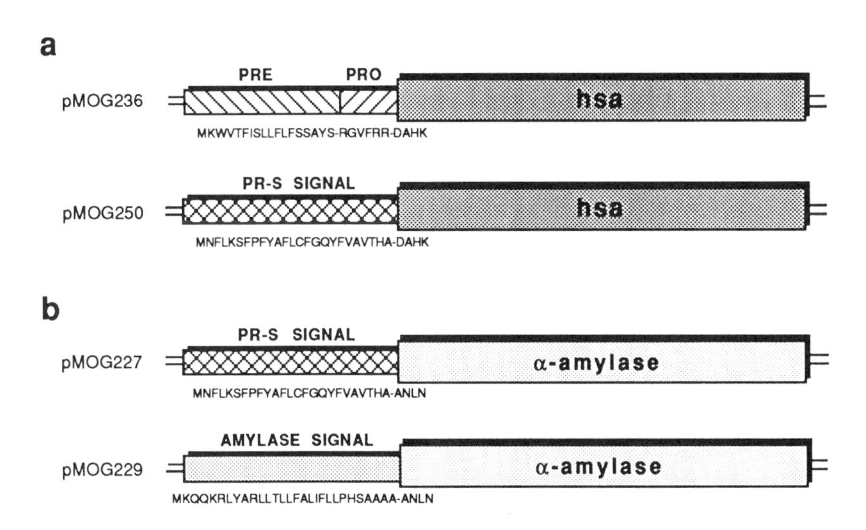

Figure 1 (a) Schematic representation of HSA chimeric gene constructs. The binary vector pMOG236 contains the pre-pro sequences from human HSA; pMOG250 contains the signal sequence of tobacco PR-S (13). (b) The chimeric α-amylase genes. The vectors pMOG227 and pMOG229 contain the signal peptide encoding sequences of tobacco PR-S and *Bacillus licheniformis* α-amylase, respectively.

We have tested both human and plant signal peptides for their ability to direct secretion of HSA. In all cases, we used a cauliflower mosaic virus 35S promoter that was modified to enhance the expression level of the transgene (14). Potato tuber discs (*Solanum tuberosum* cv. Désiré) were transformed as previously described (5) by using the *Agrobacterium* binary vector system (15). Mature transgenic plants were used for expression analysis of HSA mRNA, protein production, and processing.

The amount of HSA mRNA varied among plants, to up to 0.025% of the total mRNA. The amount of HSA protein accumulated up to 0.02% of the total protein; expression was detected in all plant parts, including the tubers. All transgenic plants were phenotypically normal. No expression was found from transcripts without signal sequences. The reason for this is unclear; the presence of a signal sequence may influence the translatability of the HSA mRNA; also, the secretion of the protein may enhance HSA stability.

Analysis of the extracellular fluid from transgenic potato leaves showed that the human and the plant signal peptides were equally capable of targeting HSA extracellularly. The two signal peptides share some characteristic properties: a basic amino-terminal domain, a stretch of apolar residues, and a more polar carboxy-terminal region followed by the cleavage site. To study the nature of the secreted products in both cases, we purified HSA from transgenic potato plants.

B. Analysis of HSA Purified from Transgenic Potato Plants

Potato HSA was extracted from homogenized plants, purified to chromatographic homogeneity, and used for N-terminal amino acid sequence analysis (14). Expression of the human pre-pro-HSA gene led to partial processing of the precursor and secretion of pro-HSA. The sequence information within the human signal peptide is sufficient for secretion in plants; the serine protease needed to remove the pro-sequence, however, may not be present in plants. The fusion between the plant signal peptide and mature HSA resulted in cleavage of the pre sequence at its natural site and in secretion of correctly processed HSA.

III. *BACILLUS LICHENIFORMIS* α-AMYLASE

A. Production

As a first example of the production of bulk industrial enzymes in plants, α-amylase from *Bacillus licheniformis* was tested. Alpha-amylases

(E.C. 3.2.1.1) from bacterial and fungal origins are widely used in industry for the hydrolysis of α-1,4-glycosidic linkages in the starch components amylose and amylopectin (16). For example, they are used in the starch processing and alcohol industries for liquefaction, in the brewing industry for producing low-calorie beer, in the baking industry for increasing bread volume, in the juice and wine industry for clarification, and in the detergent industry (17–19). The α-amylase from *Bacillus licheniformis* is the most commonly used enzyme in starch liquefaction, the first step in starch processing. This enzyme is particularly suited for this purpose because of its extreme heat stability and its activity over a wide pH range (19). Heat stability is important for the activity of the enzyme under industrial conditions because heat-gelatinized starch is a better substrate for the enzyme than is native starch.

A schematic representation of the chimeric α-amylase genes used for expression in tobacco are shown in Fig. 1b. The expression strategy was identical to the one for HSA. The modified 35S expression cassette was used; for secretion we used the *Bacillus* amylase signal peptide and the signal peptide of the tobacco PR-S protein. The enzyme accumulated to 0.5% of the soluble protein in transgenic tobacco plants. Both signal peptides were found to function equally well in secretion of the α-amylase to the extracellular spaces.

The α-amylase produced in plants consisted of two molecular forms with a molecular weight of approximately 64,000, clearly different from the *Bacillus licheniformis* enzyme, which has a molecular weight of 55,200 (19). This difference in molecular weight between the microbial α-amylase and the enzyme produced in tobacco could be fully attributed to carbohydrate chains attached to the protein. There are six potential asparagine-linked glycosylation sites in the primary structure of α-amylase (19). Comparison of the deglycosylation patterns with endo-*N*-acetylglucosaminidase H and trifluoromethanesulfonic acid shows that the carbohydrate chains found in tobacco are of the complex type. Deglycosylation of the plant enzyme reduced its molecular weight to 55 kDa (Fig. 2).

B. Activity of α-Amylase from Transgenic Plants

The enzyme produced by the transgenic plants exhibited biological activity. The glycosylation of the protein did not seem to have a major influence on the activity, although specific activities still have to be determined. In the application experiment (see Section III.C), in which starch degradation

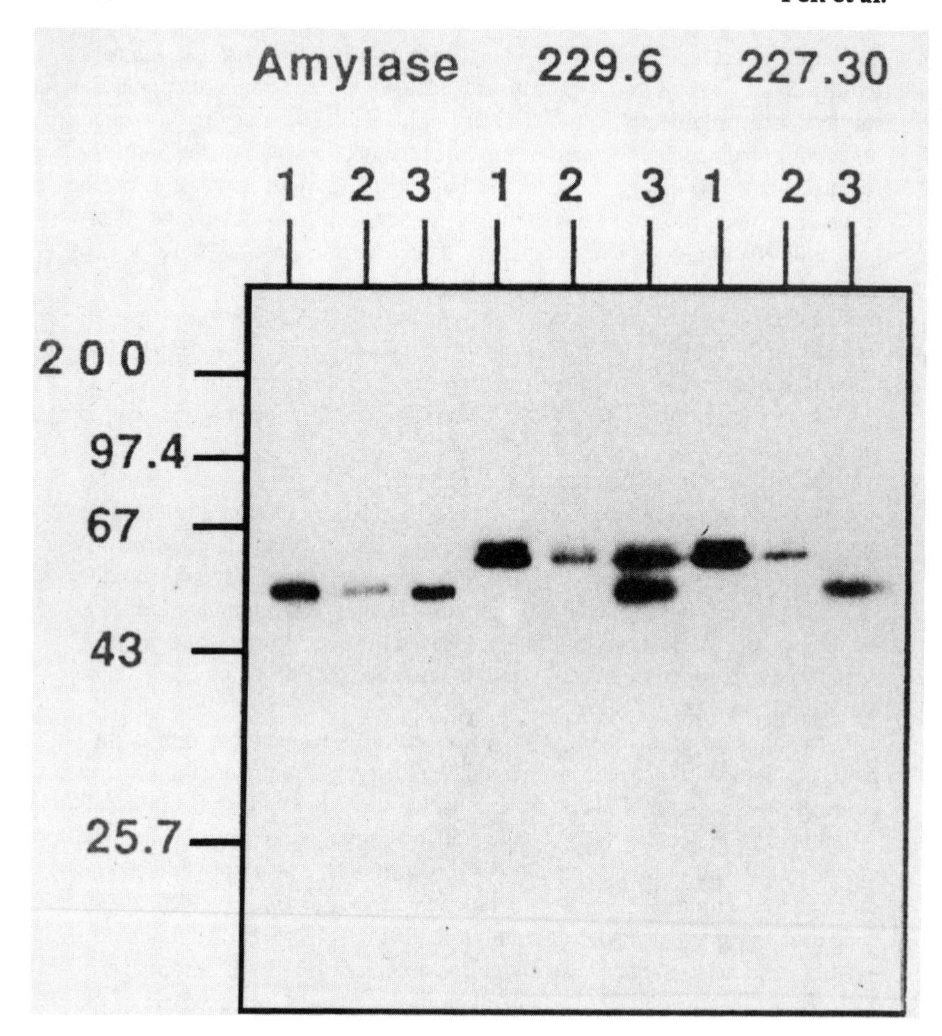

Figure 2 Deglycosylation of α-amylase produced in tobacco with endo-*N*-acetylglucosaminidase H (endoH) or trifluoromethanesulfonic acid (TFMS). Amylase lanes contain the purified microbial enzyme; extracts from transgenic lines 229.6 (amylase signal sequence) and 227.30 (PR-S signal sequence) are shown on the right. Treatments: 1, untreated; 2, EndoH; 3, TFMS. After electrophoresis on a 10% polyacrylamide gel, the gel was subjected to Western blot analysis.

was performed at 95–100°C, the thermostability of the tobacco-produced enzyme was demonstrated. These analyses revealed no differences in the biological activities between the enzyme from plants and its bacterial counterpart.

The generated transgenic plants were phenotypically indistinguishable from nontransgenic tobacco plants. Leaves from transgenic and nontransgenic plants collected halfway through the photoperiod stained similarly for starch. This indicates that the enzyme has no access to the starch present in the leaves, as is expected from the extracellular compartmentalization of α-amylase and the chloroplastic location of starch.

C. Application in Starch Liquefaction

We addressed the question whether the enzyme produced in plants would be suitable for application in starch liquefaction. Seeds were chosen as the preferred plant material, because they may provide the most stable environment for long-term storage of the enzyme. Transgenic tobacco line MOG227.3 with an α-amylase expression level of 0.2% in seeds was used. Seeds were harvested and milled, and the milled seeds were then used in liquefaction of corn and potato starches. Nontransgenic seeds were used as a control, and the microbial enzymes from *Bacillus licheniformis* and *Bacillus amyloliquefaciens* were used for comparison of activities.

High-performance liquid chromatography analysis of the degradation products obtained with the transgenic seeds showed that these were virtually identical to those obtained with *Bacillus licheniformis* α-amylase; maltopentaose (DP5) was the predominant product. The products were significantly different from those obtained with *Bacillus amyloliquefaciens* α-amylase, in which maltohexaose (DP6) was the major degradation product (Fig. 3). Control seeds showed no activity.

The hydrolyzed starches were then analyzed for their dextrose equivalent (DE) values. As shown in Table 1, the DE values obtained with transgenic seeds were found to be similar to those for the commercial enzyme preparations and within the commercially acceptable range (DE ≥ 12, preferably ≥ 16; see Ref. 20).

These data indicate that plant-borne α-amylase can be applied in starch liquefaction without detectable difference in activity from the microbial enzyme. This is the first demonstrated example of a potential industrial application of transgenic plant material expressing a heterologous enzyme (21).

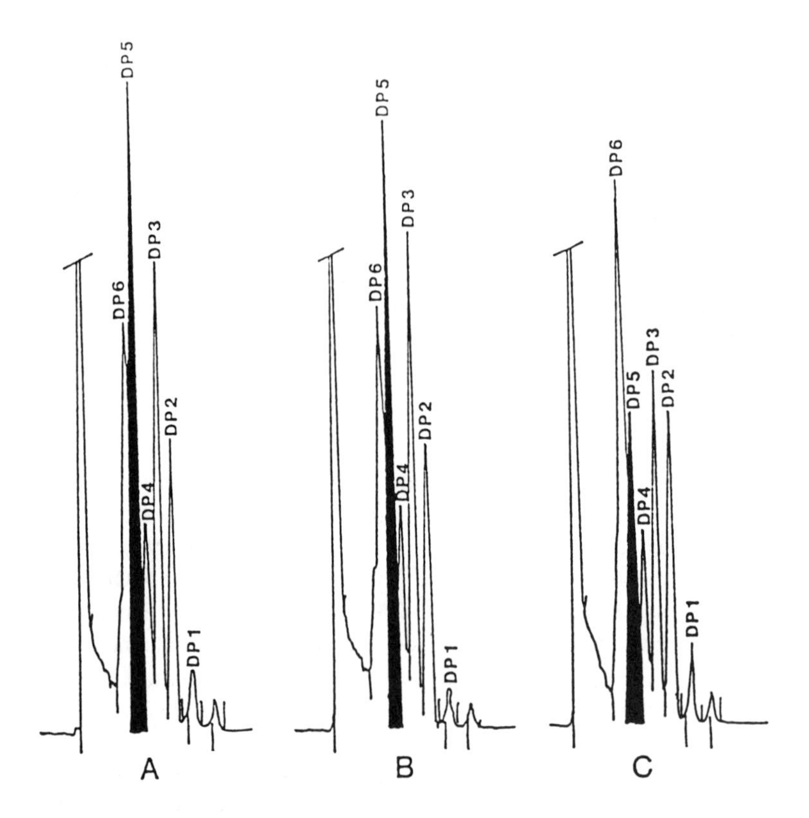

Figure 3 Comparison of oligosaccharide patterns obtained from starch hydrolysis. Corn starch was hydrolyzed by using (A) milled transgenic seeds of tobacco line MOG227.3, (B) *Bacillus licheniformis* α-amylase, and (C) *Bacillus amyloliquefaciens* α-amylase. Hydrolysis products were separated by HPLC. The degree of polymerization (DP) of glucose is indicated. The main difference between the products obtained with the seeds and *Bacillus licheniformis* amylase on the one hand and *Bacillus amyloliquefaciens* amylase on the other is the amount of DP5; this peak is indicated in black.

IV. DISCUSSION AND CONCLUSIONS

Plant genetic engineering technology may affect the applicability of crop-based resources in the industry. The modification of crops for specific industrial purposes may lead to new processes and products, ranging from the improved use of plants in the food and feed industry to new production

Table 1 Dextrose Equivalent (DE) Values Obtained from Hydrolysis of Corn and Potato Starch

Material	DE values	
	Potato starch	Corn starch
Transgenic tobacco seeds (MOG227.3)	16	13
Nontransgenic tobacco seeds	0	0
Bacillus licheniformis amylase	18	16
Bacillus amyloliquefaciens amylase	15	18

methods of pharmaceuticals (22,23), specialty chemicals, and bulk industrial products.

As a first step in studying the potential of crop modification for industrial purposes, we have chosen to produce a direct gene product in plants. In this chapter we have demonstrated the feasibility of producing a pharmaceutical protein and an industrial enzyme in plants. Ultimately, the practical success of producing proteins in plants will be determined by the ability of these proteins to compete with those produced by established production methods.

The expression level of the protein is important. There are large differences between the expression levels of various genes. In this chapter relatively high levels (max. 0.5% of total protein) were shown for α-amylase, which were at least 20 times that of HSA, obtained with an identical expression cassette. To date, maximum expression levels of heterologous proteins expressed in plants have reached about 1.5% of soluble protein (Ref. 24 and Pen et al., unpublished). The differences seem largely gene dependent and may be due to efficiencies in transcription/translation or stability of the gene products. Further improvements in the expression levels seem likely, given the modest effort that has been invested in optimization so far.

Whether dedicated growing of a crop for protein production is feasible will depend on the value of the protein. In this respect starch, potatoes, and oilseed rape could be attractive target crops, because the protein may be produced in addition to the primary products (starch and oil, respectively), thus reducing costs. The industrial process of producing the primary product must be compatible with the specific requirements for producing the protein. For many proteins production in starch potatoes will be feasible, because starch is extracted at about 14°C. Production in crops like oilseed rape will be limited to thermostable proteins, such as *Bacillus licheniformis* α-amylase, because the extraction temperature for oil of 80°C will denature other types of proteins.

Downstream processing is also an important factor in the overall cost of manufacturing proteins. To date, most research to develop industrial-scale protein extraction and purification has concentrated on microorganisms and mammalian cells. The processes and effort needed for protein purification from transgenic plants will depend on the source of material and on the required purity. For producing recombinant proteins from plants, the importance of purity is shown by a comparison of HSA and α-amylase. For pharmaceutical proteins, such as HSA, the purity requirements are very strict, because of the intended use. Therefore, the downstream processing costs will constitute a major part of the overall manufacturing costs and may thus override the advantages of a low-cost production system. The situation for α-amylase is clearly different, because the enzyme is used at relatively low purities. Here, inexpensive production in plants can be combined with moderate downstream processing costs.

To illustrate this point, we have studied two different scenarios. In one the costs for downstream processing would constitute 75% of the manufacturing costs (production and downstream processing costs), whereas in the other they would amount of 99%. To do this a rough estimate was made of growing costs for a crop, average yield, and protein content (all based on available agronomic data). We assumed that these data can be applied to estimating the costs for production of recombinant proteins in crops.

Figure 4 is a graph in which the manufacturing costs for producing recombinant proteins in starch potatoes are roughly estimated for these two cases. This clearly illustrates that the requirement of high purity, as for pharmaceutical proteins, offsets the low cost of production in transgenic plants. The potential of plant molecular farming is much greater for industrial enzymes. Even at the expression levels that can be obtained today with the available expression strategies, industrial enzymes can be manufactured at low cost. The relatively low purity at which these enzymes are used makes them good candidates for successful and competitive production from crops. In some cases, as shown in this chapter for α-amylase, seeds containing the enzyme can even be applied directly in industrial processes, without further purification or formulation of the enzyme.

In addition to the direct production of enzymes in crops, there is potential for the expression of an industrial enzyme in a crop in which the substrate is also present, thus creating a bioconversion. This may completely obviate the need for downstream processing of the enzyme. The feasibility of such an approach

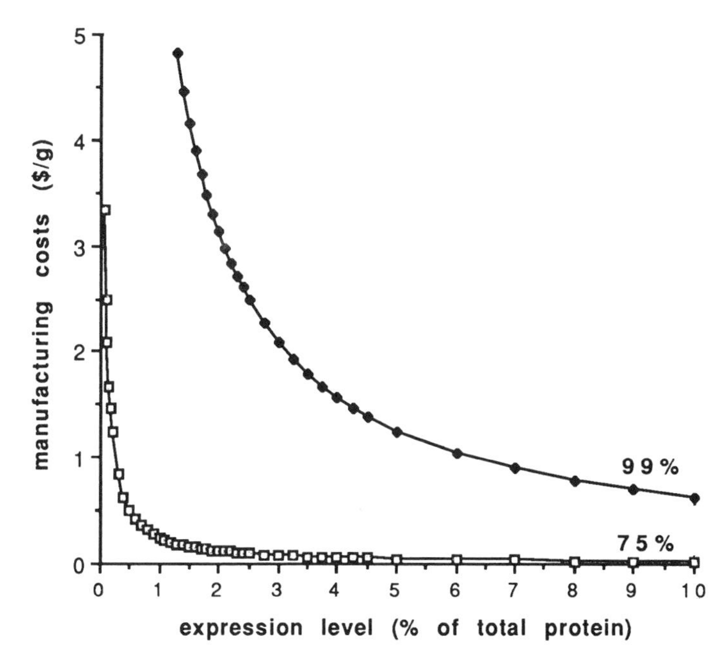

Figure 4 Estimated manufacturing costs of recombinant proteins in starch potatoes at different expression levels, in cases for which the downstream processing costs would contribute 75% (open squares) and 99% (filled squares) of the total manufacturing costs.

is exemplified in this chapter by the production of extracellular α-amylase in the model crop tobacco. The enzyme can be produced in plants without affecting endogenous starch, which shows that the enzyme and substrate are in separate locations. An industrial example of this approach may be the expression of an extracellular α-amylase in potato or corn. Because the enzyme and substrate are present in different cellular compartments, no degradation of the substrate occurs during growth. The enzymatic reaction is initiated during processing of the crop, when by homogenization enzyme and substrate are brought together. The currently obtained α-amylase expression levels would be sufficient to hydrolyze all starch present. Many applications for heterologous enzymes in crops may be found for other industrial processes, possibly leading to higher production of natural plant products, modification and bioconversion of plant materials, and new compounds.

ACKNOWLEDGMENTS

This work was supported by grant PBTS/BIO 88097 from the Dutch Ministry of Economic Affairs. We thank Theo Verwoerd, Ben Dekker, Barbara Schram-meijer, and Lucy Molendijk for their expert help during these studies. We are grateful to Dr. Peter van den Elzen, Dr. Nick Garner, and Dr. Theo de Bock for critical reading of the manuscript.

REFERENCES

1. Powell, A. P., Nelson, R. S., De, B., Hoffman, N., Rogers, S. G., Fraley, R. T., and Beachy, R. N. Delay of disease development in transgenic plants that express the tobacco mosaic virus coat protein gene, *Science, 232*: 738–743, 1986.
2. Tumer, N. E., O'Connell, K. M. O., Nelson, R. S., Sanders, P. R., Beachy, R. N., Fraley, R. T., and Shah, D. M. Expression of alfalfa mosaic virus coat protein gene confers cross-protection in transgenic tobacco and tomato plants, *EMBO J., 6*: 1181–1188, 1987.
3. Loesch-Fries, L. S., Merlo, D., Zinnen, T., Burhop, L., Hill, K., Krahn, K., Jarvis, N., Nelson, S., and Halk, E. Expression of alfalfa mosaic virus RNA 4 in transgenic plants confers virus resistance, *EMBO J., 6*: 1845–1851, 1987.
4. Van Dun, C. M. P., and Bol, J. F. Transgenic tobacco plants accumulating tobacco rattle virus coat protein resist infection with tobacco rattle virus and pea early browning virus, *Virology, 167*: 649–652, 1988.
5. Hoekema, A., Huisman, M. J., Molendijk, L., Van den Elzen, P. J. M., and Cornelissen, P. J. M. The genetic engineering of two commercial potato cultivars for resistance to potato virus X, *Biotechnology, 7*: 273–278, 1989.
6. Hilder, V. A., Gatehouse, A. M. R., Sheerman, S. E., Barker, R. F., and Boulter, D. A novel mechanism of insect resistance engineered into tobacco, *Nature, 330*: 204–207, 1987.
7. Fischoff, D. A., Bowdish, K. S., Perlak, F. J., Marone, P. G., McCormick, S. M., Niedermeyer, J. G., Kusano-Kretzmer, K., Mayer, E. J., Rochester, D. E., Rogers, S. G., and Fraley, R. T. Insect tolerant transgenic tomato plants, *Biotechnology, 5*, 807–813, 1988.
8. Vaeck, M., Reynaerts, A., Hofte, H., Jansens, H., De Beuckeleer, M. D., Dean, C., Zabeau, M., Van Montagu, M., and Leemans, J. Transgenic plants protected from insect attack, *Nature, 328*: 33–37, 1987.
9. Gasser, C. S., and Fraley, R. T. Genetically engineering plants for crop improvement, *Science, 244*: 1293–1299, 1989.
10. Etcheverry, T., Forrester, W., and Hitzeman, R. Regulation of the chelatin promoter during the expression of human serum albumin or yeast phosphoglycerate kinase in yeast, *Biotechnology, 4*: 726–730, 1986.
11. Saunders, C. W., Schmidt, B. J., Mallonee, R. L., and Guyer, M. S. Secretion of human serum albumin from *Bacillus subtilis*, *J. Bacteriol., 169*: 2917–2925, 1987.

12. Latta, M., Knapp, M., Sarmientos, P., Brefort, G., Becquart, J., Guerrier, L., Jung, G., and Mayaux, J.-F. Synthesis and purification of mature human serum albumin from *E. coli, Biotechnology, 5*: 1309–1314, 1987.
13. Hooft van Huijsduijnen, R. A. M., Cornelissen, B. J. C., Loon, L. C. V., Boom, J. H. V., Tromp, M., and Bol, J. F. Virus-induced synthesis of messenger RNAs for precursors of pathogenesis-related proteins in tobacco, *EMBO J., 4*: 2167–2171, 1985.
14. Sijmons, P. C., Dekker, B. M. M., Schrammeijer, B., Verwoerd, T. C., Van den Elzen, P. J. M., and Hoekema, A. Production of correctly processed human serum albumin in transgenic plants, *Biotechnology, 8*: 217–221, 1990.
15. Hoekema, A., Hirsch, P. R., Hooykaas, P. J. J., and Schilperoort, R. A. A binary plant vector strategy based on separation of *vir*- and T-region of the *Agrobacterium tumefaciens* Ti-plasmid, *Nature, 303*: 179–180, 1983.
16. Schwardt, E. Production and use of enzymes degrading starch and some other polysaccharides, *Food Biotechnol., 4*: 337–351, 1990.
17. Peppler, H. J., and Reed, G. Enzymes in food and feed processing, *Biotechnology*, Vol. 7A (J. F. Kennedy, ed.), VCH, Weinheim, Germany, 1987, pp. 547–603.
18. Kennedy, J. F., Cabalda, V. M., and White, C. A. Enzymic starch utilization and genetic engineering, *TIBTECH, 6*: 184–189, 1988.
19. Yuuki, T., Nomura, T., Tezuka, H., Tsuboi, A., Yamagata, H., Tsukagoshi, N., and Udaka, S. Complete nucleotide sequence of gene coding for heat- and pH-stable α-amylase of *Bacillus licheniformis*: comparison of the amino acid sequences of three bacterial liquefying α-amylase deduced from the cDNA sequences, *J. Biochem., 98*: 1147–1156, 1985.
20. Reilly, P. J. Enzymic degradation of starch, *Starch Conversion Technology* (G. M. A. Van Beynum and J. A. Roels, eds.), Marcel Dekker, New York, 1985, pp. 101–142.
21. Pen, J., Molendÿk, L., Quax, W. J., Sÿmons, P. C., Van Ooijen, A. J. J., Van den Elzen, P. J. M., Rietveld, K., and Hoekema, A. Production of active *Bacillus licheniformis* alpha-amylase in tobacco and its application in starch liquefaction, *Biotechnology, 10*: 292–296, 1992.
22. Krebbers, E., and Vandekerckhove, J. Production of peptides in plant seeds, *TIBTECH, 8*: 1–3, 1990.
23. Vandekerckhove, J., VanDamme, J., Van Lijsebettens, M., Botterman, J., De Block, M., Vandewiele, M., De Clercq, A., Leemans, J., Van Montagu, M., and Krebbers, E. Enkephalins produced in transgenic plants using modified 2S seed storage proteins, *Biotechnology, 7*: 929–932, 1989.
24. Hiatt, A., Cafferkey, R., and Bowdish, K. Production of antibodies in transgenic plants, *Nature, 342*: 76–78, 1989.

V

RIBOZYMES AND ANTISENSE RNA

13

Antisense RNA in Plants

C. F. Watson and Don Grierson

Nottingham University, Sutton Bonington, England

I. INTRODUCTION

Naturally occurring or artificially generated antisense RNA has been implicated in the control of gene expression in prokaryotes and eukaryotes. The mechanism of action involves duplex formation between antisense RNA and a complementary target nucleic acid sequence. Although the precise consequences of this may vary, this specific base pairing precludes the target sequence from participating in other molecular interactions, thus leading to a change in gene function or expression. The mechanism of action of antisense RNA has been particularly well characterized in relation to replication of the *Escherichia coli* ColE1 and R1 plasmids (for recent reviews see Refs. 1 and 2). Replication of ColE1 is initiated by the transcription of an RNA primer, the conformation of which is crucial to hybridization to DNA and priming of DNA synthesis. In forming a duplex with the RNA primer, the antisense RNA irreversibly alters the structure of the primer and hence its ability to initiate DNA replication. The presence of naturally occurring antisense RNA has also been reported, for example, in myelin-deficient (*Mld*) mice (3,4) and in *Xenopus* oocytes. In *Xenopus* oocytes, antisense RNA, after hybrid formation with a target fibroblast growth factor mRNA, is thought to cause the chemical modification and subsequent degradation of the target mRNA (5). An allelic mutation of the myelin basic protein (MBP) gene in mice can in the homozygous state lead to a 98% reduction of MBP mRNA and protein levels. This reduction in MBP leads to a severe myelin deficiency in the central nervous system characterized by the *Shiverer* phenotype. The reduction is due to a duplication of the entire MBP gene and an inversion of five out of the seven exons upstream of the normal MBP gene. Although expression of the upstream gene gives rise to antisense MBP RNA, it is not clear whether the reduction in levels of MBP RNA is due to the antisense MBP RNA (3) or transcriptional interference of the normal MBP gene (4) from the upstream duplicated gene.

Antisense RNA has also been used experimentally to modulate gene expression in animals and plants. Some of the early model systems devised to study the effect of antisense RNA used circular fragments of plasmid DNA carrying antisense genes. These genes were injected into cultured cells that were either coinjected with a target sense gene (6) or were already stably expressing a sense target gene (7). In this way thymidine kinase (TK) mRNA expression was successfully reduced in cultured mouse cell lines injected with plasmid DNA carrying an antisense TK gene. Subsequently, Ecker et al. (8) developed a transient assay system in which carrot protoplasts were infused with chimeric sense and antisense genes of bacterial origin by electroporation. Several different gene promoters were used to drive the constitutive expression of an antisense chloramphenicol acetyltransferase (CAT) gene, including the "strong" cauliflower mosaic (CaMV) 35S promoter. Simultaneous electroporation of protoplasts with sense and antisense CAT genes led to a maximum 95% inhibition of CAT activity 48 hours later; the degree of inhibition was proportional to the strength of the gene promoter driving the transcription of the antisense gene and depended on an excess of antisense plasmid DNA.

The use of *Agrobacterium tumefaciens* to mediate the stable introduction of antigenes into dicotyledonous plants has proved to be a highly versatile and efficient system for studying the effects of antisense RNA on gene expression using with reporter gene systems (9–13) and for causing the down-regulation of endogenous plant genes with a consequent measurable change in phenotype (14–18). This antisense approach has permitted the function of specific genes to be tested (14,19), has led to the identification of previously unknown genes (16,18), and has generated new plants with altered characteristics of commercial significance (20,21).

II. NATURALLY OCCURRING ANTISENSE RNA

There is some evidence that antisense RNA may function naturally to regulate gene expression in plants as it does in animals. Mobilization of starch stored in the amyloplasts of endosperm cells of germinating cereal grains requires the synthesis and secretion of hydrolytic enzymes from the aleurone layer. In barley this is stimulated by a hormonal trigger, gibberellin (GA) synthesized by the imbibing embryo and scutellum, and is inhibited by abscisic acid (ABA). Alpha amylases, the most abundant hydrolases synthesized, catalyze the hydrolysis of the starch subunits amylose and amylopectin. Two classes of α-amylase (isoenzymes A and B) are synthesized as a result of an increase in the steady-state levels of α-amylase mRNA (22).

Rogers (23) identified two different antisense RNA transcripts with sequence complementarity to barley α-amylase mRNA. By using Northern blot analysis and an S1 nuclease protection assay, in which sequence complementarity between a radiolabeled DNA fragment and the antisense RNA rendered the hybrid resistant to nuclease digestion, Rogers (23) demonstrated that the antisense transcripts have imperfect complementarity. It is likely, therefore, that the antisense RNA and the barley α-amylase mRNA are transcribed from different genes. Although the presence of antisense α-amylase RNA in the ABA-treated barley aleurone layer in amounts approximately equal to the amounts of the sense mRNA is particularly intriguing (23), a specific role for the antisense RNA in α-amylase gene expression has not been demonstrated in vivo. Isolation of the antisense gene and studies of either naturally occurring or induced mutations are needed to address further the role of this antisense RNA in the modulation of endogenous α-amylase gene expression.

The transposable element isolated from *Antirrhinum majus*, Tam3 (for a review see Ref. 24), has been used to study in vivo the gene encoding chalcone synthase (*niv*) and hence define those regions 5′ to the coding sequence involved with the spatial and temporal expression of the *niv* gene. Transposition and excision of Tam3 in the *niv* gene promoter is associated with large scale rearrangements or deletions or both, giving rise to novel patterns of *niv* gene expression. Because chalcone synthase is a key enzyme in flavonoid biosynthesis in flower petals, any significant change in *niv* gene expression can be easily identified as an alteration in flower pigmentation.

Characterization of a semidominant allele of the *niv* locus, *Niv*-525, revealed that an inverted duplication of untranslated leader sequence some 40 nucleotides long had occurred in the promoter region (25). In the heterozygous state (niv-525/Niv+) chalcone synthase protein and mRNA steady-state levels were reduced to less than 15% of normal. Coen et al. (25) postulated that transcription through the inverted duplication, and termination prior to the downstream untranslated leader sequence, could lead to the production of an antisense chalcone synthase transcript. It was suggested that this transcript could block *in trans* mRNA originating from the homologous intact *niv* gene by forming an RNA/RNA duplex. Although no such antisense transcript was detected in these plants, low concentrations of antisense RNA appear to be very efficient at down-regulating gene expression, and a similar failure to detect antisense RNA has been reported in transgenic plant systems (15).

III. MODEL ANTISENSE SYSTEMS

The transient assay system used by Ecker et al. (8) to demonstrate the artificial regulation of gene expression in plants by antisense RNA was further improved by Rothstein et al. (9), Delauney et al. (10), Sandler et al. (11), Robert et al. (12), and Cannon et al. (13). Tobacco plants previously transformed with wild-type nopaline synthase (NOS), CAT, or β-glucuronidase (GUS) reporter genes were retransformed with the respective reporter gene in the antisense orientation, and modulation of CAT, NOS, or GUS expression by antisense RNA was assayed. Rothstein et al. (9) described the constitutive expression of an antisense NOS RNA transcript 820 nucleotides long, covering the translation initiation codon, from a CaMV 35S promoter in transgenic tobacco. In greenhouse-grown plants expressing antisense NOS RNA, NOS enzyme activity was reduced by a factor of eight as a result of the substantial reduction in steady-state levels of NOS mRNA. By using a strong tissue-specific promoter from the light-regulated chlorophyll a/b-binding protein gene from petunia (*Petunia hybrida*) to drive expression of several different regions of the NOS gene in the antisense orientation, Sandler et al. (11) attained reductions in NOS enzyme activity and steady-state levels of NOS mRNA similar to those described by Rothstein et al. (9). The most effective antisense construct was one that contained a 3′ region of the NOS gene downstream from nucleotide 373 and hence overlapping that of Rothstein et al. (9). Surprisingly, however, neither a full-length nor a 373 nucleotide 5′ antisense NOS gene that included the translation start codon had a statistically significant effect on NOS gene expression compared with that in control plants. The variation in results attained by using different regions of the NOS gene was attributed to the different stabilities of the individual antisense NOS RNAs transcribed and the effect of RNA three-dimensional structure on the ability of the sense and antisense RNA molecules to form a hybrid.

Unlike Sandler et al. (11), Cannon et al. (13) demonstrated that the light-regulated expression of antisense GUS RNA from the 5′ region overlapping the translation start codon could reduce GUS expression by up to 100% in leaves of transgenic tobacco previously transformed with a constitutively expressed GUS gene. Because in this particular experiment transcription of the sense and antisense GUS genes were from promoters of similar strength and antisense RNA was undetectable, Cannon et al. (13) reasoned that an excess of antisense RNA was not required to give inhibition of gene expression. Robert et al. (12) also used the GUS reporter gene in an antisense construct but under the regulation of the CaMV 35S promoter. The relatively low levels of sense and antisense GUS gene transcripts detected in these plants was thought to indicate the rapid degradation of RNA/RNA hybrids within the cell.

Experimental evidence from these in vivo model systems is therefore consistent with one theory of antisense RNA action that postulates its effect *in trans* on gene expression by hybridizing to and forming a duplex with the target sense mRNA, thus preventing its accumulation. In principle, antisense RNA could bring about this effect by interfering with transcription, mRNA processing, transport, or translation or by enhancing the degradation of the RNA. Evidence that antisense RNA works by rendering the target RNA susceptible to degradation is discussed more fully later.

IV. DOWN-REGULATION OF ENDOGENOUS PLANT GENES WITH ANTISENSE RNA

A. Antisense Chalcone Synthase Genes

The first report of the use of antisense RNA to manipulate the expression of an endogenous plant gene was by van der Krol et al. (15). The entire coding region of the chalcone synthase (CHS) gene A from *Petunia hybrida* was cloned in the antisense orientation into the binary plant transformation vector Bin19 (15). Constitutive expression of the antisense CHS gene from a CaMV 35S promoter led to a reduction in the level of CHS mRNA and enzyme. This was scored phenotypically as a reduction in flower pigmentation. In 10 out of 20 primary transformants the CHS antisense gene resulted in a sectored pigmentation pattern or evenly reduced pigmentation of corolla tissue. Although all 20 plants contained the antisense CHS gene and expressed the antisense CHS RNA at varying steady-state levels in leaf tissue, 10 plants showed a wild-type flower pigmentation pattern of evenly colored purple flowers. Those transgenic plants expressing the antisense CHS RNA and exhibiting a phenotypic effect had between one and three copies of the antisense gene integrated into the genome. There did not appear to be a correlation between antisense gene copy number, the steady-state level of antisense RNA, or reduction in sense CHS transcript level in the plants exhibiting a change in flower pigmentation.

Steady-state levels of mRNAs of other enzymes involved in the flavonoid biosynthetic pathway of petunia, such as chalcone flavone isomerase (CHI) and dihydroflavonol reductase (DFR), were unaltered by expression of the introduced antisense CHS gene (26). However, the expression of a closely related CHS-J gene present in petunia was affected by the antisense CHS-A gene. In addition, expression of endogenous CHS genes in potato and tobacco plants transformed with an antisense petunia CHS-A gene was also down-regulated (15,21). Of the two developmentally expressed CHS genes in floral tissues of petunia, about 90% of the total CHS mRNA steady-state level is transcribed from CHS-A, whereas CHS-J contributes about 10% (27). Homology between CHS-A and CHS-J mRNA and between petunia CHS-A and tobacco or potato CHS has been estimated to be 80% (26). Van der Krol et al. were able to demonstrate, using gene-specific probes, that the expression of both CHS-A and CHS-J were down-regulated by the expression of antisense CHS-A RNA in petunia (26) and that the same antisense gene caused the down-regulation of CHS in tobacco (15) and potato (26). These observations indicate that a significant degree of sequence divergence between the antisense RNA transcript and the targeted sense RNA is tolerated (26).

Further analysis of the promoter strengths required to drive transcription of minimal antisense CHS sequences and affect flower pigmentation in transgenic petunia led van der Krol et al. (28) to conclude that a simple RNA/RNA interaction alone was not sufficient to explain the antisense effect. This conclusion was based on the observation that transcription of the antisense CHS gene from either the CaMV 35S promoter or an endogenous CHS gene promoter did not lead to an excess of antisense CHS RNA as compared with endogenous CHS mRNA.

B. Antisense Polygalacturonase Genes in Tomato

Climacteric fruits, such as apples, bananas, and tomatoes (*Lycopersicon esculentum*), show an increase in respiration during fruit ripening and an accompanying rise in ethylene production. Coincident with this rise in ethylene evolution in tomato is a series of coordinated biochemical events caused by dramatic changes in gene expression (29). These include the degradation of chlorophyll and starch, the accumulation of lycopene, sugars, and organic acids, and the production of cell-wall-degrading enzymes. The close relationship between softening and the accumulation of the cell-wall-hydrolyzing enzyme polygalacturonase (PG) (30) has attracted considerable attention in recent years. Solubilization of the middle lamella of unripe pericarp discs by PG in vitro (31) is similar to that of ripe fruit in vivo (32), and pectin depolymerization during ripening in vivo (33) correlates with the de novo synthesis of PG during tomato fruit ripening (34,35). Additionally, a strong correlation exists between the reduced accumulation of PG and the softening of two pleiotropic mutants of tomato, Neverripe (*Nr*) and ripening inhibitor (*rin*). The mutant *Nr* softens slowly and has a greatly reduced PG content (36), whereas *rin*, which does not soften, contains very little PG activity at any stage of ripening (34). These observations led to the postulation that PG was largely responsible for fruit softening (30,36,37).

Polygalacturonase is synthesized de novo during ripening (34–36). There are three PG isoenzymes, PG1, PG2A, and PG2B, which appear sequentially. Although PG1 is formed first, PG2A and PG2B accumulate to substantially higher levels in ripe fruit (19,34,38). All three isoenzymes of PG are thought to be structurally related (19,34,36), and the common catalytic fruit PG subunit is probably encoded by a single gene (38). Polygalacturonase cDNA clones have been identified in several laboratories (39–43) and have been used to show that the increase in PG protein synthesis during ripening is controlled at the level of PG mRNA accumulation. Polygalacturonase mRNA is virtually undetectable in immature green tomato fruit before any increase in ethylene evolution above a basal rate (29,40,41), increases during ripening owing to increased transcription (44), and reaches a maximum level in full-orange or firm-ripe fruit at the time of peak ethylene evolution (44).

Determining whether PG plays a role in fruit softening requires naturally occurring or induced mutations in the PG gene. Although the pleiotropic tomato ripening mutants *Nr* and *rin* have substantially reduced levels of PG activity (36) and PG mRNA (45), they are unlikely to have arisen from a single mutation in the PG gene because other ripening-associated parameters, such as ethylene synthesis and lycopene accumulation, are similarly affected. Furthermore, *Nr* and *rin* map to chromosomes 9 and 5, respectively, whereas the PG gene is on chromosome 10 (46). Our group at Nottingham, in conjunction with ICI Seeds, decided to use an antisense strategy to specifically down-regulate the endogenous PG gene. A 730-base-pair (bp) fragment from the 5′ end of the

PG cDNA clone pTOM6 (40) was cloned in reverse orientation between a 528-bp CaMV 35S promoter fragment (47) and the 3′ end of the NOS gene (48). This construction ensured that the antisense gene was expressed in leaves and fruit (14) at high levels. This chimeric gene was introduced into the binary plant transformation vector Bin19 (49) adjacent to a kanamycin resistance gene to create the plasmid JR16A (Fig. 1). This was transferred to *A. tumefaciens* strain LBA4404 by triparental mating and was used to transform tomato stem segments (14). Thirteen separate transformed plants were selected on the basis of their ability to grow and form roots on media containing kanamycin (14,39). One such transformant, GR16, has been studied extensively (14,19). Southern blot analysis of DNA extracted from leaf tissue showed that a single copy of the antisense gene was transferred intact into one site in the tomato genome. By using RNA strand-specific probes to distinguish between PG mRNA and antisense PG RNA, steady-state levels of mRNA transcribed from the endogenous PG gene and the added antisense gene were measured (Figs. 2A and 2B). Antisense PG RNA transcripts were detected in green and ripe-orange fruit of GR16 but not in untransformed fruit. Polygalacturonase mRNA was expressed in ripe but not green fruit of both the transgenic and control plants. In ripe-orange fruit of GR16, however, only 6% of the normal level of PG mRNA was found. The antisense PG RNA levels were similar in leaf tissue and orange fruit of GR16 but were only 60% that of peak PG mRNA levels in GR16 (Fig. 2B). Later in ripening, the level of antisense RNA declined (see Fig. 2A and Ref. 14).

The accumulation of the pigment lycopene occurs concomitantly with PG during ripening in both control and GR16 fruit (Fig. 3). Therefore, comparisons could be made between extractable PG activity during ripening of GR16 and the control plant. Extraction of cell-wall-associated proteins and fractionation by SDS–polyacrylamide gel electrophoresis (PAGE) revealed that PG protein accumulation in ripe fruit of GR16 was substantially less than in untransformed fruit (Fig. 3A). Polygalacturonase enzyme activity was also significantly reduced in ripening fruit of GR16, reaching only 10% of normal levels (Fig. 3B). This reduction in PG enzyme activity, which correlated well with the reduction in PG protein levels, was due to a reduction in steady-state levels of PG mRNA. Similar results were obtained for other transformants, including GR15 (Fig. 3A), which expressed only 33% of normal PG levels. Sheehy et al. (50) also transformed tomato with an antisense PG gene but used the entire PG coding region. A similar reduction in PG enzyme activity as a result of reduced PG mRNA steady-state levels was obtained in a number of different transformants that constitutively expressed the antisense PG gene.

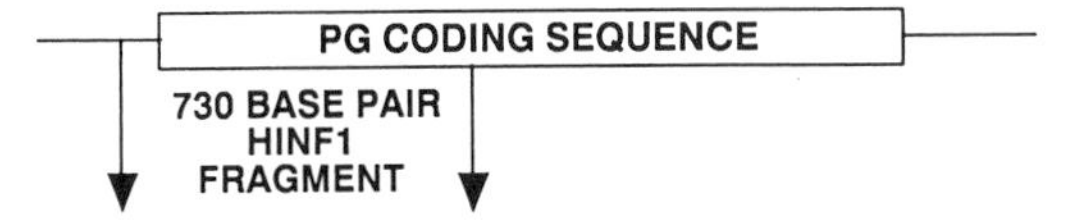

CLONE PG FRAGMENT IN REVERSE ORIENTATION BETWEEN PROMOTER AND TERMINATOR SEQUENCES

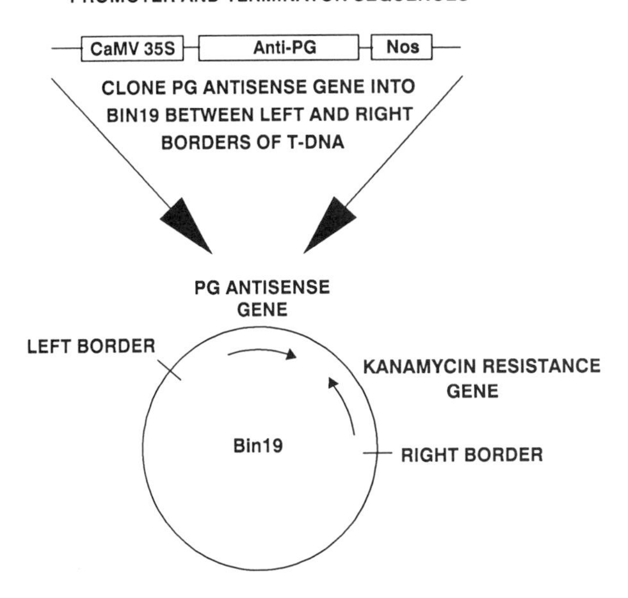

Figure 1 Construction of the PG antisense gene in Bin19. The pJR16A plasmid was transferred to *A. tumefaciens* by triparental mating and was used to transform tomato stem segments (14).

Inheritance of the antisense PG gene was studied in progeny of GR16, obtained by selfing the parent (19). Progeny plants were either homozygous (two genes) or hemizygous (one gene) for the antisense PG gene or had the antisense gene crossed out. Antisense PG RNA was undetectable in plants that did not inherit the antisense gene (GR102) or were untransformed. Progeny plants without the antisense gene also had normal levels of PG enzyme activity. This shows that the expression of the PG gene was not permanently altered and that the presence of the antisense gene was required for a reduction in PG gene activity to occur. One plant, GR105, with two antisense genes, had less than 1% PG activity throughout ripening as a result of the reduction in PG mRNA

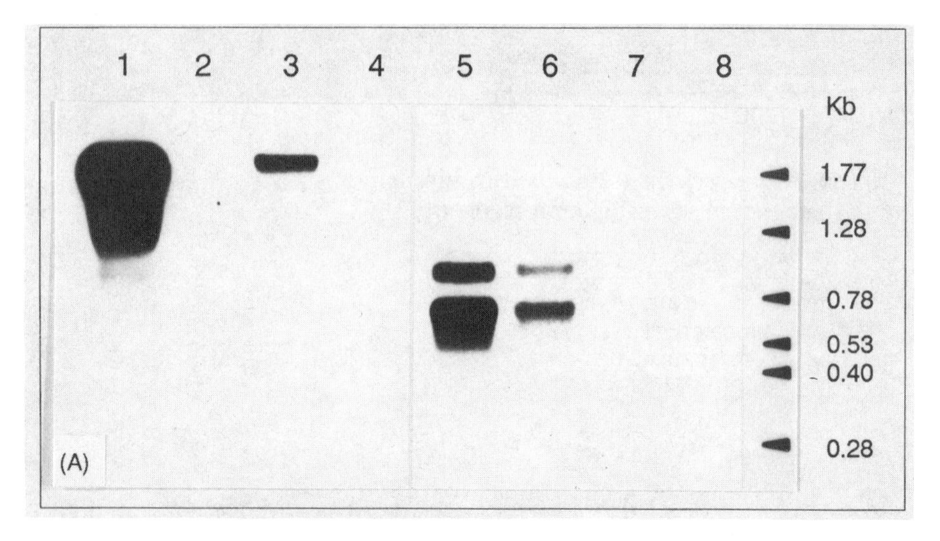

Figure 2 (A) Northern blot of poly(A)$^+$ mRNA from untransformed and transformed PG antisense fruit. RNA (1 μg per lane) from untransformed and GR16 fruit was hybridized to a PG sense-strand-specific probe (lanes 1–4) and an antisense-strand-specific (lanes 5–8): lanes 1 and 8, orange untransformed fruit; lanes 2 and 7, green untransformed fruit; lanes 3 and 6, orange transformed fruit; lanes 4 and 5, green transformed fruit.

steady-state levels. Plants with one antisense gene, including GR95, had intermediate levels of PG enzyme activity and PG mRNA (Fig. 4). There was, therefore, an obvious gene-dosage-dependent reduction in PG enzyme activity and PG mRNA levels during ripening of GR16 progeny. This relationship between antisense gene copy number and phenotypic effect is in agreement with the results of Hamilton et al. (16), who transformed *Lycopersicon esculentum* with an antisense ethylene-forming enzyme (EFE) gene, and Visser et al. (51), who transformed *Solanum tuberosum* (potato) with an antisense maize granule-bound starch synthase (GBSS) gene. However, van der Krol et al. (15) reported no such gene-dosage-dependent reduction in CHS enzyme activity in petunia transformed with an antisense CHS gene. Of crucial importance was the demonstration that the expression of the antisense PG gene did not affect other ripening-associated parameters. Measurements of ethylene evolution, lycopene accumulation, and invertase and pectin esterase enzyme activities in GR16 progeny fruit tissue were indistinguishable from those in

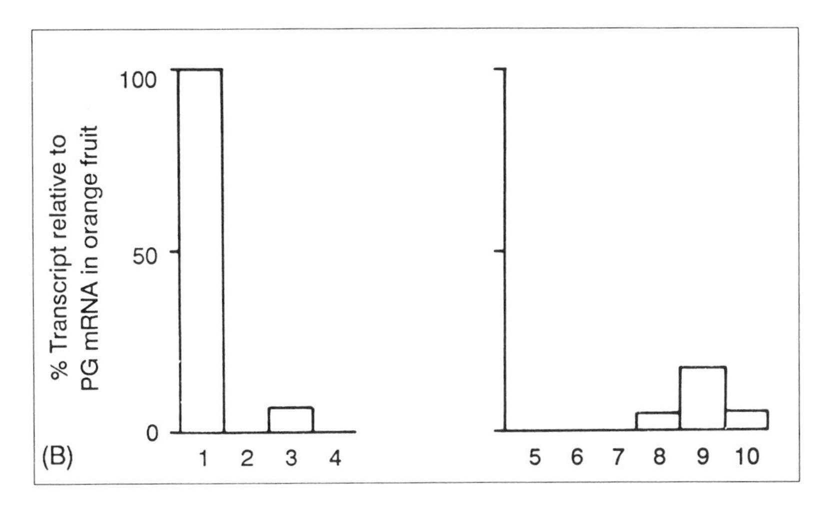

Figure 2 (continued) (B) Histograms of the dot blot analysis of poly(A)$^+$ mRNA from untransformed and transformed (GR16) plants hybridized to a sense-strand-specific probe (lanes 1–4) and an antisense-strand-specific probe (lanes 5–10): lanes 1 and 5, orange untransformed fruit; lanes 2 and 6, green untransformed fruit; lanes 3 and 8, orange transformed fruit; lanes 4 and 9, green transformed fruit; lane 7, untransformed leaf; lane 10, transformed leaf. Results are expressed as a percentage of PG mRNA level in orange untransformed fruit. (Reprinted by permission from *Nature*, Vol. 334, pp. 724–726. Copyright © 1988 by Macmillan Magazines, Ltd.)

control fruit (19). Subsequently the expression of several ripening-related genes, including pTOM13 (EFE) and pTOM5 (discussed later), was measured in mature-green and ripe-orange fruit of another antisense transformant with 1% PG activity and was found to be normal (C. F. Watson, unpublished).

Pectin breakdown during ripening in normal fruits results from an increase in the amount and a decrease in the molecular weight of soluble polyuronides (33). Plant GR105, which expressed a residual 1% PG activity in the form of PG1, had normal amounts of soluble polyuronide fragments, but pectin depolymerization was largely prevented. The weight average M_r of pectin fragments was determined and found to be similar in green fruit of antisense and normal plants (Fig. 5). As normal ripening progressed, pectin breakdown occurred, resulting in an increase in the solubility of soluble polyuronide fragments but a decrease in their size. In normal and GR102 fruit 14 days after the start of color change (breaker stage), the polyuronide fragment size had

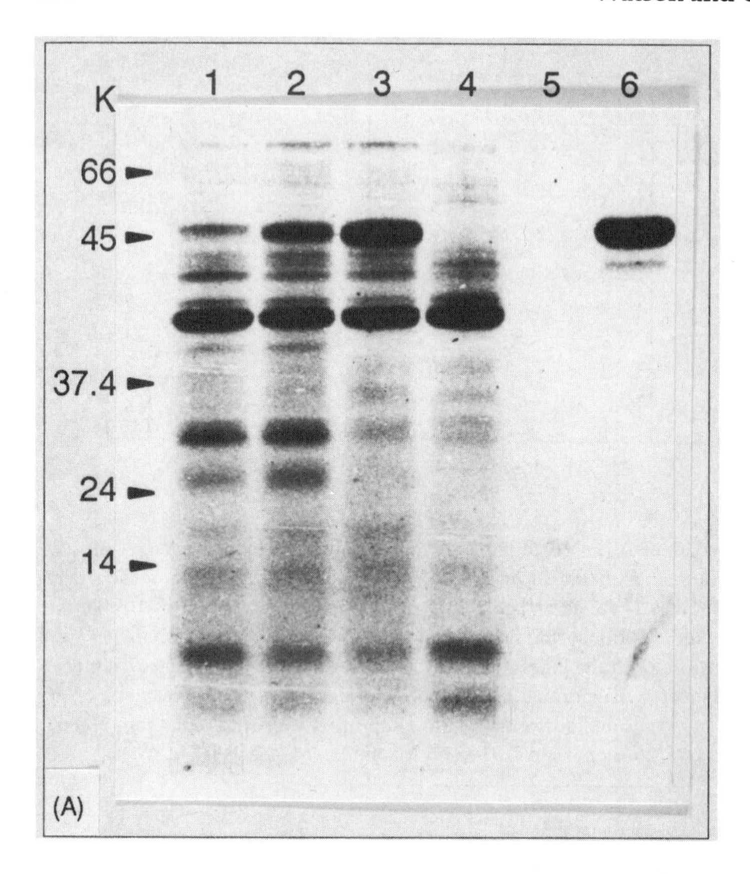

Figure 3 Polygalacturonase enzyme levels and activity during ripening of transformed PG antisense and untransformed fruit. (A) Cell-wall-associated proteins separated by SDS–PAGE: lane 1, ripe GR16; lane 2, ripe GR15; lane 3, ripe control; lane 4, mature green control; lane 5, blank; lane 6, purified PG2. Samples 1–3 were extracted 7 days after the start of ripening. Control fruit (lanes 3, 4) were from plants that had been through the transformation procedure but without antisense gene constructs.

fallen from an M_r of 158,000 at the mature-green stage to 80,000. In contrast, the M_r of fruit from GR95 and GR105 at the same ripening stage was 95,000 and 135,000, respectively.

Analysis of the transgenic fruit with reduced PG showed there was no significant reduction in softening (14,19), clearly indicating that PG is not the major determinant of fruit softening. However, PG antisense fruit are more

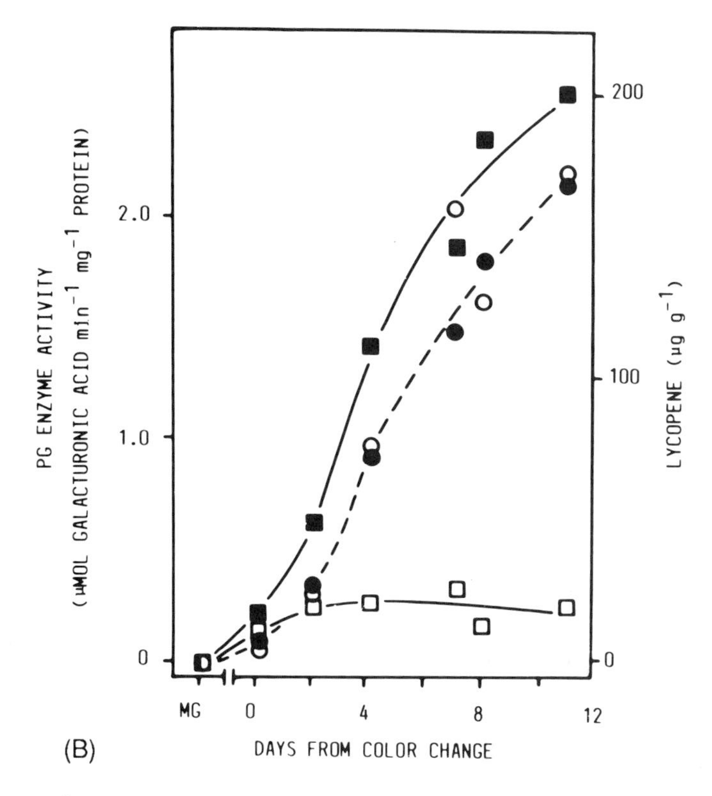

Figure 3 (continued) (B) Polygalacturonase activity of cell-wall-associated proteins and lycopene content during ripening in untransformed (■) and GR16 (□) fruit and lycopene content of untransformed (●) and GR16 (○) fruit. The MG samples were from mature-green fruit. (Reprinted by permission from *Nature*, Vol. 334, pp. 724–726. Copyright © 1988 by Macmillan Magazines, Ltd.)

robust than normal fruit and are less susceptible to splitting. They also have altered processing characteristics owing to the inhibition of pectin depolymerization (20).

C. Antisense EFE Gene in Transgenic Tomato Plants

Ethylene is an important regulator of many plant developmental processes, including fruit ripening and leaf and flower abscission and senescence. Ethylene is also synthesized in response to the wounding of plant tissue by

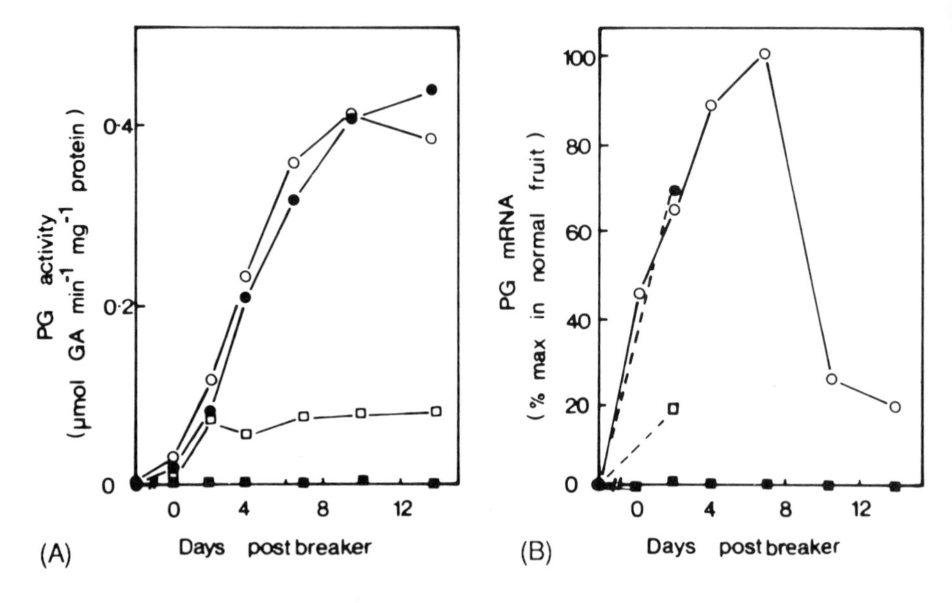

Figure 4 Polygalacturonase activity and mRNA content during ripening of selected GR16 progeny containing zero, one, or two antisense PG genes: untransformed (○); no antisense genes (100% PG) (●); one antisense gene (20% PG) (□); two antisense genes (1% PG) (■). Fruit were tagged at the first visible sign of ripening and samples picked at various stages during ripening. (A) PG activity in cell wall protein extracts. (B) PG mRNA content of pericarp tissue. (Reprinted by permission from *Plant Molecular Biology*, Vol. 14, pp. 369–379. Copyright © 1990 by Kluwer Academic Publishers.)

infection or mechanical stimuli (52,53). Two enzymes are required for ethylene production, 1-aminocyclopropane-1-carboxylic acid (ACC) synthase and EFE. Recently, genes that encode ACC synthase, which catalyzes the conversion of *S*-adenosylmethionine to ACC, have been isolated from several different plant species (54–56). However, the gene encoding EFE, which catalyzes the conversion of ACC to ethylene, proved difficult to isolate, mainly because the protein had never been purified.

The ripening-associated clone pTOM13, isolated from a tomato cDNA library, was found to hybridize to an mRNA whose accumulation was correlated with ethylene evolution during tomato fruit ripening and after wounding of leaves (57,58). Comparison of the pTOM13 nucleic acid sequence and its predicted amino acid sequence with known gene and protein sequences gave

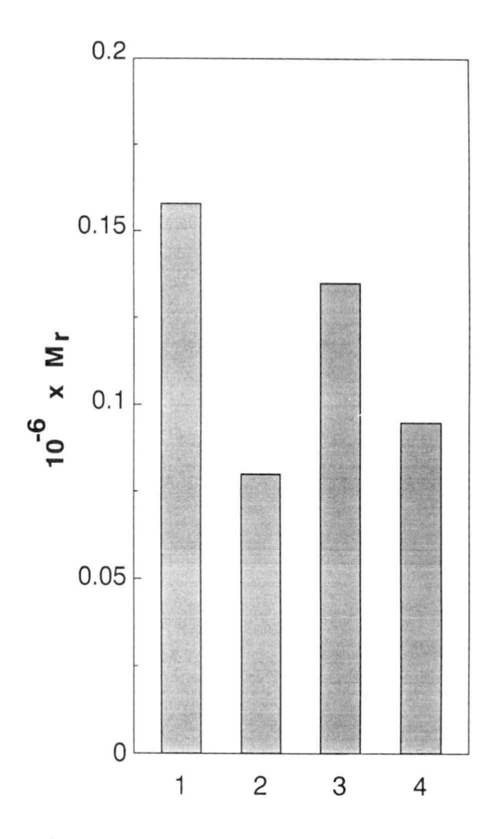

Figure 5 Change in weight average M_r of soluble polyuronides from unripe and ripe fruit of control plants and antisense plants with one (GR95) or two (GR105) antisense genes: 1, untransformed and antisense unripe fruit; 2, untransformed ripe fruit; 3, GR105 ripe fruit (1% PG); 4, GR95 ripe fruit (20% PG). Each point is the mean from three fruits assayed individually. (Reprinted by permission from *Plant Molecular Biology*, Vol. 14, pp. 369–379. Copyright © 1990 by Kluwer Academic Publishers.)

no clue as to the function of pTOM 13. In attempt to characterize the pTOM13 protein gene product, tomato plant stem segments were transformed with Bin19 containing an antisense pTOM13 gene (16). One transformant that constitutively expressed the antisense pTOM13 gene led to down-regulation of pTOM13 mRNA during ripening and after leaf wounding (Fig. 6). Subsequent selfing of this plant led to the production of offspring that had zero, one, or two antisense pTOM13 genes. In plants with two antisense genes, ethylene

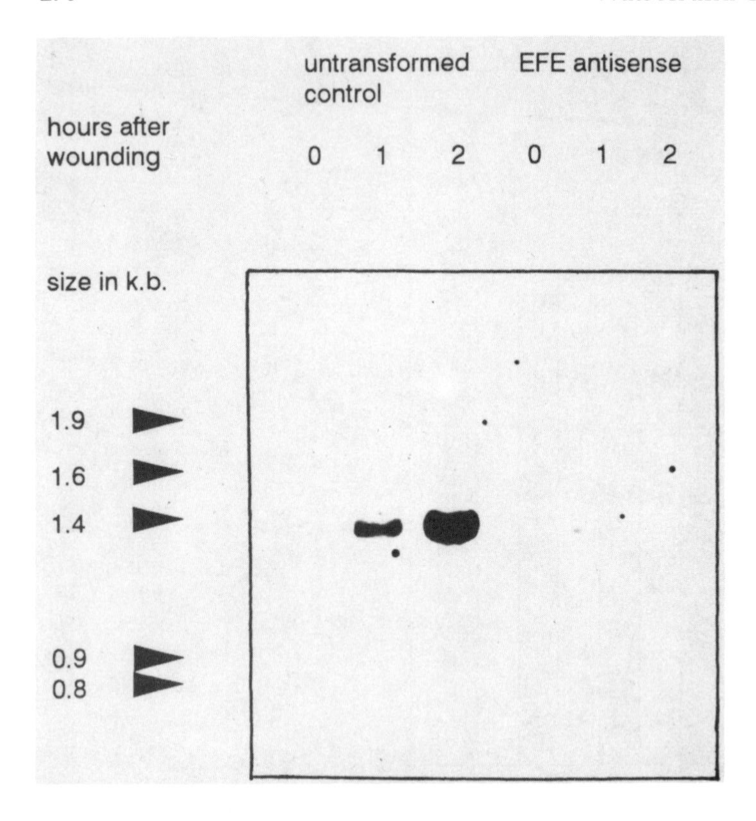

(A)

Figure 6 Expression of EFE (pTOM13) mRNA and antisense EFE RNA in normal and transgenic plants transformed with an antisense pTOM13 gene. (A) EFE (pTOM13) mRNA expression in leaves of normal and antisense plants 0, 1, and 2 hr after wounding. (B) Antisense EFE RNA expression in leaves of normal and antisense plants 0, 1, and 2 hr after wounding. (C) EFE (pTOM13) mRNA expression in normal and antisense tomatoes 0, 1, 3, and 5 days from the onset of the first color change. (Reprinted with permission from *Nature*, Vol. 346, pp. 284–286. Copyright © 1990 by Macmillan Magazines, Ltd.)

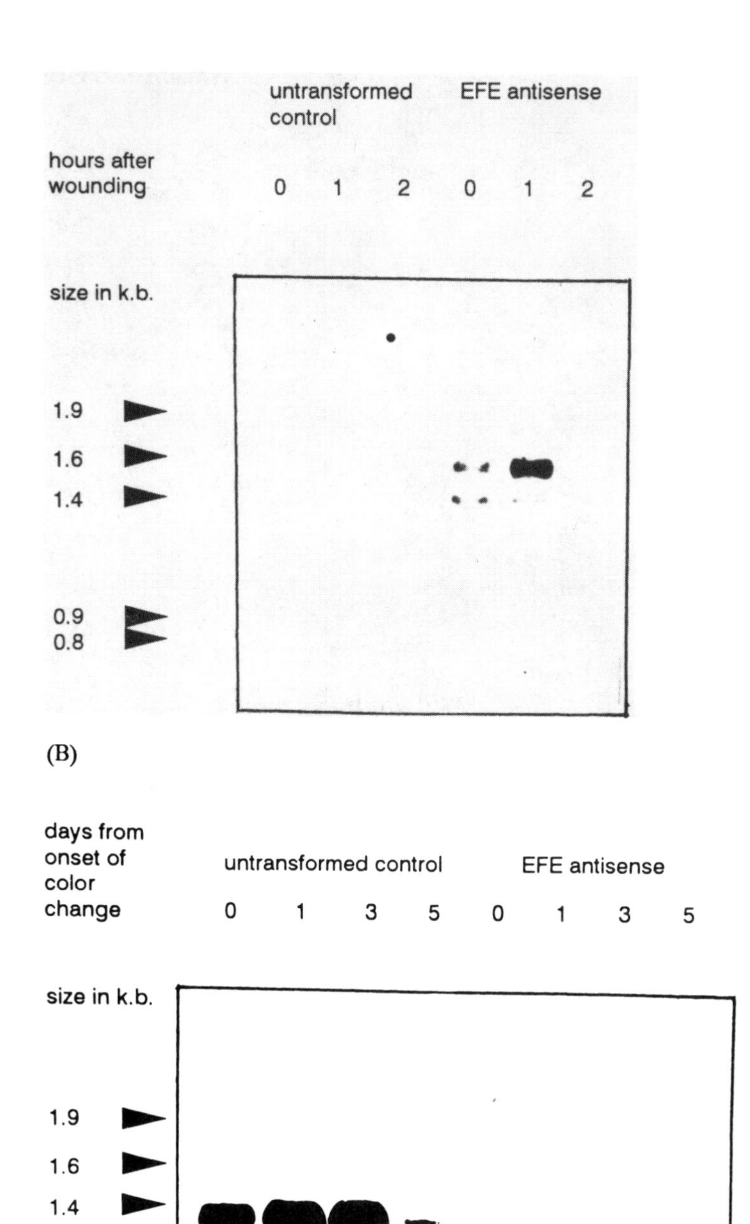
untransformed control
EFE antisense
hours after wounding
0 1 2 0 1 2
size in k.b.
1.9
1.6
1.4
0.9
0.8
days from onset of color change
untransformed control
EFE antisense
0 1 3 5 0 1 3 5
size in k.b.
1.9
1.6
1.4
0.9
0.8

(B)

(C)

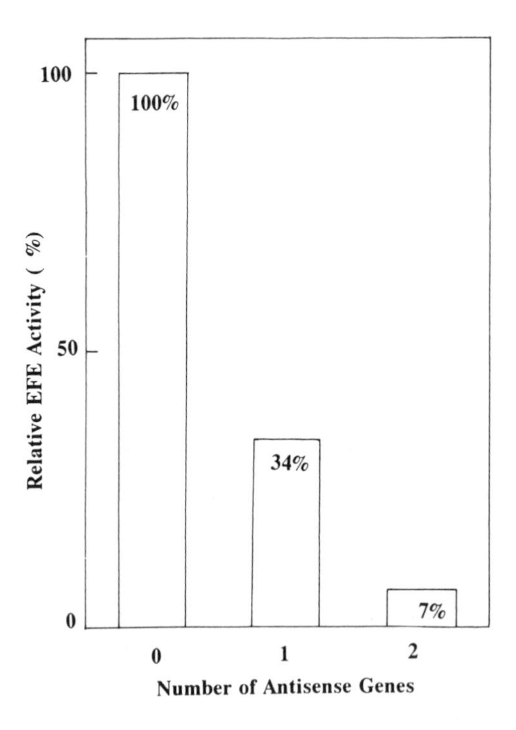

Figure 7 Ethylene-forming enzyme activity in transgenic plants with zero, one, or two pTOM13 antisense genes. (Reprinted with permission from *Nature*, Vol. 346, pp. 284–286. Copyright © 1990 by Macmillan Magazines, Ltd.)

production in detached ripening fruit and after wounding of leaf discs was inhibited by 97% and 90%, respectively. This reduction in ethylene evolution by wounded leaf discs was due to a 93% reduction in EFE activity (Fig. 7). Interestingly, the accumulation of lycopene and the extent of overripening and shriveling of detached fruit was much less in these plants than in control plants. These results strongly suggested that pTOM13 encoded EFE, or part of it, although unequivocal proof was lacking.

Comparing the pTOM13 cDNA with the 5′ sequence of the cloned pTOM13 homologous gene pETH1 (59) and pTOM13 mRNA (60) revealed several cloning artifacts. The corrected coding sequence was subsequently used to transform *Saccharomyces cerevisiae* cells to test pTOM13 gene function. Transformed yeast cells expressed pTOM13 mRNA and converted ACC to ethylene, whereas untransformed cells did not. Further characterization of the EFE activity expressed by transformed yeast has shown that it displays similarities to EFE activity in plants. These similarities include stereoselectivity for ACC analogues, inhibition by cobaltous ions, and sensitivity to iron-chelating agents (60).

D. Identification of a Gene Involved in Carotenoid Biosynthesis

Bird et al. (18) have transformed tomato plants with a previously identified ripening-related tomato clone, pTOM5 (29), in the antisense orientation. Constitutive expression of the antisense pTOM5 gene led to the production of transgenic plants with yellow ripening fruit and pale colored flowers. A dramatic reduction in pTOM5 mRNA during fruit ripening and a greater than 97% reduction in carotenoid levels were associated with this phenotype. In particular, the carotenoid lycopene, which is largely responsible for the red pigmentation in ripening fruit, was reduced to less than 0.1% of the normal level. These results clearly implicate the pTOM5 gene product in tomato fruit carotenoid biosynthesis. Interestingly, there is limited sequence homology between pTOM5 and crtB genes from bacteria involved in carotenoid biosynthesis.

V. MECHANISM OF ACTION OF ANTISENSE RNA AND POSSIBLE INVOLVEMENT OF POSTTRANSCRIPTIONAL DEGRADATION OF RNA IN TRANSGENIC PLANTS

A common mechanism of gene inactivation by antisense RNA is unlikely to occur in prokaryotes and eukaryotes in which antisense RNA is present naturally (1–5), in animal and plant model systems that transiently express antisense RNA (6–8), and in transgenic plants that express antisense RNA from stably introduced antisense genes (10–18). Kim and Wold (7) demonstrated that duplexed sense and antisense RNA did not normally enter the cytoplasm of mouse cells during transient expression of an antisense TK gene, whereas Melton (61) concluded that antisense RNA injected into the cytoplasm of frog oocytes selectively prevented translation of the target globin mRNA. In contrast, Smith et al. (14), van der Krol et al. (15), and Hamilton et al. (16) demonstrated that the down-regulation of the target gene protein product by antisense RNA in transgenic plants was due to a reduction in steady-state levels of the respective sense mRNAs. The mechanism by which this reduction in sense mRNA levels occurs is unknown. A DNA/DNA interaction between the introduced antisense gene and the endogenous gene through homologous recombination competition for *trans*-acting transcription factors could be envisaged. This is unlikely because in cases in which antisense genes have led to the down-regulation of an endogenous gene, the antisense gene has been integrated intact into the genome without any reported rearrangement of the endogenous gene (14–18). Additionally, because in most cases a viral CaMV 35S promoter was used to drive transcription of the introduced antisense genes (14–18), competition for *trans*-acting factors specific for the endogenous gene promoter are extremely unlikely. A DNA/DNA interaction *in trans* between the introduced antisense gene and the endogenous gene or an RNA/DNA interaction between the antisense RNA and the endogenous gene might lead to a reduction in the transcription rate of the endogenous gene. Currently, however, no evidence has been presented to suggest that this occurs in vivo, and, indeed, Sheehy et al. (50) have reported that in an in vitro experiment an antisense PG gene had no effect on the transcription rate of the endogenous PG gene, even though steady-state levels of PG mRNA were greatly reduced. This result may indicate that the reduction in steady-state levels of PG mRNA in ripe fruit of these transgenic plants was not due to reduced transcription of the endogenous PG gene in vivo. An RNA/RNA interaction and subsequent degradation of the RNA duplex seems the most likely explanation for the reduction in steady-state levels of the endogenous sense RNA. A mutual reduction in the amounts of antisense and sense RNA has been noted in several cases (14,16). Furthermore, Smith et al. (62) have described the appearance of possible degradation products of PG mRNA in ripe fruit of a plant constitutively expressing antisense PG RNA. Similarly, Cannon et al. (13) have shown the presence of GUS mRNA degradation products in plants expressing antisense GUS RNA. Double-stranded RNA may be specifically targeted for degradation (63) or, alternatively, may be degraded at a faster rate than single-stranded RNA. It is not clear, however, whether this occurs in the nucleus or the cytoplasm.

Smith et al. (19) showed that in RNA extracted from leaf tissue of transgenic tomato plants with one or two antisense PG genes, there was no correlation between gene copy number and the steady-state level of antisense PG RNA. An excess of antisense RNA does not seem to be required for the efficient down-regulation of gene expression, because low (15) or undetectable (13) concentrations of antisense transcripts have been shown to be equally effective. In addition, Hamilton et al. (16) correlated the almost complete disappearance of antisense RNA with the dramatic reduction in the accumulation of an inducible target mRNA in transgenic tomato plants.

Antisense genes complementary to either the 5′ (13,14) or 3′ (11,28) half of the target mRNA appear to be as effective as those covering the entire coding region of the target gene (15,50). In addition, antisense RNA as short as 41 nucleotides, including 12 bases of 5′ untranslated sequence, has been shown to reduce GUS activity by 95% (13). Significant sequence divergence between the antisense RNA and the target mRNA can also be tolerated and still produce an effect; only 80% sequence homology between an antisense RNA and a target mRNA was required to cause the down-regulation of the target gene (21).

VI. COSUPPRESSION AND POSSIBLE INVOLVEMENT OF ANTISENSE RNA

Surprisingly, a striking reduction in PG enzyme levels and activity in ripening tomato fruit was also attained in plants transformed with a sense PG gene (62) as a result of the reduction in steady-state levels of PG mRNA. Associated with this reduction in endogenous PG mRNA levels in ripe fruit was the mutual gene inactivation of the introduced PG transgene. This coordinate reduction in endogenous and homologous transgene expression has also been reported in petunia transformed with sense CHS (64,65) and DFR (65) genes, but it is not universal. For example, the expression of a full-length peroxidase cDNA in the sense orientation has been shown to increase peroxidase activity in transgenic tobacco (66).

The down-regulation of endogenous genes by sense transgenes was termed cosuppression by Napoli et al. (64). The phenomenon of cosuppression is, however, also a feature of plants transformed with antisense transgenes (13–16). The mechanism of mutual gene inactivation in plants transformed with sense transgenes is unclear. Napoli et al. (64) and Jorgensen (67) favor a model involving trans interaction between homologous DNA sequences, whereas van der Krol et al. (21,65) suggest that "excess" RNA may block transcription of the endogenous gene and the added transgene.

There are, however, a number of striking similarities between the down-regulation of gene expression by sense and antisense transgenes. These include specificity for the target gene (14–16,62,64,65), the subsequent reduction in endogenous mRNA levels (14–16,62,64,65), and the production of similar endogenous mRNA fragments (62). These factors led Grierson et al. (68) to propose that antisense RNA, generated by transcription of the "wrong" DNA strand of the inserted sense transgene, was responsible for the down-regulation of the endogenous gene in these plants. This production of antisense RNA could arise from readthrough transcription into the sense transgene from external chromosomal promoters near the site of T-DNA integration (69) or from an adjacent converging antibiotic resistance gene (62,64,65). It would not be necessary for these transcripts to be full length, because it has been shown that very short sequences are effective at causing down-regulation of gene expression (13). Interestingly, Mol et al. (70) have now reported the detection of low levels of antisense CHS RNA, as well as normal CHS transcripts, in a plant in which CHS expression was down-regulated by cosuppression.

VII. CONCLUSIONS AND PROSPECTS

The unqualified success of the antisense strategy for modifying gene expression and creating single-gene mutations in transgenic plants has been amply demonstrated. Not only has this in vivo reverse genetic approach been used to extend the characterization of known gene products, e.g., PG, but it has also elucidated the function of hitherto unidentified gene products, including pTOM5 (carotenoid biosynthesis; see Ref. 18)) and pTOM13 (EFE; see Ref. 16)). The generation of antisense EFE plants in which ethylene evolution has been drastically reduced may make it possible to study the role of this hormone in plant development. Antisense genes have also been used to modify flower color (15,21,64) and reduce the expression of the RUBISCO small subunit (71) and a 10-kDa protein of the water-splitting apparatus of photosystem II (72).

Although the mechanism of action of antisense RNA in transgenic plants is not fully understood, current evidence suggests that it exerts its effect after transcription. Further evaluation of the mode of action of antisense RNA and the factors that influence the specificity and effectiveness of antisense genes will serve to improve future design of these genes and may also shed light on the role of naturally occurring antisense RNA.

Commercial exploitation of antisense technology is limited so far to crop species amenable to transformation. Consequently, the ability to manipulate the quality and nutritional value of many plant products by antisense RNA depends on identifying efficient methods for stable gene transfer into the genome of these species.

REFERENCES

1. Cesareni, G., Helmer-Citterich, M., and Castagnoli, L. Control of ColE1 plasmid replication by antisense RNA, *Trends Genet., 7*: 230, 1991.

2. Simons, R. W. Natural antisense RNA control in bacteria, phage and plasmids, *Antisense Nucleic Acids and Proteins* (J. N. M. Mol and A. R. van der Krol, eds.), Marcel Dekker, New York, 1991, p. 7.
3. Okano, H., Ikenaka, K., and Mikoshiba, K. Recombination within the upstream gene of duplicated myelin basic protein genes of *myelin deficient shi*^{mld} mouse results in the production of antisense RNA, *EMBO J., 7*: 3407, 1988.
4. Fremeau, R. T., and Popko, B. *In situ* analysis of myelin basic protein gene expression in myelin-deficient oligodendrocytes: antisense hnRNA and readthrough transcription, *EMBO J., 9*: 3533, 1990.
5. Kimelman, D., and Kirschner, M. W. An antisense mRNA directs the covalent modification of the transcript encoding fibroblast growth factor in xenopus oocytes, *Cell, 59*: 687, 1989.
6. Izant, J. G., and Weintraub, H. Inhibition of thymidine kinase gene expression by antisense RNA: a molecular approach to genetic analysis, *Cell, 36*: 1007, 1984.
7. Kim, S. K., and Wold, B. J. Stable reduction of thymidine kinase activity in cells expressing high levels of anti-sense RNA, *Cell, 42*: 129, 1985.
8. Ecker, J. R., and Davis, R. W. Inhibition of gene expression in plant cells by expression of antisense RNA, *Proc. Natl. Acad. Sci. USA, 83*: 5372, 1986.
9. Rothstein, S. J., DiMaio, J., Strand, M., and Rice, D. Stable and heritable inhibition of the expression of nopaline synthase in tobacco expressing antisense RNA, *Proc. Natl. Acad. Sci. USA, 84*: 8439, 1987.
10. Delauney, A. J., Tabeizadeh, Z., and Verma, D. P. S. A stable bifunctional antisense transcript inhibiting gene expression in transgenic plants, *Proc. Natl. Acad. Sci. USA, 85*: 4300, 1988.
11. Sandler, S. J., Stayton, M., Townsend, J. A., Ralston, M. L., Bedbrook, J. R., and Dunsmuir, P. Inhibition of gene expression in transformed plants by antisense RNA, *Plant Mol. Biol., 11*: 301, 1988.
12. Robert, L. S., Donaldson, P. A., Ladaique, C., Altosaar, I., Arnison, P. G., and Fabijanski, S. F. Antisense RNA inhibition of β-glucuronidase gene expression in transgenic tobacco plants, *Plant Mol. Biol., 13*: 399, 1989.
13. Cannon, M., Platz, J., O'Leary, M., Sookdeo, C., and Cannon, F. Organ-specific modulation of gene expression in transgenic plants using antisense RNA, *Plant Mol. Biol., 15*: 39, 1990.
14. Smith, C. J. S., Watson, C. F., Ray, J., Bird, C. R., Morris, P. C., Schuch, W., and Grierson, D. Antisense RNA inhibition of polygalacturonase gene expression in transgenic tomatoes, *Nature, 334*: 724, 1988.
15. van der Krol, A. R., Lenting, P. E., Veenstra, J., van der Meer, I. M., Koes, R. E., Gerats, A. G. M., Mol, J. N. M., and Stuitje, A. R. An antisense chalcone synthase gene in transgenic plants inhibits flower pigmentation, *Nature, 333*: 886, 1988.
16. Hamilton, A. J., Lycett, G. W., and Grierson, D. Antisense gene that inhibits synthesis of the plant hormone ethylene in transgenic plants, *Nature, 346*: 284, 1990.

17. Lagrimini, L. M., Bradford, S., and Rothstein, S. Silencing gene expression in plants, *Oxford Surveys of Plant Molecular Biology*, Vol. 7 (B. J. Miflin, ed.), Oxford University Press, Oxford, England, 1991.
18. Bird, C. R., Ray, J. A., Fletcher, J. D., Boniwell, J. M., Bird, A. S., Teulieres, C., Blain, I., Bramley, P. M., and Schuch, W. Using antisense RNA to study gene function: Inhibition of carotenoid biosynthesis in transgenic tomatoes, *Biotechnology, 9*: 635–639, 1991.
19. Smith, C. J. S., Watson, C. F., Morris, P. C., Bird, C. R., Seymour, G. B., Gray, J. E., Arnold, C., Tucker, G. A., Schuch, W., Harding, S., and Grierson, D. Inheritance and effect on ripening of antisense polygalacturonase genes in transgenic tomatoes, *Plant Mol. Biol., 14*: 369, 1990.
20. Schuch, W., Kanczler, G., Robertson, D., Hobson, G., Tucker, G., Grierson, D., Bright, S., and Bird, C. Fruit-quality characteristics of transgenic tomato fruit with altered polygalacturonase activity. *Hort. Sci. 26*(12): 1517–1520.
21. van der Krol, A. R., Stuitje, A. R., and Mol, J. N. M. Modulation of floral pigmentation by antisense technology, *Antisense Nucleic Acids and Proteins* (J. N. M. Mol and A. R. van der Krol, eds.), Marcel Dekker, New York, 1991, p. 125.
22. Rogers, J. C. Two barley α-amylase gene families are regulated differently in aleurone cells, *J. Biol. Chem., 260*: 3731, 1985.
23. Rogers, J. C. RNA complementarity to α-amylase mRNA in barley, *Plant Mol. Biol., 11*: 125, 1988.
24. Martin, C., Lister, C., Thijs, H., Prescott, A., Jackson, D., and MacKay, S. Transposable elements from *Antirrhinum majus*: their uses in gene isolation and characterization, *Horticultural Biotechnology* (A. B. Bennett and S. D. O'Neill, eds.), Wiley-Liss, New York, 1990, p. 137.
25. Coen, E. S., and Carpenter, R. A semi-dominant allele, niv-525, acts *in trans* to inhibit expression of its wild-type homologue in *Antirrhinum majus, EMBO J., 7*: 877, 1988.
26. van der Krol, A. R., Mur, L. A., de Lange, P., Gerats, A. G. M., Mol, J. N. M., and Stuitje, A. R. Antisense chalcone synthase genes in petunia: visualisation of variable transgene expression, *Mol. Gen. Genet., 220*: 204, 1990.
27. Koes, R. E., Spelt, C. E., and Mol, J. N. M. The chalcone synthase multigene family of *Petunia hybrida* (V30): differential, light-regulated expression during flower development and UV light induction, *Plant Mol. Biol., 12*: 213, 1989.
28. van der Krol, A. R., Mur, L. A., de Lange, P., Mol, J. N. M., and Stuitje, A. R. Inhibition of flower pigmentation by antisense CHS genes: promoter and minimal sequence requirements for the antisense effect, *Plant Mol. Biol., 14*: 457, 1990.
29. Grierson, D., Maunders, M. J., Slater, A., Ray, J., Bird, C. R., Schuch, W., Holdsworth, M. J., Tucker, G. A., and Knapp, J. E. Gene expression during tomato ripening, *Philos. Trans. R. Soc. Lond. [Biol.], 314*: 399, 1986.
30. Hobson, G. E. The firmness of tomato fruit in relation to polygalacturonase activity, *J. Hort, Sci., 40*: 66, 1965.

31. Themmen, A. P. N., Tucker, G. A., and Grierson, D. Degradation of isolated tomato cell walls by purified polygalacturonase *in vitro, Plant Physiol., 69*: 122, 1980.
32. Crookes, P. R., and Grierson, D. Ultrastructure of tomato fruit ripening and the role of polygalacturonase isoenzymes in cell wall degradation, *Plant Physiol., 72*: 1088, 1983.
33. Seymour, G. B., Harding, S. E., Taylor, A. J., Hobson, G. E., and Tucker, G. A. Polyuronide solubilization during ripening of normal and mutant tomato fruit, *Phytochemistry, 26*: 1871, 1987.
34. Tucker, G. A., Robertson, N. G., and Grierson, D. Changes in polygalacturonase isoenzymes during the ripening of normal and mutant tomato fruit, *Eur. J. Biochem., 112*: 119, 1980.
35. Tucker, G. A., and Grierson, D. Synthesis of polygalacturonase during tomato fruit ripening, *Planta, 155*: 64, 1982.
36. Grierson, D., Purton, M. E., Knapp, J. E., and Bathgate, B. Tomato ripening mutants, *Developmental Mutants in Higher Plants* (H. Thomas and D. Grierson, eds.), Cambridge University Press, Cambridge, England, 1987, p. 73.
37. Grierson, D., and Tucker, G. A. Timing of ethylene and polygalacturonase synthesis in relation to the control of tomato fruit ripening, *Planta, 157*: 174, 1983.
38. Ali, Z. M., and Brady, C. J. Purification and characterisation of the polygalacturonases of tomato fruits, *Aust. J. Plant Physiol., 9*: 155, 1987.
39. Bird, C. R., Smith, C. J. S., Ray, J. A., Moureau, P., Bevan, M. W., Bird, A. S., Hughes, S., Morris, P. C., Grierson, D., and Schuch, W. The tomato polygalacturonase gene and ripening-specific expression in transgenic plants, *Plant Mol. Biol., 11*: 651, 1988.
40. Grierson, D., Tucker, G. A., Keen, J., Ray, J., Bird, C. R., and Schuch, W. Sequencing and identification of a cDNA clone for tomato polygalacturonase, *Nucleic Acids Res., 14*: 8595, 1986.
41. DellaPenna, D., Alexander, D. C., and Bennett, A. B. Molecular cloning of tomato fruit polygalacturonase: analysis of polygalacturonase mRNA levels during ripening, *Proc. Natl. Acad. Sci. USA, 83*: 6420, 1986.
42. Lincoln, J. E., Cordes, S., Read, E., and Fischer, R. L. Regulation of gene expression by ethylene during *Lycopersicon esculentum* (tomato) fruit development, *Proc. Natl. Acad. Sci. USA, 84*: 2793, 1987.
43. Sheehy, R. E., Pearson, J., Brady, C. J., and Hiatt, W. R. Molecular characterisation of tomato fruit polygalacturonase, *Mol. Gen. Genet., 208*: 30, 1987.
44. DellaPenna, D., Kates, D. S., and Bennett, A. B. Polygalacturonase gene expression in Rutgers, *rin, nor* and *Nr* tomato fruits, *Plant Physiol., 85*: 502, 1987.
45. Knapp, J., Moureau, P., Schuch, W., and Grierson, D. Organisation and expression of polygalacturonase and other ripening related agents in Ailsa Craig Neverripe and Ripening inhibitor tomato mutants, *Plant Mol. Biol., 12*: 105, 1989.
46. Mutschler, M., Guttieri, M., Kinzer, S., Grierson, D., and Tucker, G. A. Changes in ripening related processes in tomato conditioned by the alc mutant, *Theor. Appl. Genet., 76*: 285, 1988.

47. Guilley, H., Dudley, R. K., Jonard, G., Balazs, E., and Richards, K. E. Transcription of cauliflower mosaic virus: detection of promoter sequences and characterisation of transcripts, *Cell, 30*: 763, 1982.
48. Depicker, A., Stachel, S., Dehaese, P., Zambryski, P., and Goodman, H. M. Nopaline synthase: transcript mapping and DNA sequence, *J. Mol. Appl. Genet., 1*: 561, 1982.
49. Bevan, M. W. Binary *Agrobacterium* vectors for plant transformation, *Nucleic Acids Res., 12*: 8711, 1984.
50. Sheehy, R. E., Kramer, M., and Hiatt, W. R. Reduction of polygalacturonase activity in tomato fruit by antisense RNA, *Proc. Natl. Acad. Sci. USA, 85*: 8805, 1988.
51. Visser, R. G. F., Feenstra, W. J., and Jacobsen, E. Manipulation of granule-bound starch synthase activity and amylose content in potato by antisense genes, *Antisense Nucleic Acids and Proteins* (J. N. M. Mol and A. R. van der Krol, eds.), Marcel Dekker, New York, 1991, p. 141.
52. Abeles, F. B. *Ethylene in Plant Biology*, Academic Press, New York, 1973.
53. Yang, S. F., and Hoffman, N. E. Ethylene biosynthesis and its regulation in higher plants, *Annu. Rev. Plant Physiol., 35*: 155, 1984.
54. Van der Straeten, D., Van Wiermeersh, L., and Goodman, H. M. Cloning and sequence of two different cDNAs encoding 1-aminocyclopropane-1-carboxylate synthase in tomato, *Proc. Natl. Acad. Sci. USA, 87*: 4859, 1990.
55. Nakajima, N., Mori, H., Yamazaki, K., and Imaseki, H. Molecular cloning and sequence of a complementary DNA encoding 1-aminocyclopropane-1-carboxylate synthase induced by tissue wounding, *Plant Cell Physiol., 31*: 1021, 1990.
56. Sato, T., Oeller, P. W., and Theologis, A. The 1-aminocyclopropane-1-carboxylate synthase of *Cucurbita, J. Biol. Chem., 266*: 3752, 1991.
57. Smith, C. J. S., Slater, A., and Grierson, D. Rapid appearance of an mRNA correlated with ethylene synthesis encoding a protein of molecular weight 35000, *Planta, 168*: 94, 1986.
58. Holdsworth, M. J., Schuch, W., and Grierson, D. Organisation and expression of a wound/ripening-related small multigene family from tomato, *Plant Mol. Biol., 11*: 81, 1988.
59. Köck, M., Hamilton, A. J., and Grierson, D. *eth1*, a gene involved in ethylene synthesis in tomato, *Plant Mol. Biol., 17*: 141, 1991.
60. Hamilton, A. J., Bouzayen, M., and Grierson, D. Identification of a tomato gene for the ethylene forming enzyme by expression in yeast, *Proc. Natl. Acad. Sci. USA, 88*: 7434, 1991.
61. Melton, D. A. Injected anti-sense RNAs specifically block messenger RNA translation *in vivo, Proc. Natl. Acad. Sci. USA, 82*: 144, 1985.
62. Smith, C. J. S., Watson, C. F., Bird, C. R., Schuch, W., and Grierson, D. Expression of a truncated tomato polygalacturonase gene inhibits expression of the endogenous gene in transgenic plants, *Mol. Gen. Genet., 224*: 477, 1990.
63. Meegan, J. M., and Marcus, P. I. Double-stranded ribonuclease coinduced with interferon, *Science, 244*: 1089, 1989.

64. Napoli, C., Lemieux, C., and Jorgensen, C. Introduction of a chimeric chalcone synthase gene into petunia results in reversible co-suppression of homologous gene *in trans, Plant Cell, 2*: 279, 1990.
65. van der Krol, A. R., Mur, L. A., Beld, M., Mol, J. N. M., and Stuitje, A. Flavenoid genes in petunia: addition of a limited number of gene copies may lead to a suppression of gene expression, *Plant Cell, 2*: 291, 1990.
66. Lagrimini, L. M., Bradford, S., and Rothstein, S. Peroxidase-induced wilting in transgenic plants, *Plant Cell., 2*: 7, 1990.
67. Jorgensen, R. Altered gene expression in plants due to *trans* interactions between homologous genes, *Trends Biotechnol., 8*: 340, 1990.
68. Grierson, D., Fray, R. G., Hamilton, A. J., Smith, C. J. S., and Watson, C. F. Does co-suppression of sense transgenes involve antisense RNA?, *Trends Biotechnol., 9*: 122, 1991.
69. Herman, L., Jacobs, A., Van Montagu, M., and Depicker, A. *Mol. Gen. Genet., 224*: 248, 1990.
70. Mol, J., Van Blokland, R., and Kooter, J. More about co-suppression, *Trends Biotechnol., 9*: 182, 1991.
71. Rodermel, S. R., Abbott, M. S., and Bogorad, L. Nuclear–organelle interactions: nuclear antisense gene inhibits ribulose biphosphate carboxylase enzyme levels in transformed tobacco plants, *Cell, 55*: 673, 1988.
72. Stockhaus, J., Höfer, M., Renger, G., Westhoff, P., Wydrzynski, T., and Willmitzer, L. Anti-sense RNA efficiently inhibits formation of the 10 kd polypeptide of photosystem II in transgenic potato plants: analysis of the role of the 10 kd protein, *EMBO J., 9*: 3013, 1990.

14

Postharvest Evaluation of Transgenic Tomatoes with Reduced Levels of Polygalacturonase: Processing, Firmness, and Disease Resistance

Matthew Kramer and Rick A. Sanders

Calgene Fresh, Inc., Davis, California

Hassan Bolkan and Curtis M. Waters

Campbell Institute for Research and Technology, Davis, California

Raymond E. Sheehy and William R. Hiatt

Calgene Fresh, Inc., Davis, California

I. INTRODUCTION

The ripening of tomato fruit is characterized by a series of coordinated biochemical changes that result in the color, flavor, and texture of ripe tomato fruit. Among these is the solubilization of the pectin fraction of fruit cell walls. Pectin is a major component of the middle lamella, which serves as an interface between adjacent cells. Polygalacturonase (PG) is the major enzyme involved in pectin metabolism during fruit ripening and has been associated with cell wall breakdown, fruit softening, and loss of tissue integrity during ripening (1–3). Several lines of evidence support the role of PG in pectin degradation and consequent fruit softening. Large amounts of PG enzyme activity accumulate specifically during fruit ripening (4,5), resulting from the de novo synthesis of PG mRNA and protein (2,5–7). Several ripening mutations, including *rin* (ripening inhibitor), *nor* (nonripening), and *Nr* (Neverripe), have been described and are characterized by dramatically reduced levels of PG activity

(8,9) and extended shelf life. Finally, PG added to fruit cell wall preparations in vitro can induce ultrastructural changes similar to those observed during ripening (10).

Although studies aimed at defining the role of PG in tomato fruit ripening have been aided by the availability of ripening mutants, these mutations exert pleiotropic effects on the overall ripening process and do not allow a precise role for PG activity to be defined. Several groups have isolated cDNA and genomic PG clones (7,11–13) and have used these sequences to specifically manipulate PG gene expression (13–16). *Agrobacterium*-mediated transformation of tomato with PG antisense constructs has resulted in the development of lines that produce ripe fruit with dramatically reduced levels of PG activity (14–16). Transformation of plants containing the *rin* mutation with a functional PG cDNA resulted in the expression of PG activity in fruit of this mutant genotype (13). Evaluation of genetically engineered fruit with altered levels of PG activity can more clearly elucidate the role of PG in tomato fruit ripening and its effects on pectin-related characteristics and fruit quality.

We have used transformation with PG antisense gene constructs to develop tomato lines from several different varieties that have fruit expressing less than 1% of normal PG levels. These lines have been evaluated for effects on processing and fresh market tomato fruit products in a series of field trials held during 1989 and 1990. The predominant role of PG during fruit ripening is thought to be pectin metabolism (4). Because pectin is an important structural component of the fruit cell wall and middle lamella, fruit with reduced levels of PG due to antisense construct transformation may be altered with regard to a variety of parameters affecting fresh fruit and processed product quality. Here, we evaluate the effect of reduced PG levels on the product quality of processed juice and paste, firmness of fresh fruit after storage, and fruit susceptibility to fungal pathogens.

II. EXPERIMENTAL PROTOCOL

A. Plant Material

Tomato genotypes used for line development included UC82B, a processing variety, and X39, 83T, and Rutgers, all fresh market varieties. The inbred parents, X39 and 83T, were obtained from the Campbell Institute of Research and Technology (CIRT). Transgenic lines of each genotype were generated by *Agrobacterium*-mediated transformation with construct pCGN1416 (15), which contains a full-length PG cDNA (12) expressed in the antisense orientation from the cauliflower mosaic virus (CaMV) 35S promoter. Approximately 50 independent transformation events were generated for line selection from each variety. Homozygous progeny of selected transformants, identified by germination on media containing kanamycin, were produced in the greenhouse (processing) or field (fresh market) and were used to produce seed for further field testing. Tomato genotypes used for disease testing included cv. Ailsa craig expressing the *Nr* mutation (17) and Ailsa craig *Nr* transformed with construct pCGN1415. Construct pCGN1415 is identical to pCGN1416 (15) except that the full-length PG cDNA is in the sense orientation relative to the CaMV 35S promoter for expression of functional PG mRNA and enzyme activity.

B. Fungal Cultures

Cultures of *Botrytis cinerea, Rhizopus stolonifer*, and *Geotrichum candidum* were isolated from naturally infected tomato fruit and were maintained on potato-dextrose agar (PDA).

C. Yolo County, Calif., Trial (USDA Permit 88-344-07)

The Yolo field trial was performed on land leased from a commercial tomato producer in Yolo County near Woodland, Calif. The experimental design was a randomized complete block with all entries replicated five times within the field. Processing tomatoes were direct seeded in early May 1989. Plots consisted of single rows 40 ft (12.2 m) long and 60 in. (1.52 m) apart. At the third-true-leaf stage all processing plots were thinned to a normal commercial stand density of 6 in. (15.2 cm) per plant. Fresh market tomatoes were started as transplants in the greenhouse at the Calgene facility in Davis, Calif., and were transplanted to the field 30 days later, in early June 1989. Plants were transplanted in single rows 40 ft (12.2 m) long and 60 in. (1.52 m) apart with 18 in. (45.7 cm) between plants.

For processing analysis, 40-lb (10.2-kg) samples of ripe fruit from each replication of each treatment were harvested by hand when the plots had attained 95% red fruit. Fruit was processed at 185°F (85°C) and used to produce three #10 cans of both pulp and paste from each replication of each treatment for analysis. Paste was produced by evaporation of pulp and adjustment to 10.8° Brix. Laboratory analysis was carried out at the CIRT laboratory facilities in Davis, Calif. (18).

For evaluation of firmness, T_2 segregating populations of fresh market lines X39 16-11 and 83T 16-34 in the Yolo trial were progeny tested to identify plants homozygous for the antisense PG gene. Fruit of identified homozygous plants and nontransgenic controls was harvested by hand at mature-green, pink, and red stages, and approximately 100 lb (45.4-kg) of fruit for each line was sorted over a commercial packing line (DiMarie Bros., Newman, Calif.) and packed into 25-lb (11.4-kg) boxes. Because of rain late in the harvest, only red fruit from line X39 16-11 was determined to be of sufficient quality to withstand the packing process. Green fruit was subjected to ethylene treatment at 65°F (18.3°C) for a period of 5 days, after which it was stored at 65°F (18.3°C) and 65% relative humidity for an additional 5 days. Pink and red fruit were stored for 10 days after packing and without ethylene treatment. After storage, the fruit was shipped to the Calgene facility at Davis, Calif., and a subset was evaluated for firmness by using a flat-plate deformation meter. Fruit softening was measured by compressing 10 fruit from each treatment under a 500-g weight for 15 sec (19,20). The mean compression values were analyzed by Student's t test to determine the significance of their differences. Additional fruit from identified homozygous plants was collected for seed increase and transplant production for planting in the Ruskin, Fla., trial.

Tests for infectivity by each fungus involved four replicates of 25 fruit per genotype for each treatment. Fruit were washed and sorted for uniformity, surface sterilized in 0.52% sodium hypochlorite, rinsed twice in deionized water, and air dried overnight at room temperature. Material used for infection by *B. cinerea* was chilled overnight at 4°C before inoculation. Three 2-mm-deep wounds were made near the stem end of each fruit with the tip of a 2-mm-diameter nail. Each wound was inoculated with 15 μl of conidia suspensions (1×10^5 spores/ml) obtained from cultures grown on PDA for 6–15 days under continuous fluorescent light. After inoculation, fruit were incubated at room temperature for 4–7 days on wire mesh in produce crispers containing water to ensure high humidity. The results are reported as lesion area on the surface of each fruit.

D. Ruskin, Fla., Trial (USDA Permit 89-320-01)

The Florida trial was carried out in cooperation with Voegle Farms of Ruskin, Fla., at their Hillsborough County production site. Transplants were sown in early January 1990 at the Sun City, Fla., production facility of Speedling, Inc., and were planted to the field in mid-February 1990. Lines selected from the Yolo trial were planted in a randomized complete block design with each treatment and nontransgenic control replicated three times within the field.

Plots consisted of 40 plants spaced 28 in. (71.1 cm) apart in single rows 92 ft (28.0 m) long and 60 in. (152 cm) apart.

At 30% coloration of the field, four 25-lb (11.4-kg) boxes of fruit from each stage (mature-green, pink, and red) and from each replication of each treatment were harvested by hand and packed. One 25-lb (11.4-kg) box of each stage of each treatment was shipped to the Calgene facility in Davis, Calif., under USDA- approved conditions. The remaining boxes were transported to a commercial packing house (Stake Tomato of Ruskin, Fla.), where green fruit was exposed to ethylene gas at 65°F (18.3°C) and 65% relative humidity for a period of 4 days. After ethylene treatment, the fruit were transferred to a holding room and stored an additional 6 days at 65°F (18.3°C) and 65% relative humidity. Pink and red fruit were stored for 10 days at 65°F (18.3°C) and 65% relative humidity without ethylene treatment. The fruit were then transported to the field, evaluated for firmness after 10 days of storage as already described, and disposed of on site.

E. Winters, Calif., Trial (USDA Permit 90-019-01)

The 1990 Winters field trial was performed on land leased from a commercial tomato producer in Yolo county near Winters, Calif. The experimental design was a randomized complete block design with all entries replicated five times within the field. Fresh market transplants were sown in the greenhouses at the Calgene facility in Davis, Calif., and transplanted to the field 30 days later, in late May 1990. Plants were transplanted in a single row 40 ft (12.2 m) long with 18 in. (45.7 cm) between plants and 60 in. (1.52 m) between rows.

At 40% coloration in the field, approximately 300 lb (136-kg) of fruit per entry was harvested by hand, sorted over a commercial packing line (DiMarie Bros., Newman, Calif.), and packed into 25-lb (11.4-kg) boxes. Green fruit was treated with ethylene at 56–60°F (10–15.6°C) and 65% relative humidity for a period of 5 days, after which it was stored at 56–60°F (10–15.6°C) for 9 days more. A subset of green fruit was stored for an additional 20 days under the conditions described. Red fruit was stored for 14 days after packing without ethylene treatment. After storage, fruit was returned to the Calgene facility in Davis, Calif., and a subset of the fruit was evaluated for firmness as already described.

F. Analytical Methods

Polygalacturonase assays were performed as previously described (15), with the following modifications. Following centrifugation of initial homogenates, the supernatant was precipitated with 80% ammonium sulfate and the protein pellet was suspended in 150 m*M* NaCl, 10 m*M* sodium acetate, and 1 m*M* dithiothreitol; it was then dialyzed against the same buffer. Enzyme preparations were then further concentrated 10-fold by lyophilization and resuspension in water. Polygalacturonase activity was determined by using 3,5-dintrosalicylate to measure reducing sugars (21). One unit of PG activity was determined as an increase in absorbance at 540 nm of 0.01 per minute. Protein concentrations were measured by using Bradford reagent (Bio-Rad Laboratories) with bovine serum albumin as a standard. RNA preparations, gel blotting, and hybridization were performed as previously described (15).

G. Statistical Methods

Data on processed pulp and paste and on lesion area were compared by analysis of variance and Duncan's multiple range test (22). Fruit firmness data were analyzed by the student's t test method.

III. RESULTS

A. Tomato Line Development

Tomato lines were transformed with the PG antisense construct pCGN1416 (15) and were selected for further evaluation on the basis of whole-plant morphology, reduced levels of PG activity in ripe fruit, and segregation ratios of the kan^r gene, which indicates a single functional insertion event. A summary of line selection and development is shown in Table 1. All transformed lines had approximately 99% or greater reductions in PG activity. Processing tomato lines were derived from transformation of UC82B cv. mill. Line 16-1 was originally tested in the winter of 1989 in a field trial carried out in Guasave, Sinaloa, Mexico (18), and again in a Yolo County, Calif., field trial during the summer of 1989. The Yolo trial contained 14 additional lines of UC28B transformed with pCGN1416, from which two lines (16-14 and 16-44) were selected for further analysis.

Fresh market tomato lines were derived from transformation of the varieties Rutgers, 83T, and X39. A progeny test of T_2 segregating populations of the three transformed varieties was performed in the Yolo trial. A total of 26 lines, each representing a separate transformation event, were screened, and four lines were selected on the basis of the aforementioned criteria. This included one line each from Rutgers (16-21) and 83T (16-34) and two from X39 (16-11 and 16-51). Individual homozygous progeny from these four selected lines were identified and advanced for evaluation in a winter 1990 field trial in

Table 1 Transformed Tomato Lines Evaluated in Field Trials

Variety	Line	PG activity[a] (unit/mg protein)	Trial (generation)[b]
UC82B	Control	62.9	Yolo (T_3)
	16-1	0.37	
	16-14	0.26	
	16-44	0.19	
Rutgers	Control	28.1	Florida (T_3)
	16-21	0.02	
X39	Control	38.8	Yolo (T_2), Florida (T_3)
	16-11	0.11	
	16-51	0.06	
83T	Control	21.6	Yolo (T_2), Florida (T_3), Winters (T_3)
	16-34	0.31	

[a]PG activity was measured as described in Section II.
[b]T_1 generation represents the initial transformation event, T_2 is the selfed progeny of the original transformant, and T_3 represents the selfed progeny of T_2 individuals selected as homozygous for the antisense PG gene.

Ruskin, Fla. In addition, 83T 16-34 was further evaluated in a Winters, Calif., trial in the summer of 1990.

B. Influence of Reduced PG Levels on Processed Juice and Paste

Processed juice from 16-1 fruit grown in the Guasave, Mexico, trial (18) had higher viscosity and better consistency than juice from UC82B controls. To extend this analysis, fruit from lines 16-1, 16-14, and 16-44 were harvested from the Yolo trial and were processed into juice as described. A subset of the juice samples were adjusted by evaporation to 10.8° Brix to produce paste samples from each replication. Both juice and paste were subjected to standard processing analyses, and the results are summarized in Table 2. At a processing temperature of 185°F (85°C), no significant differences were observed between transformed and control fruit in any processing characteristic except serum viscosity (Ostwald); a significant increase in viscosity was observed in fruit from all three lines with reduced PG activity. Transgenic fruit with

Table 2 Processing Characteristics of Transformed and Control Tomato Fruit[a]

UC82B selection	Juice[b]							Paste[b]	
	Total solids	Soluble solids	pH	TA	Color	Ostwald (sec)	Bostwick (cm)	Ostwald (sec)	Bostwick (cm)
16-1									
Mean	6.430	5.780	4.41	4.84	1.92	384.2[c]	11.52	339[c]	5.50
STD	0.219	0.239	0.02	0.22	0.03	49.7	0.44		
16-14									
Mean	6.234	5.620	4.41	4.61	1.94	374.2[c]	12.94	310[c]	5.26
STD	0.960	0.912	0.05	0.41	0.11	110.5	2.35		
16-44									
Mean	6.22	5.60	4.37	4.96	1.96	431.2[c]	12.34	351[c]	5.04
STD	0.90	0.89	.070	0.47	0.04	101.5	1.63		
Control									
Mean	6.11	5.50	4.37	4.96	1.94	227.2	14.52	205	5.88
STD	0.72	0.71	0.02	0.47	0.07	61.5	1.69		
LSD	n.s.	n.s.	n.s.	n.s.	n.s.	84.2	n.s.	46.4	n.s.

[a]Abbreviations: STD, standard deviation; LSD, least significant difference ($P < 0.05$); n.s., not significant.
[b]Juice and paste were prepared as described in Section II. Processing characteristics were measured as described in Ref. 18.
[c]Significant at 95% ($P < 0.05$).

reduced PG activity did not exhibit any of the pleiotropic effects normally associated with the ripening mutations. Ripening of the fruit was not delayed, and color development was normal. No differences were observed in levels of the glycoalkaloid tomatine or β-carotene (provitamin A) and ascorbic acid (vitamin C) between transgenic and control fruit (data not shown).

C. Effects of Reduced PG on Fruit Softening During Storage

Table 3 contains the combined data for firmness evaluations from three separate trials, and the results of each trial are discussed separately. In the Yolo trial, fruit at both the green and pink stages from transformed lines 16-11 and 16-34 were significantly firmer than nontransgenic controls. In addition, 16-11 fruit that were harvested red were firmer than X39 control fruit harvested either green or pink. These fruit were purposely sorted over a commercial packing line to simulate industrial handling and storage.

Homozygous lines of the three fresh market genotypes were further evaluated for firmness in the Florida trial to confirm the results obtained in the Yolo trial with a larger sample of fruit. When approximately 30% of the fruit in the field had attained color, fruit at the green, pink, and red stages were harvested from each line, packed by hand in the field, and stored as described in Section II. After storage, fruit firmness was again compared with that of nontransgenic controls. The values shown in Table 3 for each line represent the mean of three replications of 15 individual measurements each. Except for one stage of one genotype (16-11 and 16-51, harvested green), the transgenic fruit were again significantly firmer than the nontransgenic controls at all stages of harvest after 10 days of storage. In particular, red fruit harvested from transformed lines were as firm as their nontransformed counterparts harvested green and treated with ethylene. Although compression values were within a narrower range than in the Yolo trial, the differences in firmness were easily discernible to the touch.

Line 16-34 was again evaluated in the Winters trial to confirm the effect of the antisense PG gene on fruit firmness and quality. When 40% of the fruit in the field had attained color, fruit was harvested across replications, again sorted over a commercial packing line for size and color (green and red), and packed into 25-lb (11.4-kg) boxes. All fruit was treated as previously described, and two replications of 15 fruit from each stage packed were measured for firmness 14 days after storage. Significant differences were observed between 16-34 fruit and nontransformed controls for both stages tested (Table 3). In addition,

Table 3 Evaluation of Tomato Fruit Firmness after Storage[a]

	Fruit compression[b]					
	Green stage		Pink stage		Red stage	
Trial and Line	X	t	X	t	X	t
Yolo						
X39						
16-11	1.18	6.18[c]	1.06	15.53[c]	1.31	
Control	2.07		2.52		n.d.	
83T						
16-34	1.85	8.36[c]	1.78	5.03[c]	n.d.	
Control	3.69		3.14			
Florida						
X39						
16-11	0.44	n.s.	0.65	4.00[c]	0.64	11.00[c]
16-51	0.46	n.s.	0.66	3.98[c]	0.77	9.14[c]
Control	0.57		0.93		1.41	
83T						
16-34	0.79	14.41[c]	1.03	3.38[d]	1.09	6.16[c]
Control	1.28		1.27		2.20	
Rutgers						
16-21	1.40	9.65[c]	1.32	14.56[c]	1.79	5.66[c]
Control	1.95		2.15		2.47	
Winters						
83T						
16-34	0.98	2.37[d]			1.71	3.03[c]
Control	1.42				2.41	

[a]Abbreviations: n.s., not significantly different; n.d., not determined owing to lack of fruit from the field.
[b]Deformation analysis was performed as described in Section II, and results are reported as the mean deformation (X) in millimeters. The t values reported are calculated values of t for determination of differences between means (22).
[c]Significant at 99% ($P < 0.01$).
[d]Significant at 95% ($P < 0.05$).

as a measure of the ability of the 16-34 fruit to remain intact for an extended period, fruit that had been harvested green and treated with ethylene were allowed to remain in storage for a total of 25 days. As shown in Fig. 1, the overall quality of 16-34 fruit was substantially higher than that of the 83T control material.

D. Effects of Reduced PG on Susceptibility to Fungal Pathogens

A component of improved fruit quality appeared to be a lower incidence of fungal infection (Fig. 1). An attempt was made to correlate levels of PG activity in fruit to the severity of infection by specific fungi, particularly fungi

Figure 1 Quality of transgenic and control fresh market tomatoes after extended storage. Mature green 83T control (left) and 83T 16-34 (right) fruit from the Winters trial were harvested and packed into 25-lb (11.4-kg) boxes, treated with ethylene, and stored for a total of 25 days as described in Section II. After storage, fruit were removed from randomly selected boxes for comparison as shown.

that preferentially infect ripe fruit, in which PG is normally abundant. Spores of two such fungi, *Rhizopus stolonifer*, the causal agent of rhizopus soft rot, and *Geotrichum candidum,* the causal agent of sour rot, were used in laboratory experiments to inoculate ripe fruit from the Yolo trial. *Botrytis cinerea*, the causal agent of gray mold, was used as a control because of its ability to infect green fruit, a stage at which PG is not expressed. Infection of 16-1 fruit by *B. cinerea* and *R. stolonifer* was compared with infection of fruit from nontransformed UC28B and a line [16-1 (-/-)] that had lost the antisense PG gene through segregation. No differences were observed in the area of lesions caused by *B. cinerea* after inoculation (data not shown). In the case of *R. stolonifer* (Fig. 2A), the lesion area of infected 16-1 fruit was significantly smaller than that of either nontransformed UC82B controls or 16-1 (-/-) fruit. A similar result was obtained when infections of 83T control and transformed fruit (16-34) by *G. candidum* were compared (Fig. 2B).

In an effort to further correlate levels of PG activity and susceptibility to fungal infections, fruit with reduced PG due to the presence of the *Nr* mutation and resistance to infection by *G. candidum* (Fig. 2B) were complemented with functional PG activity. The tomato variety Ailsa craig with the *Nr* mutation was transformed with a full-length PG cDNA expressed constitutively from the CaMV 35S promoter (pCGN1415), and several transformants that expressed active PG in leaves and green fruit were identified.

Line AC-44 was selected for further analysis. As shown in Fig. 3, mature *Nr* fruit (lane 6) contained lower levels of PG mRNA and activity than did wild-type fruit (lane 3), which is in agreement with previous results (9). Constitutive expression of the PG cDNA in line AC-44 resulted in substantial levels of PG mRNA in leaf tissue (lane 7), developing fruit tissue (lane 8), and mature fruit tissue (lane 9). Approximately 85% of normal PG activity was observed in mature transgenic *Nr* fruit, and similar levels of activity were also observed in leaf and developing fruit (data not shown). In spite of the high levels of PG activity, growth and development of AC-44 and similar transformants appeared to be unaffected. When mature fruit from line AC-44 was infected with *G. candidum*, the *Nr* fruit expressing the PG gene developed lesions approximately twice as large as those of the *Nr* fruit with low levels of PG activity (Fig. 2B).

IV. DISCUSSION

Pectin composition is known to influence the consistency and viscosity of processed tomato products. High processing temperatures (170°–205°F;

A

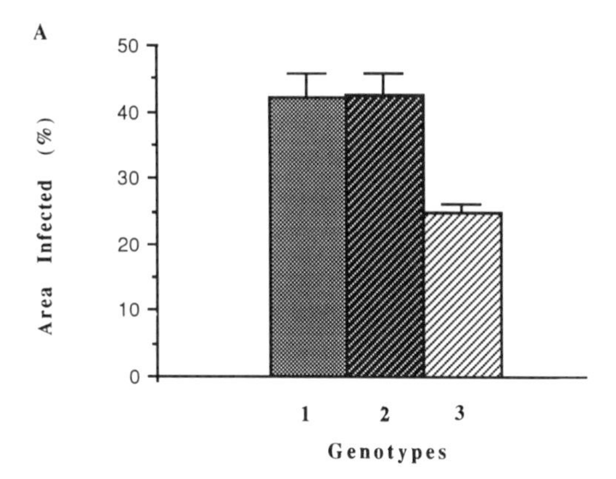

B

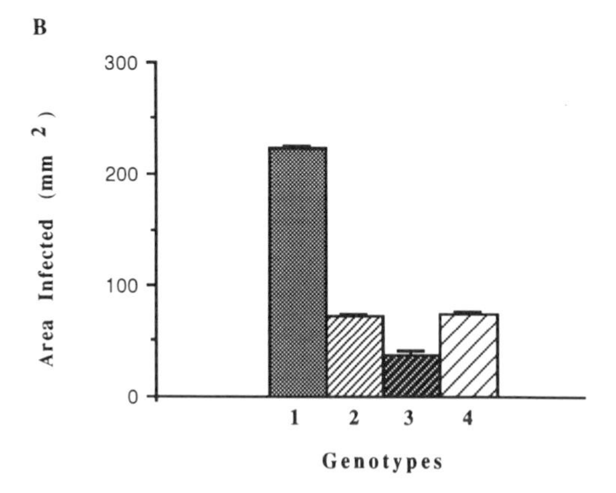

Figure 2 Infection of transgenic and control fruit by *R. stolonifer* and *G. candidum*. Fruit were inoculated as described in Section II, and the results are reported as the area (either actual or percent of total) of fruit infected. (A) Infection by *R. stolonifer* of fruit: 1, UC82B control; 2, UC82B 16-1 (-/-); 3, UC82B 16-1. (B) Infection by *G. candidum* of fruit: 1, 83T control; 2, 83T 16-34; 3, Ailsa craig *Nr*; 4, Ailsa craig *Nr* transformed with pCGN1415 (line AC-44). Standard errors are denoted by bars.

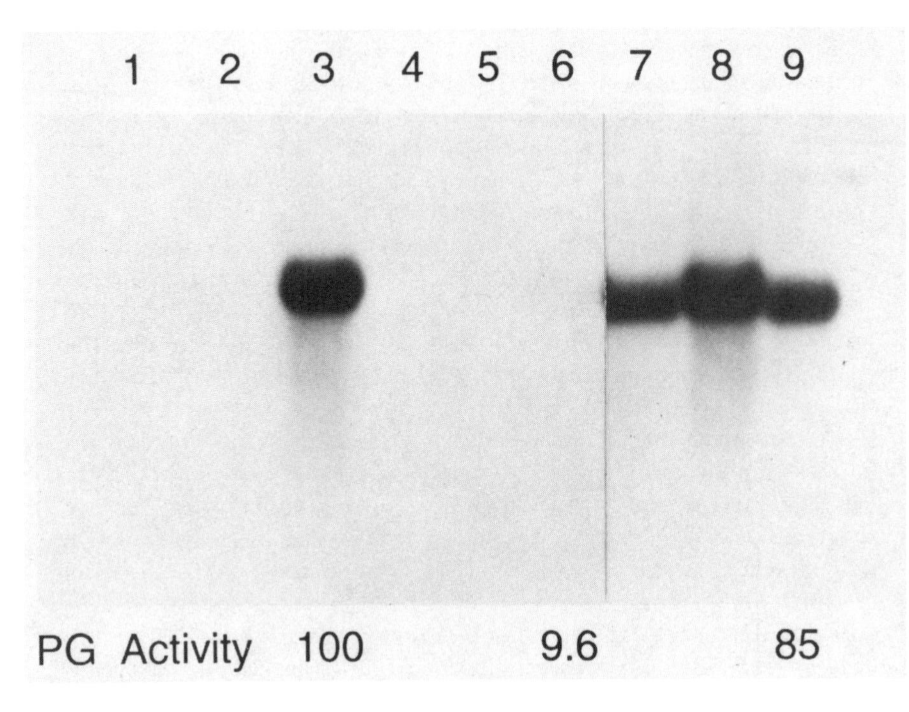

Figure 3 Expression of PG mRNA and enzyme activity in transgenic Ailsa craig *Nr*. Total RNA was prepared from leaf (lanes 1, 4, 7), green fruit (lanes 2, 5, 8), and ripe fruit (lanes 3, 6, 9) from varieties Ailsa craig (lanes 1–3), Ailsa craig *Nr* (lanes 4–6), and line AC-44 resulting from transformation of Ailsa craig *Nr* with pCGN1415 (lanes 7–9). The RNA (20 μg) was separated on a 0.8% agarose gel containing formamide, transferred to nitrocellulose, and probed with a nick-translated PG cDNA prepared according to the instructions of the manufacturer (Bio-Rad). Polygalacturonase activity was determined as described in Section II and is expressed as the percent activity in ripe Ailsa craig fruit (lane 3).

77°–96°C) are used to destroy the activity of pectolytic enzymes, and the retention of intact pectin by high break temperatures has been shown to have a major influence on viscosity (23,24) and, to a lesser degree, consistency (23). The results presented here demonstrate that a reduction in PG activity resulted in a significant increase in serum viscosity of paste and juice processed at 185°F (85°C). This is in agreement with our previous results from the Guasave, Mexico, trial (18), in which juice from 16-1 had increased serum viscosity and improved consistency. Whether this effect on serum viscosity is observed at higher break temperatures is currently unknown. Smith et al. (16) have demonstrated that reduced PG levels result in less pectin depolymerization and an increase in the molecular weight of soluble pectin in ripe fruit. Our observed increase in serum viscosity of processed products may be attributable to this increase in pectin molecular weight. Other processing characteristics, as measured by a variety of methods, were essentially unchanged.

The role of PG in the softening of tomato fruit is controversial. Under the production conditions described here, transgenic fruit from three different fresh market varieties remained firmer after storage when harvested at the mature-green, pink, or red stages of development. However, Smith et al. reported no differences in compression between control and transgenic fruit with PG levels reduced by 90% (14) or 99% (16) with a similar antisense construct. This apparent contradiction in results is difficult to reconcile because no actual data or measurement methods have been reported to date (14,16). However, harvest and handling methods, genotype, growing environment, and residual PG level may influence measurements of tomato firmness. For example, fruit from the Yolo trial was sorted over a commercial packing line, whereas fruit from the Florida trial was packed by hand in the field. Differences in compressibility between control and transgenic fruit from the Florida trial were less dramatic and were probably due to these differences in handling. The importance of genotype can be seen from the differences between Rutgers and X39 in final firmness after storage. In addition, PG is extremely abundant in ripe tomato fruit, and even reductions of 90–99% result in substantial levels of remaining activity. This activity is present as PG1 (16), an isozyme of PG that may be responsible for the initial solubilization of pectin (13,16,19,25).

There is a temporal correlation between the onset of PG activity and softening in tomato. However, our results demonstrate that fruit lacking 99% of normal PG activity will soften, although more slowly than nontransgenic controls. Conversely, functional PG activity in a *rin* background resulted in an increase in chelator-soluble pectin, but not in softening (13). We have obtained similar results in that Ailsa craig (*Nr*) fruit transformed with pCGN1415 showed no increase in softening (data not shown). Polygalacturonase activity appears to be highly regulated with regard to access to the pectin substrate, as is evidenced by the normal growth and morphology of plants with constitutive PG expression. These results strongly suggest that PG activity is not solely responsible for softening but is instead regulated as part of a more complex process that occurs during normal ripening. It is during this process that pectin degradation appears to contribute to the overall loss of fruit firmness and integrity.

Pectin is abundant in the middle lamella, the region between adjacent cells, and may lend structural support though adhesion and interaction with ionic calcium (26). The decrease in PG activity may allow the middle lamella to remain intact for an extended period and may slow the rate at which adjacent cells are allowed to move relative to one another. This increase in cellular integrity may improve texture and also slow the disease process of certain fungi. Both *R. stolonifer* and *G. candidum* cause fruit rots and preferentially infect ripe tomato fruit that are wounded or damaged. Many fungi that cause fruit rots, including *R. stolonifer*, secrete pectolytic enzymes (27,28), and the infectivity of *G. candidum* on lemon fruit tissue can be enhanced by the application of pectin-degrading enzymes (29). Thus, it is possible that these organisms are exploiting PG produced by the host as part of the infective process.

Pectin composition is known to influence the quality of processed tomato products, and manufacturing processes have been tailored to eliminate pectin-degrading enzymes. However, the effect of inhibiting pectin degradation in fresh tomato products in an effort to improve postharvest fruit quality has been more difficult to assess. Although many PG-deficient mutants have been characterized, the pleiotropic nature of these mutations has limited their use as both experimental models and commercial products. We have demonstrated that tomato fruit with dramatically reduced levels of PG exhibit improved postharvest fruit quality. Enhanced firmness and increased resistance to post-harvest fungal disease processes appear to be major components of this overall improvement in postharvest quality. The improved firmness and reduced post-harvest disease are pertinent to commercial producers of fresh market tomatoes, who stand to gain from improved fruit quality and reduced losses due to disease and rot. In addition, these same characteristics are pertinent to producers of vine-ripe tomato products, because the enhanced firmness and reduced susceptibility to postharvest disease may allow for the harvest of a greater proportion of fruit with color while avoiding the usual losses attributed the packing and shipping of vine-ripe fruit.

ACKNOWLEDGMENTS

The authors would like to thank DiMarie Bros. of Newman, Calif., for assistance in packing and evaluating fruit; the CIRT facility of Davis, Calif., for juice and paste processing data, C. Facciotti for help in manuscript preparation, and K. Redenbaugh and C. Shewmaker for critical reading of the manuscript.

REFERENCES

1. Hobson, G. The firmness of tomato fruit in relation to polygalacturonase activity, *J. Hort. Sci., 40*: 66–72, 1965.
2. Brady, C., MacAlpine, G., McGlasson, W., and Ueda, Y. Polygalacturonase in tomato fruit and the induction of ripening, *Aust. J. Plant Physiol., 9*: 171–178, 1982.
3. Brady, C., McGlasson, W., Pearson, J., Meldrum, S., and Kopeliovitch, E. Interaction between the amount and molecular forms of polygalacturonase, calcium, and firmness in tomato fruit, *J. Am. Soc. Hort. Sci., 110*: 254–258, 1985.
4. Huber, D. Polyuronide degradation and hemicellulose modifications in ripening tomato fruit, *J. Am. Soc. Hort. Sci., 108*: 405–409, 1983.
5. Bennett, A., and DellaPenna, D. Polygalacturonase gene expression in ripening tomato fruit, *Tomato Biotechnology* (D. Nevins and R. Jones, eds.), Alan R. Liss, New York, 1987, pp. 399–308.
6. Tucker, G. A., and Grierson, D. Synthesis of polygalacturonase during tomato fruit ripening, *Planta, 155*: 64–67, 1982.
7. DellaPenna, D., Alexander, D., and Bennett, A. Molecular cloning of tomato fruit polygalacturonase: analysis of polygalacturonase mRNA levels during ripening, *Proc. Natl. Acad. Sci. USA, 83*: 6420–6424, 1986.
8. Tigchelaar, E., McGlasson, W., and Buescher, R. Genetic regulation of tomato fruit ripening, *Hort. Sci., 13*: 508–513, 1978.
9. DellaPenna, D., Kates, D., and Bennett, A. Polygalacturonase gene expression in Rutgers, *rin, nor*, and *Nr* tomato fruits, *Plant Physiol., 85*: 502–507, 1987.
10. Crookes, P., and Grierson, D. Ultrastructure of tomato fruit ripening and the role of polygalacturonase isoenzymes in cell wall degradation, *Plant Physiol., 72*: 1088–1093, 1983.
11. Bird, C. R., Smith, C. J. S., Ray, J., Moureau, P., Bevan, M. J., Bird, A. S., Hughes, S., Morris, P. C., Grierson, D., and Schuch, W. The tomato polygalacturonase gene and ripening specific expression in transgenic plants, *Plant Mol. Biol., 11*: 651–662, 1988.
12. Sheehy, R. E., Pearson, J., Brady, C., and Hiatt, W. R. Molecular characterization of tomato fruit polygalacturonase, *Mol. Gen. Genet., 208*: 30–36, 1987.
13. Giovannoni, J., DellaPenna, D., Bennett, A., and Fischer, R. Expression of a chimeric polygalacturonase gene in transgenic *rin* (ripening inhibitor) tomato fruit results in polyuronide degradation but not fruit softening, *Plant Cell, 1*: 53–63, 1989.
14. Smith, C., Watson, C., Ray, J., Bird, C., Morris, P., Schuch, W., and Grierson, D. Antisense RNA inhibition of polygalacturonase gene expression in transgenic tomatoes, *Nature, 334*: 724–726, 1988.
15. Sheehy, R. E., Kramer, M., and Hiatt, W. R. Reduction of polygalacturonase activity in tomato fruit by antisense RNA, *Proc. Natl. Acad. Sci. USA, 85*: 8805–8809, 1988.
16. Smith, C. J. S., Watson, C. F., Morris, P. C., Bird, C. R., Seymour, G. B., Grey, J. E., Arnold, C., Tucker, G. A., Schuch, W., Harding, S., and Grierson, D. Inheritance and effect on ripening of antisense polygalacturonase genes in transgenic tomatoes, *Plant Mol. Biol., 14*: 369–379, 1990.

17. Tucker, G., Robertson, N., and Grierson, D. Changes in polygalacturonase isozymes during the 'ripening' of normal and mutant tomato fruit, *Eur. J. Biochem., 112*: 119–124, 1980.
18. Kramer, M., Sanders, R. A., Sheehy, R. E., Melis, M., Kuehn, M., and Hiatt, W. R. Field evaluation of tomatoes with reduced polygalacturonase by antisense RNA, *Horticultural Biotechnology* (A. Bennett and S. O'Neill, eds.), Wiley-Liss, New York, 1990, pp. 347–355.
19. Brady, C., Meldrum, S. D., McGlasson, W. B., and Ali, Z. M. Differential accumulation of the molecular forms of polygalacturonase in tomato mutants, *J. Food Biochem., 7*: 7–14, 1983.
20. Jackman, R. L., Marangoni, A. G., and Stanley, D. W. Measurement of tomato fruit firmness, *Hort. Sci., 25*: 781–783, 1990.
21. Ali, Z., and Brady, C. Purification and characterization of polygalacturonases of tomato fruits, *Aust. J. Plant Physiol., 9*: 155–169, 1982.
22. Steele, R. G. D., and Torrie, J. H. *Principles and Procedures of Statistics: A Biometrical Approach*, 2nd ed., McGraw-Hill, New York, 1980.
23. Luh, B., and Daoud, H. Effect of break temperature and holding time on pectin and pectin enzymes in tomato pulp, *J. Food Sci., 36*: 1039–1043, 1971.
24. McColloch, R., Nielsen, B., and Beavens, E. Factors influencing the quality of tomato paste, 2: pectic changes during processing, *Food Technol., 4*: 339–343, 1950.
25. Brady, C. J., McGlasson, W. B., Pearson, J. A., Meldrum, S. K., and Kopeliovitch, E. Interactions between the amount and molecular forms of polygalacturonase, calcium and firmness in tomato fruit, *J. Am. Soc. Hort. Sci., 110*: 254–258, 1985.
26. Nevins, D. J. The primary cell wall, *Models in Plant Physiology and Biochemistry*, Vol. 1 (D. W. Newman and K. C. Wilson, eds.), CRC Press, Chicago, 1987, pp. 75–77.
27. Keon, J. P. R., Byrde, R. J. W., and Cooper, R. M. Some aspects of fungal enzymes that degrade plant cell walls, *Fungal Infections of Plants* (G. F. Pegg and P. G. Ayres, eds.), Cambridge University Press, Cambridge, England, 1987, pp. 133–157.
28. Collmer, A., and Keen, N. T. The role of pectic enzymes in plant pathogenesis, *Annu. Rev. Phytopathol., 24*: 383–409, 1986.
29. Baudoin, A. B. A. M., and Eckert, J. W. Influence of preformed characteristics of lemon peel on susceptibility to *Geotrichum candidum, Physiol. Plant Pathol., 26*: 151–163, 1985.

15

Ribozymes: Descriptions and Uses

Brent V. Edington,* Richard A. Dixon, and Richard S. Nelson

The Samuel Roberts Noble Foundation, Ardmore, Oklahoma

I. INTRODUCTION

The RNA molecules that catalyze the cleavage of RNA substrates intra- or intermolecularly have been termed ribozymes. Ribozymes are involved in the maturation of RNA molecules in a diverse group of organisms and organellar structures, including *Tetrahymena*, newt, a bacteriophage, a satellite virus of an animal virus, satellite RNAs of plant viruses, plant virusoids and viroids, and yeast mitochondria (1,2). Such molecules present possibilities for altering gene expression in all organisms. This chapter describes the various known ribozymes and how some of these have been modified to work *in trans*. The requirements for *trans* functioning (transactivity) are described and related to the use of ribozymes in plant biological research. Specific areas in which ribozymes could immediately affect plant research include the prevention of plant disease and the manipulation of plant metabolism. Reviews that provide further details regarding ribozymes are available (2–6).

**Present affiliation*: Ribozyme Pharmaceuticals, Inc., Cleveland, Ohio

II. SELF-PROCESSING RNAs

The maturation of some RNA transcripts and the processing of many small pathogenic RNAs occur in vitro and possibly in vivo by the activity of self-splicing or self-cleaving RNA molecules, respectively. Self-splicing activity was first identified in the Cech laboratory during studies of the maturation of ribosomal RNA precursors in *Tetrahymena* (7). Introns that possess self-splicing capacity have been separated into two groups on the basis of conserved secondary structure and mechanism of cleavage. Group I introns (Fig. 1A) have an approximately 70-nucleotide core, which is a region containing sequences and structures necessary for catalytic functions (9,10). This type of splicing unit is characterized by its requirements for the 3′ hydroxyl of a free guanosine molecule (GTP, GDP, GMP, or guanosine), which acts as a nucleophile and is added to the 5′ end of the intervening sequence during splicing. Also, a U·G wobble base pair must be adjacent to the cleavage site. Products of this reaction include the spliced transcript and a circular intron, produced by a second transesterification reaction involving the hydroxyl of the 3′ nucleotide of the intron. The circular intron can be linearized in a reversible reaction (1). Group II self-splicing introns have no requirement for a free guanosine molecule because the 2′ hydroxyl of a specific nucleotide in the RNA chain acts as a nucleophile. Site-specific cleavage and a transesterification reaction produce the mature transcript and a lariat structure held together by a branch containing a 2′,5′ phosphodiester bond. Group I and II self-splicing introns are reviewed in more detail in Refs. 1, 9, and 10.

Several small pathogenic RNAs that infect plants (viroids, virusoids, and satellite RNAs) and one that infects animals (hepatitis delta virus) exist as single-stranded circular RNAs during at least a portion of their life cycles. All are considered to replicate by a rolling-circle mechanism that produces strands of multimer-length genomes (2,11,12). These long single-stranded concatamers then undergo specific cleavage to produce monomers of the pathogenic RNAs. The maturation of these small RNAs can proceed in vitro in the absence of proteins through self-cleaving reactions (13–15). Conserved sequences that form a secondary structure termed a hammerhead (Fig. 1B) have been identified around many of the cleavage sites in the plant pathogenic RNAs. A hammerhead structure is also believed to function in the cleavage of synthetic satellite 2 transcripts from the newt (16,17). The hammerhead structure consists of three base-paired stems around an open single-stranded region. In some cases, a double hammerhead structure consisting of two catalytic units connected by a stem has been shown to form; this can lead to greater efficiency

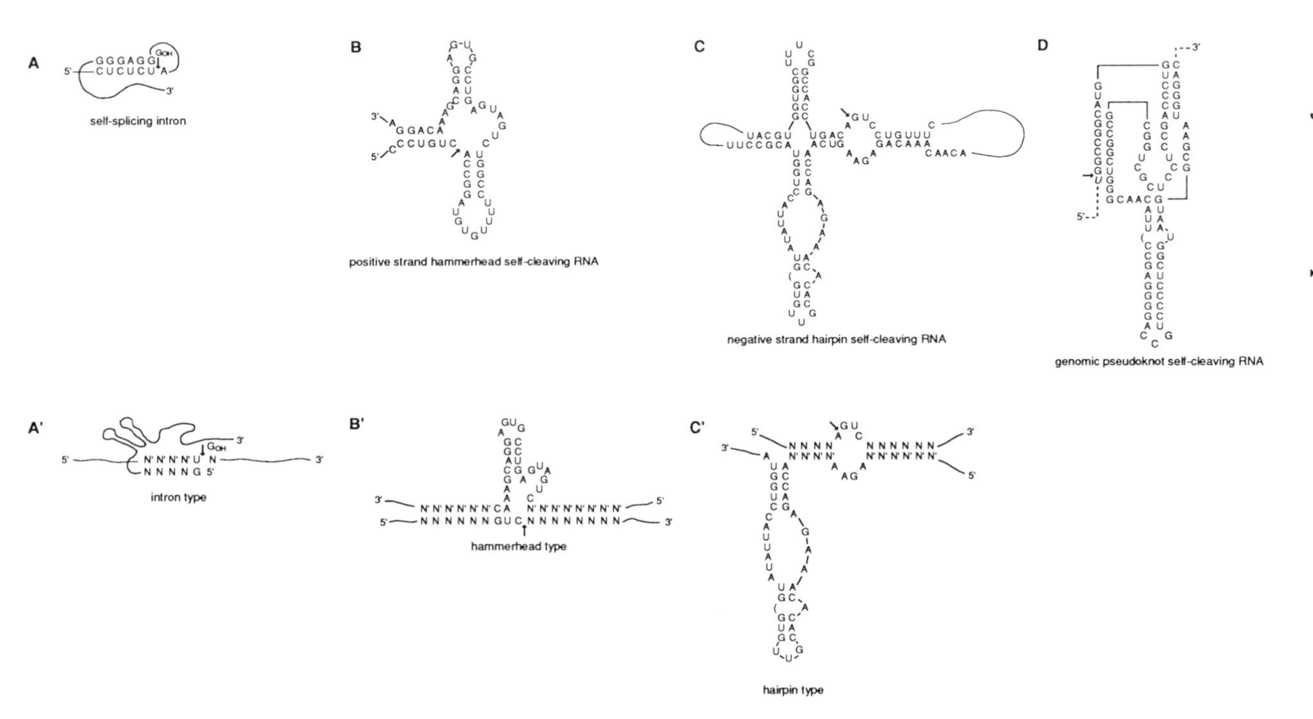

Figure 1 Structures of naturally occurring and engineered ribozymes. The top portion of the diagram represents self-cleaving structures from a group I intron (A), the positive-strand hammerhead of the satellite RNA of tobacco ringspot virus (B), the negative-strand hairpin of the satellite RNA of tobacco ringspot virus (C), and the genomic-strand pseudoknot of hepatitis delta virus (D). The bottom depicts engineered ribozymes of the intron type (A′), hammerhead type (B′), and hairpin type (C′). An arrow designates the cleavage point in each representation. (Adapted from Ref. 8.)

in RNA processing than can a single hammerhead structure (18,19). Unlike the cleavage catalyzed by self-splicing introns, the cleavage catalyzed by the hammerhead structure results in a free 5′ hydroxyl and a 2′,3′-cyclic phosphate at the 3′ terminus. These reaction products are identical to those produced during base hydrolysis of RNA, although the rate-limiting step in ribozyme-mediated cleavage apparently does not involve hydroxide ions (20).

Another type of self-cleavage structure, often referred to as a hairpin, has been identified in the negative strand of the satellite RNA of tobacco ringspot virus [(-)sTRSV] (14,21,22). Secondary structure analysis indicates that two intramolecular helices form on the sides of the site where transesterification occurs (Fig. 1C) (21). A conserved region in U6 small nuclear RNA, a molecule involved in pre-mRNA splicing, forms a hairpin structure that is remarkably similar to the catalytic hairpin of (-)sTRSV (23). Adjacent to the hairpin are nine bases of sequence identity between the U6 RNA and (-)sTRSV (23). Site-specific mutagenesis and deletions or insertions within these regions of U6 RNA disrupt splicing activity (24–26) and indicate the importance of this RNA structure for the excision of introns.

The self-catalytic region of hepatitis delta virus (HDV) differs from all the previously mentioned self-splicing structures in that a pseudoknot-like structure may be required for efficient self-cleavage of genomic and antigenomic sequences (see Ref. 27 and Fig. 1D). Self-cleavage of HDV RNA occurs at a faster rate than that of other known ribozymes, even at low Mg^{++} concentrations, which indicates that these cleavage sequences may represent a distinct class of ribozymes (28). It should be noted, however, that the putative HDV cleavage structures have a U·G wobble base pair in the position next to the cleavage site, which is similar to the group I intron structure (3,27). As with both the hammerhead- and the hairpin-catalyzed cleavages in sTRSV, the pseudoknot-catalyzed cleavage produces a free 5′ hydroxyl and a 2′,3′ cyclic phosphate. Recently, a hammerhead structure that may be modified to contain a pseudoknot has been identified in the satellite RNA of barley yellow dwarf virus (29).

Bacterial RNase P is the only naturally occurring *trans*-acting ribozyme described to date. This enzyme is responsible for the cleavage that produces the mature 5′ terminus of tRNA molecules. Unlike other ribozymes, bacterial RNase P interacts with a protein subunit during the catalytic reaction (30); however, the capacity to cleave RNA resides only in the RNA moiety (31,32). In contrast, a chloroplastic RNase P is thought to mediate RNA cleavage through a protein component (33). Many eukaryotic RNase Ps also contain both protein and RNA components (34–36), and the moiety responsible for RNA cleavage has not yet been determined.

Ribozymes are thus a diverse group of enzymes involved in the processing of RNA. They do, however, share some characteristics. None of the known naturally occurring ribozymes requires energy from the hydrolysis of nucleoside triphosphates for cleavage to occur. All but one use Mg^{++} during cleavage, although Ca^{++} and Mn^{++} can, in some cases, be substituted (1,20,28). The single ribozyme that does not use Mg^{++} was isolated from a *Tetrahymena* intervening sequence and requires Mn^{++} for activity (37). Further studies will probably reveal other natural systems in which RNA processing is catalyzed by RNA enzymes.

III. SELF-CLEAVING RNAs DESIGNED FOR TRANSACTIVITY

Several self-cleaving RNA enzymes have been adapted for cleavage of a target *in trans* (transactivity). The group I self-splicing structure of the *Tetrahymena* ribosomal RNA intron (Fig. 1A), the hammerhead self-cleaving structure from the positive strand of the satellite RNA of tobacco ringspot virus (Fig. 1B), and the hairpin self-cleaving structure of the minus strand of the satellite RNA of tobacco ringspot virus (Fig. 1C) have all been modified for transactivity (Figs. 1A′, 1B′, and 1C′). Each type of adapted ribozyme has been shown to cleave substrate RNA in a catalytic and site-specific manner when incubated under the appropriate conditions in vitro (9,38–40). The *trans*-acting hammerhead ribozyme has the smallest catalytic domain, which may allow easier access to cleavage sites on target RNAs. Also, when compared with the other cleavage structures, the structure of the hammerhead catalytic domain allows for greater variability in both complementary flanking sequences that target the ribozyme to the substrate (compare Figs. 1A′, 1B′, and 1C′). Thus, the most promising cleavage structure appears to be the hammerhead type. Naturally occurring hammerhead ribozymes cleave target RNAs after the sequence GUX (X = A, C, or U), although certain other sequences are tolerated (41 and 44).

These catalytic structures have been analyzed for transactivity during in vitro and in vivo studies. The catalytic nature of the *trans*-acting ribozymes has most often been assessed by observing the in vitro or in vivo cleavage of short RNA transcripts having limited secondary structure (20,38–40,42–45). When some of these ribozymes were challenged with large and sometimes protein-associated targets, either less efficient cleavage or no cleavage was observed

(40,42,44). Thus, ribozyme-mediated cleavage of natural substrates appears to be more complex than the cleavage of short transcripts.

In addition to understanding the effects of substrate size on cleavage kinetics, greater knowledge of the properties of ribozymes themselves will lead to the production of ribozymes having greater practical application in vivo. By using U7 small nuclear ribonucleoprotein–mediated histone pre-mRNA processing as an assay, Birnstiel et al. compared antisense DNA, antisense DNA oligonucleotides, antisense RNA, and ribozymes for their ability to inhibit processing activity in nuclear extracts (42). In this assay the ribozyme was the least effective inhibitor, requiring a 1000-fold molar excess of ribozyme to U7 RNA to achieve the same inhibition as a 600-fold excess of antisense DNA oligonucleotides, a 60-fold excess of antisense DNA, or a 6-fold excess of antisense RNA. This comparison may not have reflected the optimum cleaving ability of the ribozyme because of differences in stability between it and the antisense RNA and DNA sequences in nuclear extracts. This study has been extended to an in vivo system in which the ribozyme targeted to U7 RNA and the U7 RNA substrate were co-injected into the nucleus of *Xenopus* oocytes (45). It was found that ribozyme RNA, expressed within a tRNA, remained localized mainly in the nucleus, whereas the U7 RNA substrate exited rapidly into the cytoplasm. In spite of this compartmentalization, selective ribozyme-mediated degradation of U7 RNA was demonstrated. Efficiency of the reaction was low, however, in that an estimated 200-fold excess of cytoplasmic ribozyme to substrate was necessary to destroy the U7 RNA.

The size of a ribozyme transcript can affect the stability of the ribozyme and thus its cleavage activity. When a construct producing a short ribozyme transcript directed against chloramphenical acetyltransferase (CAT) and driven by the SV40 early promoter was cotransfected into COS cells with a construct producing a CAT transcript, no inhibition of CAT activity was detected (46). The short ribozyme transcript appeared to be unstable and unable to accumulate to levels that could affect CAT activity. In our own work we have seen that a small ribozyme transcript (350 nt) expressed from behind a strong promoter (i.e., the cauliflower mosaic virus 35S) in plant protoplasts accumulated to very low steady-state levels of transcripts (unpublished results). This may be because of the rapid degradation of this small transcript.

An advantageous way to address the stability problem may be to embed the ribozyme sequence in a larger transcript. Cameron et al. (46) and Cotten et al. (45) attempted this when they placed their ribozymes in the 3′ untranslated region of a luciferase transcript (~1700 bp) and in the anticodon loop of a tRNA, respectively. Cameron et al. obtained a 60% inhibition of CAT activity in the presence of a 1000-fold molar excess of ribozyme (46). This apparent low efficiency may reflect an inhibition of activity due to an interaction between the ribozyme and luciferase sequences, but this has not been proven. The efficiency of the ribozyme in the anticodon loop of a tRNA may have been compromised owing to the compartmentalization of this transcript (45).

Although interactions between ribozyme and nonribozyme sequences can be detrimental for cleavage activity, in some cases, they can be beneficial. Analysis of a ribozyme derived from HDV has shown that certain cotranscribed vector sequences at the 5′ end of the transcript increased the in vitro activity of the ribozyme (47). It was suggested that this result may be explained by the interaction of the 5′ vector-derived sequences with 3′ virus-derived sequences that, in the absence of the 5′ vector sequences, may have interacted with nucleotides adjacent to the cleavage site and inhibited the cleavage reaction. In other in vitro studies single nucleotide alterations changing the content but not the number of complementary bases in the flanking regions altered cleavage rates 50-fold (48). Cech (9) suggested and reviewed four possible explanations for such an observation, including the need for (a) a specific helical geometry, (b) helix stability, (c) the ability to produce alternate pairing partners, and (d) the ability to form base triplets.

Although these results indicate a lack of optimization for ribozyme-mediated cleavage in vivo, the potential usefulness of ribozyme technology is being demonstrated. When a hammerhead ribozyme targeted to cleave the human immunodeficiency virus (HIV-1) *gag* transcript was expressed in human cells infected with HIV-1, less of the *gag* transcript was observed in these cells than in infected cells not expressing the ribozyme (49). Cleavage was inferred from results showing a greater accumulation of a polymerase chain reaction (PCR) product that does not span the cleavage site than of one that does. The molar ratio of ribozyme to target was unknown because neither species was quantitated.

IV. USES OF RIBOZYMES IN PLANT BIOLOGY

Ribozymes can be used as an alternative to antisense inhibition in every instance, assuming that a cleavage site is present in the target sequence. Because sequence requirements for cleavage are minimal, i.e., three or fewer bases for group I and hammerhead- or hairpin-type ribozymes (3,38,39), a cleavage site will be found on average at least every 64 nucleotides in any target gene or noncoding sequence with nucleotides arranged randomly. Areas near the AUG translation initiation codon, splice sites, and 3′ end of the coding regions have been successfully targeted for antisense inhibition (50,51). Thus, for both antisense and ribozyme-mediated inhibition, no position within a nucleotide sequence can be assumed to be inaccessible. Limitations of ribozyme-mediated cleavage, such as inhibitory secondary or tertiary structures of the target RNA, can also apply to antisense-mediated inhibition.

Although antisense and ribozyme constructs can be used in similar situations, results may vary for similar size constructs because of their different modes of action. Disassociation of antisense transcripts from their targets is not desirable, and therefore the length of their complementary sequences is limited only by the length of the target-site sequence. Ribozymes are limited in the length of their complementary flanking region sequences because turnover (i.e., their catalytic function) is critical for maximum effectiveness. Determining the best length for the complementary flanking regions of ribozymes is difficult at this time. For RNA–RNA interactions the rate constant of association is relatively independent of nucleotide sequence and structure, compared with the rate constant of dissociation which is very much affected by these factors (52). Because in many cases RNA secondary structures have not yet been predicted with great certainty, it is difficult to determine the strength of an RNA–RNA interaction in large helices. Shortcomings also exist in our understanding of tertiary interactions of large RNA molecules (52). For these reasons and those mentioned in the previous section, the optimum lengths for the flanking regions of ribozymes have not been determined (53) and must now be determined empirically.

If the preceding constraints are kept in mind, ribozyme targets within plants could include both pathogen- and plant-derived transcripts. Preventing plant disease and manipulating plant metabolism for applied and basic purposes are desirable objectives. The ease with which large numbers of transgenic plants can be produced (54) will aid in the use of ribozymes in agriculture. The following sections discuss various target molecules and the potential for cleaving them with ribozymes.

V. RIBOZYMES FOR THE PREVENTION OF PLANT DISEASE

A. Fungi and Bacteria as Targets

Fungi and bacteria pose a significant problem for ribozyme-mediated protection strategies. Because membrane barriers always separate the cytoplasm of the plant cell from the cytoplasm of the fungal or bacterial cell (55), it is difficult to envisage a mechanism by which a ribozyme transcribed in a plant cell could cleave a pathogen-derived transcript. This problem may be solved if ribozymes, transcribed on wounding, are directed against essential host messages. These wound-induced ribozymes must be transcribed from nonleaky promoters to avoid nonspecific cell death. Destruction of these host messages would kill the host cells at the wound site. This approach may be useful against biotrophic fungi, but for the necrotrophic fungi (e.g., *Botrytis* sp.) host transcript cleavage may simply aid the pathogen in killing the plant cell. Another approach may be to direct ribozymes against host transcripts that, if allowed to express, result in a cell that is more susceptible to pathogen infection. Polygalacturonase helps to loosen cell walls, and thus its expression may aid fungi in colonizing the tissue. Expression of an antisense construct against the mRNA of polygalacturonase resulted in greater fungal resistance in tomato fruit (B. Martineau, oral report at the 1991 annual meeting of the American Phytopathological Society).

If a system could be envisioned for uptake of host cell transcripts by a pathogen, ribozymes targeted to pathogen transcripts could be used. Recently, Bryngelsson et al. (56) have shown that tandemly repeated sequences of apparent host origin were present in the parasitic fungus *Plasmodiophora brassicae*. The function of this DNA is unknown, but intact ribozymes potentially could be taken up by a similar mechanism. The extent of host DNA uptake into fungi requires further investigation. If DNA uptake is more common in fungi than is currently known, ribozyme specificity is critical to avoid cleavage of transcripts in nonpathogenic fungi that have taken up the DNA.

B. Viruses as Targets

Plant viruses rely on host machinery for replication. Although replication may take place in a specific area of the cell, such as the viroplasm in a cell infected with tobacco mosaic virus (TMV) (57), this area is theoretically accessible to host constituents because no membrane surrounds this structure. Even plant viruses that are enveloped, e.g., the plant rhabdoviruses and tomato spotted wilt virus, or have viral constituents associated with membranes, e.g., the potyviruses, are in intimate contact with host constituents during the replication and translation phases of their life cycles. Thus, viruses should be susceptible to cleavage by ribozyme at some point in their life cycles.

Positive-stranded RNA viruses make up 77% of the known plant viruses (58) and are generally simple in that none are membrane bound. The nucleotide sequences of more than 20 of these viruses were known as of 1989 (59), and findings of new sequences are being published at a rapid pace (for examples, see Refs. 60–67). These sequences provide the information necessary for ribozyme construction. Although protection against plant viruses has been achieved through viral coat protein expression in transgenic plants (reviewed in Ref. 68), attempts to use antisense transcripts to inhibit infection by RNA viruses have not been very successful (69–72) until recently (73). This lack of effectiveness may be the result of viral replication taking place in the cytoplasm where viral proteins with helicase activity could unwind antisense-viral RNA duplexes. Helicase domains have been found encoded in many positive-stranded RNA viruses, and a viral protein containing such a conserved sequence has recently been shown to unwind RNA duplexes in vitro (74). Ribozymes directed against plant viruses avoid some of the problems associated with antisense transcripts because they need to be bound for only a short time to allow cleavage and subsequent inhibition of virus replication.

Although the majority of plant viruses are composed of plus-stranded RNA, a significant number is composed of minus-stranded RNA, double-stranded RNA, single-stranded DNA, or double-stranded DNA. The replication of all of these viruses may be inhibited by targeting ribozymes against their genomic RNA, their intermediate transcripts in replication, or their mRNAs.

1. Selection of a Target Sequence

In selecting a target site within a viral sequence, the life cycle of that virus should be taken into account. For many positive-stranded RNA viruses, intact RNA is necessary for infection (Ref. 75 and references therein). Positive-stranded RNA viruses that produce polyproteins such as potyviruses and comoviruses are excellent targets because they produce no subgenomic RNA that could compete with genomic RNA for cleavage. The life cycles of viruses that contain multipartite genomes or produce subgenomic RNAs should be studied to determine which sequences of the viral genome do not appear in the subgenomic RNA or are essential for replication. These sequences would be excellent candidates for targeting. The possibility of multiple cleavage structures per transcript or multiple transcripts each with a separately targeted ribozyme should be considered.

An objective of any protection strategy is to prevent virus replication early in infection. Targeting ribozymes against the complementary RNA of a plus-stranded RNA virus is potentially useful because in many cases the complementary strand is produced in much lower quantities than the genomic RNA. For TMV, the molar ratio of plus- to minus-strand RNA was approximately 200:1 in tobacco protoplasts inoculated 12 hr previously (76). The drawback of targeting only the minus strand is that it allows the genomic RNA to escape cleavage and thereby allows both translation and replication to again be initiated on this RNA. Thus, it is best to target the genomic RNA or early RNA transcripts of DNA viruses. In this way infection may be stopped before the replicase can be translated and allowed to replicate the virus. For TMV specifically, after the rod-shaped particle enters the cytoplasm the rod is thought to swell, starting with the 5′ end of the RNA, allowing exposure of the TMV genomic RNA (77). The 5′-proximal open reading frame of the genomic RNA is then translated, producing the putative RNA-dependent RNA polymerase (RdRp) (Fig. 2). The RdRp, possibly with the involvement of host factors, is then able to produce the negative strand of the TMV genome and, subsequently, genomic positive-strand RNA and subgenomic transcripts. Cleavage of the genomic RNA within the region encoding the RdRp should prevent translation of a full-length polymerase and inhibit production of progeny viral RNA. In our studies, we therefore chose to target a ribozyme against a sequence in the open reading frame for the putative RdRp. For identifying a target site within this reading frame we used computer programs to search for regions lacking excessive RNA secondary structure. We used the program devised by Zuker et al. (78), but a program is now available that recognizes additional secondary structures, such as pseudoknots (79). Regions encoding conserved amino acid motifs for viral or eukaryotic polymerases (80) and NTP-binding motifs (81–83) were avoided. Ribozymes targeted to sequences encoding such conserved amino acid motifs may result in cleavage of endogenous host transcripts and death of the cell if conservation extends to the nucleic acid level. Avoidance of conserved sequences in target RNAs is not a good approach if these sequences are known to occur only in plant virus sequences. The conserved region may be required for virus replication and thus may be less likely to mutate. Ribozymes targeted against such regions may remain effective over longer periods because the possibility of virus RNA escaping cleavage due to mutation at the site is lessened (84). In addition, if the target sequence is conserved by many virus strains within a virus family, protection may increase horizontally when this strategy is used (84).

By taking the preceding factors into consideration, a hammerhead-type ribozyme was designed to cleave the positive strand of the genomic TMV RNA after the GUC sequence at position 2467. Flanking the catalytic domain on either side are 20 nucleotides complementary to the TMV genomic RNA sequence (Fig. 2). These ribozyme sequences have been cloned into a vector enabling in vitro transcription of the ribozyme and into a binary vector allowing integration into the plant genome. The latter transformation vector

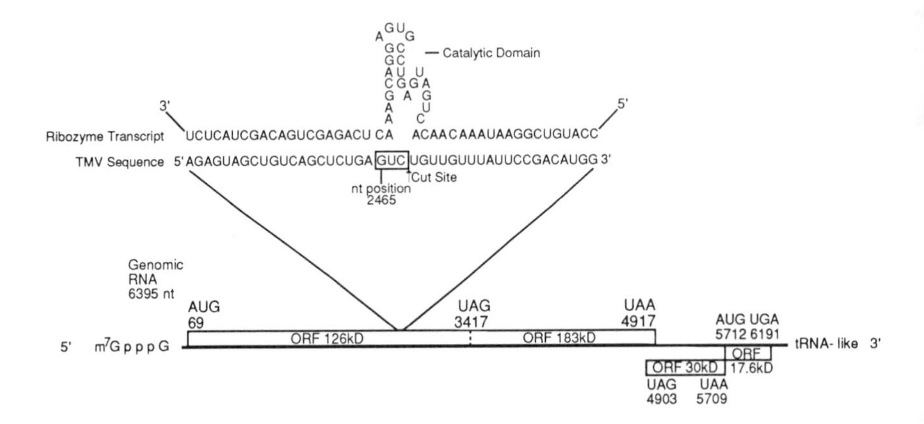

Figure 2 Base pairing of a ribozyme and its target site within the positive-strand genomic RNA of TMV. The ribozyme is targeted to the GUC (boxed) at nucleotides 2465–2467 of TMV RNA. The ribozyme catalytic domain and the cleavage site are also designated. The genomic organization of TMV and the position of the ribozyme target site in that organization are shown. Open reading frames (ORFs) in the TMV genome are designated by open bars. The 126-kDa and 183-kDa ORFs code for proteins putatively considered part of the RNA-dependent RNA polymerase. The 30-kDa ORF codes for a protein associated with cell-to-cell movement of the virus, and the 17.6-kDa ORF codes for the coat protein of the virus. A tRNA-like structure is at the 3′ end of the virus genome. The dashed line in the open bar between ORF 126 kDa and ORF 183 kDa indicates the site of the UAG leaky termination codon. (Adapted from Ref. 8.)

contains a cauliflower mosaic virus (CaMV) 35S promoter for high constitutive expression in plants or plant protoplasts and a ribulose-1,5-bisphosphate carboxylase small subunit (rbcS) 3′ polyadenylation signal to ensure polyadenylation of the transcript (85). Two other groups that are developing ribozymes as antiviral agents in plants have targeted regions coding for viral proteins. Lamb et al. have targeted one ribozyme to the predicted RdRp coding region and another to the coat protein coding region of potato leafroll virus (PLRV) (84). These ribozymes are flanked by 19 or 20 nucleotides (11, 10, or 9 nucleotides on each side of the cleavage site) complementary to the positive-strand genomic RNA of PLRV. Gerlach et al. have constructed a ribozyme containing three hammerhead structures targeted to three successive GUC sites at the 5′ end of the coding sequence for the RdRp of TMV. The cleavage domains are embedded in an extensive antisense sequence (1 kb) to the 5′ end of the polymerase coding region (W. Gerlach, personal communication).

Because of the predicted length of duplex formation between the target sequence and this ribozyme, it appears that there can be no significant catalytic turnover of this enzyme. The degree to which the RNA secondary structure of this large antisense ribozyme inhibits the proper formation of the cleavage domains is not known.

2. Cleavage of Viral RNAs In Vitro

In our work, transcribed ribozyme RNA was able to cleave full-length genomic RNA of both the masked and U_1 strains of TMV in a site-specific manner (Ref. 8 and unpublished results). The masked strain of TMV has 55 nucleotide substitutions with respect to the U_1 strain (60), but has none in the ribozyme flanking regions. The in vitro catalysis appears to be very inefficient. A ribozyme molar excess of approximately 20-fold was used to cleave the masked strain genomic RNA, and complete cleavage was not achieved in 70 min. This apparent inefficiency may be due to secondary or tertiary structures in either the 6.4-kb target RNA or the ribozyme. Alternative secondary or tertiary structures can significantly affect ribozyme activity, as described in Section III.

3. Protection Against Virus Infection In Vivo

Using tobacco protoplasts has allowed us to assess the ability of the ribozyme targeted against TMV RNA to inhibit replication of TMV in vivo. Cleavage of the viral genomic RNA in vivo may proceed differently from the in vitro cleavage, because of differences in protein–RNA associations or TMV genomic RNA secondary structure in the two situations. More specifically, during infection TMV cotranslationally disassembles, which frees its RNA from the 5′ end to the 3′ end (77). Thus, secondary structure of the genomic RNA may be more limited in the in vivo than the in vitro situation until full disassembly is achieved. Such a limit on secondary structure may expose, for a brief period, the site to which the ribozyme is targeted. In our assays, protoplasts were treated with polyethylene glycol (PEG) to induce the simultaneous uptake of TMV and constructs containing a ribozyme or antisense sequence. The antisense construct was used to demonstrate that any inhibition of viral replication was due to the ribozyme and not simply to an antisense interaction with the TMV target sequence. Both the ribozyme and antisense transcripts were produced in protoplasts, although at levels detectable only after PCR amplification of cDNAs from the transcripts. Cotreatment with TMV and either vector or antisense construct had no significant effect on TMV replication and accumulation after 24 hr of incubation (8). However, expression of the antisense construct slightly inhibited TMV accumulation in the early stages of replication, according to results seen after 15 hr of incubation. In contrast, expression of the ribozyme construct appeared to prevent detectable TMV accumulation at 15 hr and 90% of the TMV accumulation seen with vector controls at 24 hr (8). Increasing the viral inoculum by 5- or 10-fold did not overcome the ribozyme-mediated protection in protoplasts (unpublished results). A problem with these transient protoplast assays is that the protoplast must allow both replication of the introduced virus and transcription of the introduced vector containing the ribozyme construct. We have obtained results showing that virus replication was suppressed in the presence of various vector DNAs not containing ribozyme constructs. A similar nonspecific inhibitory effect has been observed in transient assays in which plasmids containing antisense constructs specific or not specific (i.e., anti-CAT) for HIV-1 were comicroinjected with HIV-1 into human cells (86). In an attempt to avoid this variability, protoplasts from transgenic plants expressing the ribozyme will be challenged with virus. Assuming that adequate levels of ribozyme transcript are present in the transgenic protoplasts, the variability may decrease because only virus is being introduced by PEG treatment.

The ribozyme designed by Gerlach et al. and targeted to TMV has also demonstrated an antiviral capacity in vivo. This ribozyme has been cloned into an appropriate vector and used to transform tobacco plants that allow systemic spread of the virus. Transgenic plants expressing the ribozyme display TMV-associated symptoms significantly later than do controls (W. Gerlach, personal communication). The extent of this delay in symptom development is positively correlated with the presence of ribozyme transcript in succeeding generations. We have also produced transgenic systemic host tobacco plants (*N. tabacum* cv. Xanthi nn) that contain our ribozyme construct, and these plants are being analyzed for resistance to TMV.

VI. RIBOZYMES FOR THE MANIPULATION OF PLANT METABOLISM

In the last several years there has been a rapid increase in the number of genes cloned from plants. Whereas earlier reports were primarily concerned with genes encoded by highly abundant transcripts (e.g., seed storage proteins, Rubisco), a wide range of enzymes (of primary and secondary metabolism), structural proteins and regulatory proteins have now been cloned and sequenced. There are strong practical and fundamental reasons for wanting to inhibit expression of specific plant genes by the use of molecular genetic techniques, such as antisense or ribozyme expression. A useful application would be inhibition of the formation of unfavorable secondary metabolites in crop plants. Specific examples are the synthesis of nicotine in tobacco, gossypol in cotton, glucosinolates in *Brassica*, and alkaloids in cocoa. Our understanding of the biochemistry and molecular biology of the biosynthesis of plant secondary products is gradually increasing to a point where such pathways may be targetable. In more fundamental studies, preventing expression of a specific transcript may in some cases be the only direct way of assessing the function of its gene product; this is especially important because some plant genes are now being cloned without any biochemical information on their products (e.g., by PCR, homology probing, or gene tagging) (87).

Antisense RNA has been shown to successfully inhibit expression of endogenous or introduced genes in plants. Such studies either have used readily assayable reporter genes, such as β-glucuronidase (88) or CAT (89,90), or have been directed at specific plant transcripts (91–93). Expression of an antisense chalcone synthase (CHS) gene in transgenic *Petunia* or tobacco results, with high frequency, in a reduction in CHS transcript and enzyme levels, which lead to altered flower pigmentation (94). The pigmentation pattern of flowers is highly variable, and this variation does not correlate with the chromosomal position of the antisense transgene (94). Contrary to some reports in other systems (reviewed in Ref. 51), antisense CHS genes encoding the 3′ 50% or 25% of CHS mRNA affect *Petunia* flower pigmentation, whereas 5′ antisense sequences do not (95). Furthermore, it appears unlikely that the amount of antisense CHS transcripts produced is greater than the amount of endogenous CHS transcripts. It has been suggested that the mechanism for antisense inhibition of CHS expression in *Petunia* does not simply involve the formation of a duplex between sense and antisense transcripts (95). The situation is further complicated by the demonstration that overexpression of CHS sense transcripts in transgenic *Petunia* likewise results in altered flower pigmentation arising from an overall reduction in CHS transcript levels (96,97). A similar phenomenon has been observed for overexpression of dihydroflavonol reductase, phenylalanine ammonia-lyase (PAL), and nopaline synthase transgenes (97–99). Possible mechanisms for sense and antisense inhibition of plant gene expression have been reviewed (100,101). Clearly, these phenomena are complex and vary with gene sequence and developmental status of the plant.

The ability to target a ribozyme to a specific site on a specific mRNA transcript suggests a potentially powerful strategy for manipulating plant metabolism that has clear advantages over the use of antisense RNA. If the antisense sequences used for targeting the ribozyme are short, it may be possible to avoid any form of antisense inhibition, which might, as previously discussed, lead to variable and complex results. Experiments in which a ribozyme is targeted separately to two different areas, one of which is susceptible to antisense inhibition and one of which is not (e.g., the 3′ and 5′ halves of *Petunia* CHS, respectively), could be instructive in this respect. However, the major potential advantage of ribozymes over the antisense strategy is the specificity of their targeting. It might, for example, be possible to target a ribozyme to disrupt a transcript encoding a single member of a multigene family, while leaving the other members unaffected. This could be of value in elucidating the functions of individual isoforms of plant enzymes, such as PAL or CHS, whose isogenes are differentially activated in response to different developmental and environmental stimuli (102) but for which no specific differential metabolic functions have yet been ascribed. This specificity could also assist attempts to genetically engineer new traits into plants. For example, in cases in which modified gene product is being expressed, the product of the introduced transgene might compete unfavorably with the product of the corresponding endogenous gene (e.g., have a higher K_m value for its substrate), and it may therefore be essential to selectively eliminate the expression of the endogenous gene. Antisense RNA may lack the necessary discrimination between the endogenous gene and transgene, whereas this discrimination may be achieved with a ribozyme. If both transgene and endogenous gene contain identical target sites and there is extensive sequence similarity around these sites, it would be possible to alter the transgene by a single-base mutation by using site-directed mutagenesis to remove the target sequence. In this way specificity for the endogenous gene might be restored. This method requires that the mutagenesis does not significantly alter the amino acid sequence and therefore the activity of the transgene. Such strategies clearly warrant investigation, at least until methods for gene replacement have been developed for plants.

VII. CONCLUSIONS

The extent to which ribozymes can be useful in altering in vivo gene expression is being assessed. In mammalian tissue culture (46,49) and *Xenopus* oocytes (45), ribozymes can selectively inhibit gene expression in vivo. The use of ribozymes to inhibit fungal or bacterial infection of plants, although plausible, will require much more basic work, whereas prevention of virus infection in plant protoplasts and plants has been demonstrated and may be more effective than antisense inhibition. Besides contributing to virus resistance, ribozymes could be used (a) to inhibit production of biosynthetic enzymes that lead to undesirable secondary products in plants or (b) to prevent expression of plant transcripts to assess their function. Many of the problems with this technology arise because of our lack of knowledge regarding the rules governing RNA structure and stability. In addition, further work on increasing and localizing ribozyme expression levels in specific cells through the use of enhanced or tissue-specific promoters and specific cell or tissue delivery systems will be important to realizing the full potential of ribozymes in modifying gene expression. Widespread use of ribozymes in plant systems is not likely in the immediate future, but existing evidence supports further investigation and development of this technology.

ACKNOWLEDGMENTS

The authors wish to thank Drs. Brandt Cassidy, Abraham Oommen, and Michael Shintaku for reviewing the chapter. In addition, Cuc Ly is thanked for her excellent graphics work and Allyson Wilkins for rapid typing and editing.

REFERENCES

1. Cech, T. R., and Bass, B. L. Biological catalysis by RNA, *Annu. Rev. Biochem., 55*: 599–629, 1986.
2. Symons, R. H. Self-cleavage of RNA in the replication of small pathogens of plants and animals, *TIBS, 14*: 445–450, 1989.
3. Cech, T. R. Self-splicing of group I introns, *Annu. Rev. Biochem., 59*: 543–568, 1990.
4. Rossi, J. J., and Sarver, N. RNA enzymes (ribozymes) as antiviral therapeutic agents, *Trends Biotechnol., 8*: 179–183, 1990.
5. Cotten, M. The *in vivo* application of ribozymes, *Trends Biotechnol., 8*: 174–178, 1990.
6. Bruening, G. Compilation of self-cleaving sequences from plant virus satellite RNAs and other sources, *Methods Enzymol., 180*: 546–558, 1989.
7. Kruger, K., Grabowski, P. J., Zaug, A. J., Sands, J., Gottschling, D. E., and Cech, T. R. Self-splicing RNA: autoexcision and autocyclization of the ribosomal RNA intervening sequence of *Tetrahymena, Cell, 31*: 147–157, 1982.
8. Edington, B. V., and Nelson, R. S. Utilization of ribozymes in plants: plant viral resistance, *Gene Regulation: Biology of Antisense RNA and DNA* (R. P. Erickson and J. Izant, eds.), Raven Press, New York, pp. 209–221, 1992.
9. Cech, T. R. Conserved sequences and structures of group I introns: building an active site for RNA catalysis—a review, *Gene, 73*: 259–271, 1988.

10. Burke, J. M. Molecular genetics of group I introns: RNA structures and protein factors required for splicing—a review, *Gene, 73*: 273–294, 1988.
11. Branch, A. D., and Robertson, H. D. A replication cycle for viroids and other small infectious RNAs, *Science, 223*: 450–455, 1984.
12. van Tol, H., Buzayan, J. M., and Bruening, G. Evidence for spontaneous circle formation in the replication of the satellite RNA of tobacco ringspot virus, *Virology, 180*: 23–30, 1991.
13. Prody, G. A., Bakos, J. T., Buzayan, J. M., Schneider, I. R., and Bruening, G. Autolytic processing of dimeric plant virus satellite RNA, *Science, 231*: 1577–1580, 1986.
14. Buzayan, J. M., Gerlach, W. L., and Bruening, G. Non-enzymatic cleavage and ligation of RNAs complementary to plant virus satellite RNA, *Nature, 323*: 349–353, 1986.
15. Hutchins, C. J ., Rathjen, P. D., Forster, A. C., and Symons, R. H. Self-cleavage of plus and minus RNA transcripts of avocado sunblotch viroid, *Nucleic Acids Res., 14*: 3627–3640, 1986.
16. Epstein, L. M., and Gall, J. G. Self-cleaving transcripts of satellite DNA from the newt, *Cell, 48*: 535–543, 1987.
17. Epstein, L. M., and Pabón-Peña, L. M. Alternative modes of self-cleavage by newt satellite 2 transcripts, *Nucleic Acids Res., 19*: 1699–1705, 1991.
18. Forster, A. C., Davies, C., Sheldon, C. C., Jeffries, A. C., and Symons, R. H. Self-cleaving viroid and newt RNAs may only be active as dimers, *Nature, 334*: 265–267, 1988.
19. Sheldon, C. C., and Symons, R. H. RNA stem stability in the formation of a self-cleaving hammerhead structure, *Nucleic Acids Res., 17*: 5665–5677, 1989.
20. Uhlenbeck, O. C. A small catalytic oligoribonucleotide, *Nature, 328*: 596–600, 1987.
21. Hampel, A., and Tritz, R. RNA catalytic properties of the minimum (-)sTRSV sequence, *Biochemistry, 28*: 4929–4933, 1989.
22. Feldstein, P. A., Buzayan, J. M., and Bruening, G. Two sequences participating in the autolytic processing of satellite tobacco ringspot virus complementary RNA, *Gene, 82*: 53–61, 1989.
23. Tani, T., and Ohshima, Y. mRNA-type introns in U6 small nuclear RNA genes: implications for the catalysis in pre-mRNA splicing, *Genes Dev., 5*: 1022–1031, 1991.
24. Fabrizio, P., McPheeters, D. S., and Abelson, J. *In vitro* assembly of yeast U6 snRNP: a functional assay, *Genes Dev., 3*: 2137–2150, 1989.
25. Fabrizio, P., and Abelson, J. Two domains of yeast U6 small nuclear RNA required for both steps of nuclear precursor messenger RNA splicing, *Science, 250*: 404–409, 1990.
26. Vankan, P., McGuigan, C., and Mattaj, I. W. Domains of U4 and U6 snRNAs required for snRNP assembly and splicing complementation in *Xenopus* oocytes, *EMBO J., 9*: 3397–3404, 1990.

27. Perrotta, A. T., and Been, M. D. A pseudoknot-like structure required for efficient self-cleavage of hepatitis delta virus RNA, *Nature, 350*: 434–436, 1991.
28. Wu, H.-N., Lin, Y.-J., Lin, F.-P., Makino, S., Chang, M.-F., and Lai, M. M. C. Human hepatitis δ virus RNA subfragments contain an autocleavage activity, *Proc. Natl. Acad. Sci. USA, 86*: 1831–1835, 1989.
29. Miller, W. A., Hercus, T., Waterhouse, P. M., and Gerlach, W. L. A satellite RNA of barley yellow dwarf virus contains a novel hammerhead structure in the self-cleavage domain, *Virology, 183*: 711–720, 1991.
30. Kole, R., Baer, M. F., Stark, B. C., and Altman, S. *E. coli* RNase P has a required RNA component *in vivo, Cell, 19*: 881–887, 1980.
31. Guerrier-Takada, C., Gardiner, K., Marsh, T., Pace, N., and Altman, S. The RNA moiety of ribonuclease P is the catalytic subunit of the enzyme, *Cell, 35*: 849–857, 1983.
32. Guerrier-Takada, C., and Altman, S. Catalytic activity of an RNA molecule prepared by transcription *in vitro, Science, 223*: 285–286, 1984.
33. Wang, M. J., Davis, N. W., and Gegenheimer, P. Novel mechanisms for maturation of chloroplast transfer RNA precursors, *EMBO J., 7*: 1567–1574, 1988.
34. Nichols, M., Söll, D., and Willis, I. Yeast RNase P: catalytic activity and substrate binding are separate functions, *Proc. Natl. Acad. Sci. USA, 85*: 1379–1383, 1988.
35. Krupp, G., Cherayil, B., Frendewey, D., Nishikawa, S., and Söll, D. Two RNA species co-purify with RNase P from the fission yeast *Schizosaccharomyces pombe, EMBO J., 5*: 1697–1703, 1986.
36. Gold, H. A., Craft, J., Hardin, J. A., Bartkiewicz, M., and Altman, S. Antibodies in human serum that precipitate ribonuclease P. *Proc. Natl. Acad. Sci. USA, 85*: 5483–5487, 1988.
37. Dange, V., Van Atta, R. B., and Hecht, S. M. A Mn^{2+}-dependent ribozyme, *Science, 248*: 585–588, 1990.
38. Haseloff, J., and Gerlach, W. L. Simple RNA enzymes with new and highly specific endoribonuclease activities, *Nature, 334*: 585–591, 1988.
39. Hampel, A., Tritz, R., Hicks, M., and Cruz, P. Hairpin catalytic RNA model: evidence for helices and sequence requirement for substrate RNA, *Nucleic Acids Res., 18*: 299–304, 1990.
40. Zaug, A. J., Been, M. D., and Cech, T. R. The *Tetrahymena* ribozyme acts like an RNA restriction endonuclease, *Nature, 324*: 429–433, 1986.
41. Koizumi, M., Iwai, S., and Ohtsuka, E. Construction of a series of several self-cleaving RNA duplexes using synthetic 21-mers, *FEBS Lett., 228*: 228–230, 1988.
42. Cotten, M., Schaffner, G., and Birnstiel, M. L. Ribozyme, antisense RNA, and antisense DNA inhibition of U7 small nuclear ribonucleoprotein-mediated histone pre-mRNA processing *in vitro, Mol. Cell. Biol., 9*: 4479–4487, 1989.
43. Koizumi, M., Iwai, S., and Ohtsuka, E. Cleavage of specific sites of RNA by designed ribozymes, *FEBS Lett., 239*: 285–288, 1988.
44. Saxena, S. K., and Ackerman, E. J. Ribozymes correctly cleave a model substrate and endogenous RNA *in vivo, J. Biol. Chem., 265*: 17106–17109, 1990.

45. Cotten, M., and Birnstiel, M. L. Ribozyme mediated destruction of RNA *in vivo, EMBO J., 8*: 3861–3866, 1989.
46. Cameron, F. H., and Jennings, P. A. Specific gene suppression by engineered ribozymes in monkey cells, *Proc. Natl. Acad. Sci. USA, 86*: 9139–9143, 1989.
47. Belinsky, M. G., and Dinter-Gottlieb, G. Non-ribozyme sequences enhance self-cleavage of ribozymes derived from Hepatitis delta virus, *Nucleic Acids Res., 19*: 559–564, 1991.
48. Ruffner, D. E., Dahm, S. C., and Uhlenbeck, O. C. Studies on the hammerhead RNA self-cleaving domain, *Gene, 82*: 31–41, 1989.
49. Sarver, N., Cantin, E. M., Chang, P. S., Zaia, J. A., Ladne, P. A., Stephens, D. A., and Rossi, J. J. Ribozymes as potential anti-HIV-1 therapeutic agents, *Science, 247*: 1222–1225, 1990.
50. Goodchild, J., Agrawal, S., Civeira, M. P., Sarin, P. S., Sun, D., and Zamecnik, P. C. Inhibition of human immunodeficiency virus replication by antisense oligodeoxy-nucleotides, *Proc. Natl. Acad. Sci. USA, 85*: 5507–5511, 1988.
51. Green, P. J., Pines, O., and Inouye, M. The role of antisense RNA in gene regulation, *Annu. Rev. Biochem., 55*: 569–597, 1986.
52. Eguchi, Y., Itoh, T., and Tomizawa, J. Antisense RNA, *Annu. Rev. Biochem., 60*: 631–652, 1991.
53. Gerlach, W. L., Haseloff, J. P., Young, M. J., and Bruening, G. Use of plant virus satellite RNA sequences to control gene expression, *Viral Genes and Plant Pathogenesis* (T. P. Pirone and J. G. Shaw, eds.), Springer-Verlag, Berlin, 1990, pp. 177–184.
54. Gasser, C. S., and Fraley, R. T. Genetically engineering plants for crop improvement, *Science, 244*: 1293–1299, 1989.
55. Agrios, G. N. *Plant Pathology*, Academic Press, San Diego, 1988.
56. Bryngelsson, T., Gustafsson, M., Gréen, B., and Lind, C. Uptake of host DNA by the parasitic fungus *Plasmodiophora brassicae, Physiol. Mol. Plant Pathol., 33*: 163–171, 1988.
57. Saito, T., Hosokawa, D., Meshi, T., and Okada, Y. Immunocytochemical localization of the 130K and 180K proteins (putative replicase components) of tobacco mosaic virus, *Virology, 160*: 477–481, 1987.
58. Zaitlin, M., and Hull, R. Plant virus–host interactions, *Annu. Rev. Plant Physiol., 38*: 291–315, 1987.
59. Daubert, S. Sequence determinants of symptoms in the genomes of plant viruses, viroids, and satellites, *Mol. Plant–Microbe Interact., 1*: 317–325, 1988.
60. Holt, C. A., Hodgson, R. A. J., Coker, F. A., Beachy, R. N., and Nelson, R. S. Characterization of the masked strain of tobacco mosaic virus: identification of the region responsible for symptom attenuation by analysis of an infectious cDNA clone, *Mol. Plant–Microbe Interact., 3*: 417–423, 1990.
61. Ugaki, M., Tomiyama, M., Kakutani, T., Hidaka, S., Kiguchi, T., Nagata, R., Sato, T., Motoyoshi, F., and Nishiguchi, M. The complete nucleotide sequence of cucumber green mottle mosaic virus (SH strain) genomic RNA, *J. Gen. Virol., 72*: 1487–1495, 1991.

62. Solis, I., and Garcia-Arenal, F. The complete nucleotide sequence of the genomic RNA of the tobamovirus tobacco mild green mosaic virus, *Virology, 177*: 553–558, 1990.
63. van der Wilk, F., Huisman, M. J., Cornelissen, B. J. C., Huttinga, H., and Goldbach, R. Nucleotide sequence and organization of potato leafroll virus genomic RNA, *FEBS Lett., 245*: 51–56, 1989.
64. Le Gall, O., Candresse, T., Brault, V., Bretout, C., Hibrand, L., and Dunez, J. Cloning full-length cDNA of grapevine chrome mosaic nepovirus, *Gene 73*: 67–75, 1988.
65. Maiss, E., Timpe, U., Brisske, A., Jelkmann, W., Casper, R., Himmler, G., Mattanovich, D., and Katinger, H. W. D. The complete nucleotide sequence of plum pox virus RNA, *J. Gen. Virol., 70*: 513–524, 1989.
66. Rochon, D. M., and Tremaine, J. H. Complete nucleotide sequence of the cucumber necrosis virus genome, *Virology, 169*: 251–259, 1989.
67. Veidt, I., Lot, H., Leiser, M., Scheidecker, D., Guilley, H., Richards, K., and Jonard, G. Nucleotide sequence of beet western yellows virus RNA, *Nucleic Acids Res., 16*: 9917–9932, 1988.
68. Beachy, R. N., Loesch-Fries, S., and Tumer, N. E. Coat protein-mediated resistance against virus infections, *Annu. Rev. Phytopathol., 28*: 451–474, 1990.
69. Rezaian, M. A., Skene, K. G. M., and Ellis, J. G. Anti-sense RNAs of cucumber mosaic virus in transgenic plants assessed for control of the virus, *Plant Mol. Biol., 11*: 463–471, 1988.
70. Hemenway, C., Fang, R.-X., Kaniewski, W. K., Chua, N.-H., and Tumer, N. E. Analysis of the mechanism of protection in transgenic plants expressing the potato virus X coat protein or its antisense RNA, *EMBO J., 7*: 1273–1280, 1988.
71. Cuozzo, M., O'Connell, K. M., Kaniewski, W., Fang, R.-X., Chua, N.-H., and Tumer, N. E. Viral protection in transgenic tobacco plants expressing the cucumber mosaic virus coat protein or its antisense RNA, *Bio/Technology, 6*: 549–557, 1988.
72. Powell, P. A., Stark, D. M., Sanders, P. R., and Beachy, R. N. Protection against tobacco mosaic virus in transgenic plants that express tobacco mosaic virus antisense RNA, *Proc. Natl. Acad. Sci. USA, 86*: 6949–6952, 1989.
73. Kawchuk, L. M., Martin, R. R., and McPherson, J. Sense and antisense RNA-mediated resistance to potato leafroll virus in Russet Burbank potato plants, *Mol. Plant–Microbe Interact. 4*: 247–253, 1991.
74. Lain, S., Riechmann, J. L., and Garcia, J. A. RNA helicase: a novel activity associated with a protein encoded by a positive strand RNA virus, *Nucleic Acids Res., 18*: 7003–7006, 1990.
75. Matthews, R. E. F. *Plant Virology*, Academic Press, San Diego, 1991.
76. Ishikawa, M., Meshi, T., Takeshi, O., and Okada, Y. Specific cessation of minus-strand RNA accumulation at an early stage of tobacco mosaic virus infection, *J. Virol., 65*: 861–868, 1991.

77. Wilson, T. M. A., Plaskitt, K. A., Watts, J. W., Osbourn, J. K., and Watkins, P. A. C. Signals and structures involved in early interactions between plants and viruses or pseudoviruses, *Recognition and Response in Plant–Virus Interactions* (R. S. S. Fraser, ed.), NATO ASI Series, Vol. 441, Springer-Verlag, Berlin, 1990, pp. 123–145.
78. Zuker, M., and Stiegler, P. Optimal computer folding of large RNA sequences using thermodynamics and auxiliary information, *Nucleic Acids Res., 9*: 133–148, 1981.
79. Abrahams, J. P., van den Berg, M., van Batenburg, E., and Pleij, C. Prediction of RNA secondary structure, including pseudoknotting, by computer simulation, *Nucleic Acids Res., 18*: 3035–3044, 1990.
80. Argos, P. A sequence motif in many polymerases, *Nucleic Acids Res., 16*: 9909–9916, 1988.
81. Wierenga, R. K., and Hol, W. G. J. Predicted nucleotide-binding properties of p21 protein and its cancer-associated variant, *Nature, 302*: 842–844, 1983.
82. Hunter, T. A thousand and one protein kinases, *Cell, 50*: 823–829, 1987.
83. Gorbalenya, A. E., and Koonin, E. V. Viral proteins containing the purine NTP-binding sequence pattern, *Nucleic Acids Res., 17*: 8413–8440, 1989.
84. Lamb, J. W., and Hay, R. T. Ribozymes that cleave potato leafroll virus RNA within the coat protein and polymerase genes, *J. Gen. Virol., 71*: 2257–2264, 1990.
85. Schardl, C. L., Byrd, A. D., Benzion, G., Altschuler, M. A., Hildebrand, D. F., and Hunt, A. G. Design and construction of a versatile system for the expression of foreign genes in plants, *Gene, 61*: 1–11, 1987.
86. Rittner, K., and Sczakiel, G. Identification and analysis of antisense RNA target regions of the human immunodeficiency virus type 1, *Nucleic Acids Res., 19*: 1421–1426, 1991.
87. Dixon, R. A., and Paiva, N. L. Prospects for accessing DNA banks for the isolation of genes encoding biologically active proteins, *Conservation of Plant Genes: DNA Banking and In Vitro Biotechnology* (R. P. Adams and J. E. Adams, eds.), Academic Press, New York, pp. 99–118, 1992.
88. Robert, L. S., Donaldson, P. A., Ladaique, C., Altosaar, I., Arnison, P. G., and Fabijanski, S. F. Antisense RNA inhibition of β-glucuronidase gene expression in transgenic tobacco plants, *Plant Mol. Biol., 13*: 399–409, 1989.
89. Ecker, J. R., and Davis, R. W. Inhibition of gene expression in plant cells by expression of antisense RNA, *Proc. Natl. Acad. Sci. USA, 83*: 5372–5376, 1986.
90. Delauney, A. J., Tabaeizadeh, Z., and Verma, D. P. S. A stable bifunctional antisense transcript inhibiting gene expression in transgenic plants, *Proc. Natl. Acad. Sci. USA, 85*: 4300–4304, 1988.
91. van der Krol, A. R., Lenting, P. E., Veenstra, J., van der Meer, I. M., Koes, R. E., Gerats, A. G. M., Mol, J. N. M., and Stuitje, A. R. An anti-sense chalcone synthase gene in transgenic plants inhibits flower pigmentation, *Nature, 333*: 866–869, 1988.
92. Smith, C. J. S., Watson, C. F., Ray, J., Bird, C. R., Morris, P. C., Schuch, W., and Grierson, D. Antisense RNA inhibition of polygalacturonase gene expression in transgenic tomatoes, *Nature, 334*: 724–726, 1988.

93. Stockhaus, J., Höfer, M., Renger, G., Westhoff, P., Wydrzynski, T., and Willmitzer, L. Anti-sense RNA efficiently inhibits formation of the 10 kd polypeptide of photosystem II in transgenic potato plants: analysis of the role of the 10 kd protein, *EMBO J., 9*: 3013–3021, 1990.
94. van der Krol, A. R., Mur, L. A., de Lange, P., Gerats, A. G. M., Mol, J. N. M., and Stuitje, A. R. Antisense chalcone synthase genes in petunia: visualization of variable transgene expression, *Mol. Gen. Genet., 220*: 204–212, 1990.
95. van der Krol, A. R., Mur, L. A., de Lange, P., Mol, J. N. M., and Stuitje, A. R. Inhibition of flower pigmentation by antisense CHS genes: promoter and minimal sequence requirements for the antisense effect, *Plant Mol. Biol., 14*: 457–466, 1990.
96. Napoli, C., Lemieux, C., and Jorgensen, R. Introduction of a chimeric chalcone synthase gene into petunia results in reversible co-suppression of homologous genes in trans, *Plant Cell, 2*: 279–289, 1990.
97. van der Krol, A. R., Mur, L. A., Beld, M., Mol, J. N. M., and Stuitje, A. R. Flavonoid genes in petunia: addition of a limited number of gene copies may lead to a suppression of gene expression, *Plant Cell, 2*: 291–299, 1990.
98. Elkind, Y., Edwards, R., Mavandad, M., Hedrick, S. A., Ribak, O., Dixon, R. A., and Lamb, C. J. Abnormal plant development and down-regulation of phenylpropanoid biosynthesis in transgenic tobacco containing a heterologous phenylalanine ammonia-lyase gene, *Proc. Natl. Acad. Sci. USA, 87*: 9057–9061, 1990.
99. Goring, D. R., Thomson, L., and Rothstein, S. J. Transformation of a partial nopaline synthase gene into tobacco suppresses the expression of a resident wild-type gene, *Proc. Natl. Acad. Sci. USA, 88*: 1770–1774, 1991.
100. van der Krol, A. R., Mol, J. N. M., and Stuitje, A. R. Modulation of eukaryotic gene expression by complementary RNA or DNA sequences, *Biotechniques, 6*: 958–976, 1988.
101. Jorgensen, R. Altered gene expression in plants due to trans interactions between homologous genes, *Trends Biotechnol., 8*: 340–344, 1990.
102. Dixon, R. A., and Harrison, M. J. Activation, structure and organization of genes involved in microbial defense in plants, *Adv. Genet., 28*: 165–234, 1990.

16

The Directed Evolution of Antisense Ribozymes in Plants

Debra L. Robertson

The Scripps Research Institute, La Jolla, California

I. INTRODUCTION

Gene expression can be regulated in both prokaryotic and eukaryotic cells by the interaction of RNA and DNA repressor molecules (1–3). The regulatory molecule contains a sequence that is complementary to the target RNA, and inhibition of expression occurs by the base pairing of the antisense oligonucleotide with the target gene. Antisense molecules are tremendously versatile tools for addressing several biologically relevant questions. These complementary oligonucleotides can be used to define the role of previously defined genes in different tissues or related species, to interfere with the structure or activities of untranslated RNAs found in the cell and determine their function, or to determine the cellular activities of randomly generated sequences.

The antisense inhibition of the expression of specific cellular or viral genes in plants has become the focus of several laboratories whose interests range from normal plant development to the study of viral pathogens. Complementary oligonucleotides introduced into plants have been shown to suppress expression of endogenous and introduced genes with high specificity (4). In plants, antisense molecules have been successfully used for achieving viral resistance in RNA and DNA viral strains (5).

Antisense oligonucleotides and oligonucleotide analogues are small, stable molecules designed to bind single-stranded RNAs by Watson–Crick base pairing. The oligonucleotides specifically bind to the target RNA sequence to form a RNA–DNA heteroduplex structure. This structure interferes with either the normal processing of the RNA or the translation of the target RNA, thereby inhibiting specific protein synthesis in the cell. The goal of antisense technology is to design a oligodeoxynucleotide or oligonucleotide analogue that accurately and efficiently binds and cleaves a target sequence without interfering with normal cellular function.

The discovery of catalytic RNA has expanded the antisense field by providing a molecule that can physically bind and cleave in a specific sequence within a target gene. The hammerhead self-cleaving RNA has been used to induce specific gene expression in a variety of systems, including mammalian cells (6,7) and higher plants (4).

Although the use of catalytic RNAs as antisense molecules is still in its infancy, the limitations of this technology are becoming obvious. The catalytic efficiency and substrate selection of RNA cleavage by the hammerhead ribozyme and other RNA enzyme in vivo are directly influenced by several factors endogenous to the intracellular microenvironment. Using the group 1 ribozyme from *Tetrahymena*, our laboratory has approached this problem by developing an in vitro evolution system that allows the identification of variant ribozymes that have a greater catalytic efficiency inside the cell (8). This method allows in vitro identification of novel ribozymes that are catalytically active under reaction conditions that mimic the intracellular conditions. This approach will potentially allow investigators to develop antisense ribozymes that will be able to catalyze the cleavage of a target substrate within any cell type. The purpose of this chapter is to introduce the group 1 ribozyme as an effective antisense ribozyme and then describe the in vitro evolution of a system that allows the selective amplification of novel group 1 ribozymes with new catalytic functions.

II. ANTISENSE STRATEGIES

The antisense inhibition of gene expression uses two general strategies in the cell. In prokaryotes, antisense genes have been shown to regulate several aspects of bacterial and bacteriophage development and replication. The hybridization of small antisense transcripts to the *Escherichia coli* Tn 10 mRNA regulates the transposition of the element (9). Tomizawa et al. showed that a short antisense RNA forms a duplex structure with the 5′ end of the replication dimer of the Col E1. This interaction inhibits the tertiary structure required for primer activity during plasmid replication (10).

The discovery of antisense-mediated gene regulation in prokaryotes suggested that gene expression in eukaryotes may also be inhibited by antisense genes. By using antisense transcripts in the cell and synthetic antisense oligodeoxynucleotides, investigators have been successful in down-regulating gene expression in higher plants and mammalian cells (11,12).

Antisense oligonucleotides also have provided a genetic tool for analyzing the cellular function of specific DNAs. These molecules bind to double-stranded RNA in a sequence-specific manner to form a triple helix (13). These oligonucleotide analogues bind to the major groove of the duplex structure of DNA and selectively interrupt gene expression. Targeting of DNA with antisense molecules offers several advantages over targeting RNA. A transcriptionally active gene can be represented in the cell more than 1000 copies of the corresponding mRNA. Messenger RNA is continually being synthesized in vivo and requires a continual supply of antisense molecules for the inhibition of gene expression. In contrast, the DNA encoding the target gene only exists as a few copies per cell.

Antisense strategies are limited by the reversible nature of the inhibitory duplex structure and the degradation of the complementary oligonucleotide by intracellular nucleases. Several approaches are currently being used to design an antisense molecule that binds its target sequence and directs the degradation of the duplex or heteroduplex formed by the interaction of complementary nucleotides. The first approach is based on evidence that the DNA–RNA heteroduplex formed by the binding of the antisense oligodeoxynucleotide can serve as a substrate for cellular ribonuclease H (14–16). The problem now becomes to create a nuclease-resistant antisense molecule that directs the degradation of the heteroduplex by RNase H faster than the oligonucleotide is degraded by intracellular enzymes. A second approach is to attach a reagent with endonuclease activity to the antisense oligonucleotide. The binding of the oligonucleotide delivers the cleavage agent. An Fe–EDTA complex added to the 5′ end of an antisense DNA has been used to generate hydroxyl radicals that result in the cleavage of the complementary target (17–21). The third approach involves the use of antisense ribozymes targeted toward specific mRNAs. The discovery of the RNA enzymes has expanded the potential of RNA-mediated antisense inhibition by providing a molecule that both binds and cleaves the target sequence.

III. RNA ENZYMES

A. RNA Catalysis

In 1982, Cech et al. demonstrated in separate experiments that some RNA molecules had both catalytic and informational properties. By using the ciliated protozoa *Tetrahymena pigmentosa*, it was demonstrated that an intervening sequence (intron) within the ribosomal RNA (rRNA) could facilitate the processing of the precursor RNA by self-splicing (22). Self-splicing is an intramolecular event in which the intron acts upon itself to alter its own covalent structure through two consecutive transesterification reactions (23). The accurate excision of the intron and the concomitant ligations of the exons are catalyzed by a single activity that resides within the intron. The requirements for self-splicing have been defined by in vitro studies, and it has been demonstrated that this catalytic activity occurs in the absence of proteins and requires only a guanosine cofactor and Mg^{2+} (22).

B. RNA-Catalyzed *trans*-Splicing

It was shown later that a modified form of the *Tetrahymena* ribozyme (L-21) is able to carry out two intermolecular reactions with an external oligonucleotide substrate in a reaction that is analogous to the first step of self-splicing (24). During these "*trans*-splicing" reactions the intron remains unchanged after cleavage of the substrate and is able to turn over and catalyze additional *trans*-splicing events. The high level of sequence specificity exhibited by the *Tetrahymena* ribozyme is determined by the binding of a pyrimidine sequence within the substrate to the internal guide sequence (IGS) of the intron. Cleavage occurs at the phosphodiester bond located downstream from the target sequence (Fig. 1) (24–26).

IV. GROUP 1 RIBOZYMES AS ANTISENSE MOLECULES

Four distinct classes of RNA enzymes have been shown to cleave an RNA substrate in a sequence-specific manner: RNase P group 1 and group 2 self-splicing introns and the hammerhead and hairpin self-cleaving RNAs (22, 27–33). Each category is distinguished from the others by secondary and tertiary structural features and the chemical mechanism used in RNA catalysis (34,35). Among these RNA enzymes, the group 1 self-splicing intron from *Tetrahymena* and the hammerhead self-cleaving RNA have been studied most extensively as antisense molecules.

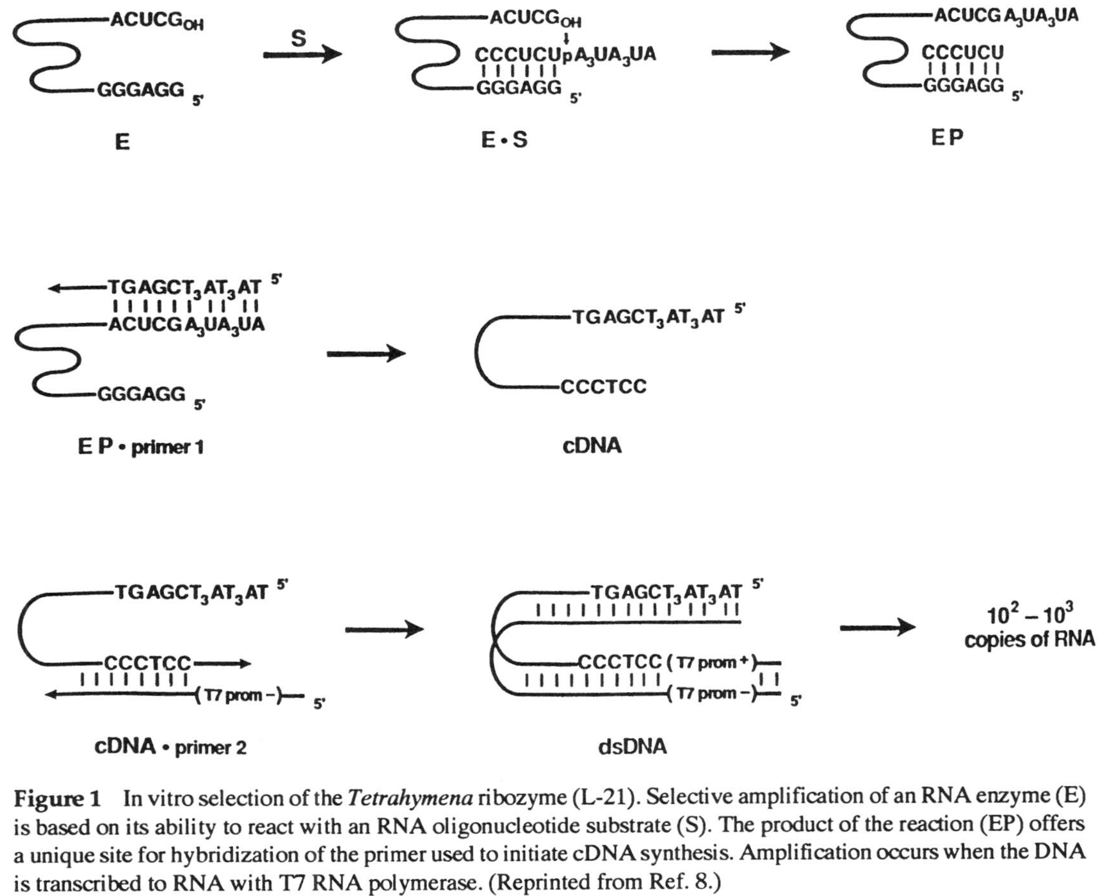

Figure 1 In vitro selection of the *Tetrahymena* ribozyme (L-21). Selective amplification of an RNA enzyme (E) is based on its ability to react with an RNA oligonucleotide substrate (S). The product of the reaction (EP) offers a unique site for hybridization of the primer used to initiate cDNA synthesis. Amplification occurs when the DNA is transcribed to RNA with T7 RNA polymerase. (Reprinted from Ref. 8.)

The high level of sequence specificity and catalytic efficiency of the *Tetrahymena* ribozyme can be generalized to include any sequence by making the compensatory changes in the IGS of the intron (36,37). By optimizing the reaction conditions, it is possible to achieve a high level of catalytic efficiency with almost any IGS–substrate combination.

The substrate repertoire of the ribozymes has been expanded to include single-stranded DNA (8,38). An RNA-catalyzed DNA cleavage occurs by the same successive transesterification reactions used in RNA cleavage. Although a detailed kinetic analysis of DNA cleavage by the wild-type *Tetrahymena* ribozyme showed that the reaction is less efficient by five orders of magnitude than is cleavage of an RNA substrate, this reaction rate is 10^9-fold greater than spontaneous DNA cleavage (38). The *Tetrahymena* ribozyme can be used as an antisense agent specific for either RNA or DNA target substrates.

The efficiency of the transesterification reaction catalyzed by the ribozyme is directly influenced by the local environment of the cell. It is known that in vitro RNA catalysis requires the availability of a divalent cation and the correct folding of the ribozyme into a specific secondary or tertiary structure.

V. SELECTED EVOLUTION OF THE *TETRAHYMENA* RIBOZYME

A. Selective Amplification of RNA Enzymes

The development of RNA enzymes as useful tools for the specific inactivation of RNA and DNA targets ultimately depends on the catalytic activity of these molecules in vivo. Although the results of the in vitro studies are promising, transferring technologies developed in vitro to the milieu of the intracellular compartment presents persistent problems. Our laboratory has developed a system for optimizing ribozyme function by selecting variant ribozymes that are most reactive under reaction conditions that mimic the microenvironment of the cell (8,39).

The first catalytic molecules that can be manipulated at the levels of genotype and phenotype are RNA enzymes. Using the *Tetrahymena* ribozyme, our laboratory has developed an in vitro system that allows the generation of RNA enzymes with novel catalytic functions. This system integrates the processes of mutation, selection, and amplification to identify catalytic RNAs that are active within new environmental conditions or with new substrates (8).

Several techniques based on a reciprocal primer technology have been developed for RNA and DNA amplification. Reciprocal primer technology involves the hybridization of an initial primer to the 3′ end of the RNA. Using this primer, reverse transcriptase copies the RNA template to yield a complementary DNA molecule, cDNA. The second primer carrying the T7 RNA polymerase promoter hybridizes at the 3′ end of the cDNA and is extended to yield a double-stranded DNA (dsDNA). The dsDNA molecule contains a functional T7 RNA polymerase promoter that produces 200–1200 copies of the RNA transcript per copy of DNA template (40). It has been shown that this protocol can proceed continuously as a single reaction in which the RNA is copied to cDNA, the cDNA is converted to dsDNA, and then the dsDNA is transcribed into hundreds of copies of RNA. Each RNA becomes a new template for additional rounds of amplification. During a 1-hr incubation, this system results in a 10^3- to 10^4-fold amplification of the input RNA (Fig. 1).

The catalytic versatility of the ribozyme can best be studied by generating a population of RNA enzymes carrying random mutations and then selectively amplifying the RNA molecule that is active under the newly defined conditions. Random mutations are introduced into the ribozyme with oligodeoxynucleotides that are synthesized with a 5% degeneracy per nucleotide position (41). This generates a complex population of ribozymes containing 10^{12} mutants with a starting population of 1 pmol of RNA. Selective amplification is based on the RNA enzyme's ability to react with an oligonucleotide substrate (either DNA or RNA). The cleavage–ligation reaction catalyzed by the ribozyme involves the attack of the phosphodiester bond after a sequence of pyrimidines (CCCUCU) located within the substrate by the 3′-terminal guanosine of the ribozyme. This intermolecular phosphate exchange reaction results in the cleavage of the substrate. The substrate sequences located 3′ from the target phosphodiester bond at the cleavage site are joined to the 3′ end of the ribozyme. Selection, which occurs when the highly specific oligonucleotide primer is hybridized across the ligation junction between the ribozyme and substrate sequences, is used to initiate the synthesis of a DNA molecule complementary to the catalytic RNA molecule. The first primer is designed to hybridize only to the substrate sequences found at the 3′ end of any ribozyme capable of catalyzing the cleavage–ligation reaction. Selection occurs regardless of whether an RNA or DNA substrate is used. Amplification occurs when a second primer carrying the T7 RNA polymerase promoter then hybridizes to the 3′ end of the cDNA molecules and the DNA is transcribed to RNA by the T7 RNA polymerase, as described previously (Fig. 1).

B. Selection of Targeted Sequences

The wild-type *Tetrahymena* ribozyme cleaves single-stranded RNA and DNA molecules at the phosphodiester bond immediately following the sequence that is complementary to the IGS of the intron. Changing the IGS has yielded several variant ribozymes that recognize novel RNA sequences. It is possible to extend this work to develop ribozymes that recognize all 64 IGS–substrate combinations (24,32,33). Cleavage of single-stranded DNA substrates has been limited to substrates containing the sequence CTCT. By analogy to the work done with RNA substrates, a variety of DNA sequences can be recognized and cleaved by the ribozyme. It is important to note that altering the specificity of the ribozyme requires factors in vitro that affect duplex stability. It is not known whether these requirements exist in vivo among the complex microenvironment of the cell.

The cleavage activity of antisense ribozymes depends on the secondary structure of the target RNA substrate. Because the cleavage–ligation activity of the ribozyme is limited to single-stranded RNAs, the structure of the target RNA is an important consideration in the selection of target substrates. The RNA sequences buried in the stem of a hairpin or pseudoknot structure are not accessible for binding by the IGS of the ribozyme and are not a substrate for cleavage. There are several computer programs that predict the lowest-energy RNA structure for a given RNA sequence. These programs are useful as a first step in evaluating potential RNA cleavage sites. It is important to extend this analysis to include chemical probing of the RNA to guarantee the single-stranded nature of the target sequence.

VI. CONCLUSIONS

Antisense ribozymes offer several advantages over complementary oligonucleotides as tools for investigating the role of specific gene expression in a normal system and for use as antiviral therapeutic agents. First, ribozymes have a highly specific inhibitory function because of their intrinsic ability to bind and cleave an RNA substrate in a sequence-specific manner. Second, the enzymatic properties of these RNAs allow them to turn over and catalyze the cleavage of several molecules. Finally, the group 1 ribozyme from *Tetrahymena* provides a molecule that can cleave a DNA substrate in addition to an RNA substrate. The catalytic efficiency of the RNA enzymes (including the *Tetrahymena* ribozyme) has been measured in vitro under reaction conditions that are dramatically different from those found in cells. In designing a realistic strategy to inhibit gene function in normal or virus-infected cells, it is necessary to consider the effects cellular components may have on the catalytic activity of the ribozyme. Little is known about the proteins that interact with the *Tetrahymena* ribozyme and their effects on its *trans*-splicing ability. Our approach suggests that given the undefined cellular environment, it is possible to select in vitro RNA enzymes with greater catalytic efficiency in vivo. Once identified, these variant ribozymes can be introduced into the cell and expressed in a constitutive or inducible manner. The extensive secondary structure of the ribozyme stabilizes the enzyme from nuclease degradation within the cell. The integration of in vitro selection with in vivo expression of the ribozyme for the first time provides a system in which a ribozyme with known catalytic efficiency and specificity can be used as an antisense agent in vivo.

REFERENCES

1. Zamecnik, P. C., and Stephenson, M. L. Inhibition of rous sarcoma virus replication and cell transformation by a specific oligodeoxynucleotide, *Proc. Natl. Acad. Sci. USA, 75*: 280, 1978.
2. Goodchild, J., Agrawal, S., Civeira, M. P., Sarin, P. S., Sun, D., and Zamecnik, P. C. Inhibition of human immunodeficiency virus replication by antisense oligodeoxynucleotides, *Proc. Natl. Acad. Sci. USA, 85*: 5506, 1988.
3. Taniguchi, T., and Weissman, C. Inhibition of Qb RNA 70S ribosome initiation complex formation by an oligonucleotide complementary to the 3′ terminal region of E. coli 16S ribosomal RNA, *Nature, 275*: 770, 1978.
4. Haseloff, J., and Gerlach, W. L. Simple RNA enzymes with new and highly specific endoribonuclease activities, Nature, 334: 585, 1988.
5. Day, A. G., Bejarano, E. R., Buck, K. W., Burrell, M., and Lichtenstein, C. P. Expression of an antisense viral gene in transgenic tobacco confers resistance to the DNA virus tomato golden mosaic virus, *Proc. Natl. Acad. Sci. USA, 88*: 6721, 1991.
6. Cameron, F. H., and Jennings, P. A. Specific gene suppression by engineered ribozyme in monkey cells, *Proc. Natl. Acad. Sci. USA, 86*: 9139, 1989.
7. Sarver, N., Cantin, E. M., Chang, P. S., Zara, J. A., Ladne, P. A., Stephens, D. A., and Rossi, J. J. Ribozymes as potential anti-HIV-1 therapeutic agents, *Science, 247*: 1222, 1990.
8. Robertson, D. L., and Joyce, G. F. *In vitro* selection of an RNA enzyme that specifically cleaves single-stranded DNA, *Nature, 344*: 467, 1990.
9. Simons, R. W., Hoopes, B. C., McClure, W. R., and Kleckner, N. Three promoters near the termini of IS10: pIN, pOUT, and pIII, *Cell, 34*: 673, 1983.
10. Tomizawa, J. I., and Itoh, T. The importance of RNA secondary structure in Col E1 primer formation, *Cell, 31*: 575, 1982.

11. Izant, J. G., and Weintraub, H. Inhibition of thymidine kinase gene expression by antisense RNA: a molecular approach to genetic analysis, *Cell, 36*: 1007, 1984.
12. Melton, D. A. Injected antisense RNAs specifically block messenger RNA translation *in vivo, Proc. Natl. Acad. Sci. USA, 82*: 144, 1985.
13. Riordan, M., and Martin, J. C. Oligonucleotide-based therapeutics, *Nature, 350*: 442, 1991.
14. Minshull, J., and Hunt, T. The use of single-stranded DNA and RNase H to promote quantative "hybrid arrest of translation" of mRNA/DNA hybrids in reticulocyte lysate cell-free translation, *Nucleic Acids Res., 14*: 6433, 1986.
15. Dash, P., Lotan, I., Knapp, M., Kandel, E. R., and Goelet, P. Selective elimination of mRNAs in vivo: complementary oligonucleotides promote RNA degradation by an RNase H-like activity, *Proc. Natl. Acad. Sci. USA, 84*: 7896, 1987.
16. Walder, R. Y., and Walder, J. A. Role of RNase H in hybrid-arrest translation by antisenses oligonucleotides, *Proc. Natl. Acad. Sci. USA, 85*: 5011, 1988.
17. Matsukura, M., Shinozuka, K., Zon, G., Mitsuya, H., Reitz, M., Cohen, J. S., and Broder, S. Phosphorothioate analogs of oligodeoxynucleotides: inhibitors of replication and cytopathic effects of immunodeficiency virus, *Proc. Natl. Acad. Sci. USA, 84*: 7706, 1987.
18. Agrawal, A., Goodchild, J., Civeira, M. P., Thornton, A. H., Sarin, P. S., and Zamecnik, P. C. Oligodeoxynucleoside phosphoramidates and phosphorothioates as inhibitors of human immunodeficiency virus, *Proc. Natl. Acad. Sci. USA, 85*: 7079, 1988.
19. Matsukura, M., Zon, G., Shinozuka, K., Robert-Guroff, M., Shimada, T., Stein, C. A., Mitsuya, H., Wong-Staal, F., Cohen, J. S., and Broder, S. Regulation of viral expression of human immunodeficiency virus *in vitro* by an antisense phosphorothioate oligodeoxynucleotide against *rev* (*art/trs*) in chronically infected cells, *Proc. Natl. Acad. Sci. USA, 86*: 4244, 1989.
20. Dreyer, G. B., and Dervan, P. B. Sequence specific cleavage of single-stranded DNA: oligodeoxynucleotide-EDTA-Fe(II), *Proc. Natl. Acad. Sci. USA, 82*: 986, 1985.
21. Chu, B. C. F., and Orgel, L. E. Nonenzymatic sequence specific cleavage of single-stranded DNA, *Proc. Natl. Acad. Sci. USA, 82*: 963, 1985.
22. Kruger, K., Grabowski, P. J., Zang, A. J., Sands, J., Gottschling, D. E., and Cech, T. R. Self-splicing RNA: auto excision and autocyclization of ribosomal RNA intervening sequence of *Tetrahymena, Cell, 31*: 147, 1982.
23. Zaug, A. J., Grabowski, P. J., and Cech, T. R. Autocyclization of an excised intervening sequence is a cleavage-ligation reduction, *Nature, 310*: 578, 1983.
24. Zaug, A. J., Been, M. D., and Cech, T. R. The *Tetrahymena* ribozyme acts like an RNA restriction endonuclease, *Nature, 324*: 429, 1986.
25. Waring, R. B., Towner, P., Minter, S. J., and Davies, R. W. Splice site selection by a self-splicing RNA of *Tetrahymena, Nature, 321*: 133, 1983.
26. Been, M. D., and Cech, T. R. One binding site determines sequence, *Cell, 47*: 207, 1986.

27. Guerrier-Takada, C., Gardiner, K., Marsh, T., Pace, N., and Altman, S. The RNA moiety of ribonuclease P is the catalytic subunit of the enzyme, *Cell, 35*: 849, 1983.
28. Peebles, C. L., Perlman, P. S., Mecklenburg, K. L., Petrillo, M. L., Tabor, J. H., Jarrell, K. A., and Cheng, H. L. A self-splicing RNA excises an intron lariat, *Cell, 44*: 213, 1986.
29. Uhlenbeck, O. C. A small catalytic oligoribonucleotide, *Nature, 328*: 596, 1987.
30. McClain, W. H., Guerrier-Takada, C., and Altman, S. Model substrates for an RNA enzyme, *Science, 238*: 527, 1987.
31. Sharmeen, L., Kuo, M. Y. P., Dinter-Gottlieb, G., and Taylor, J. Antigenomic RNA of human hepatitis delta virus can undergo self-cleavage, *J. Virol., 62*: 2674, 1988.
32. Zaug, A. J., Grosshans, C. A., and Cech, T. R. Sequence-specific endoribonuclease activity of the Tetrahymena ribozyme: enhanced cleavage of certain oligonucleotide substrates that form mismatched ribozyme-substrate complexes, *Biochemistry, 27*: 8924, 1988.
33. Hampel, A., Tritz, R., Hicks, M., and Cruz, P. "Hairpin" catalytic RNA model: evidence for helices and sequence requirements for substrate RNA, *Nucleic Acids Res., 18*: 299, 1990.
34. Davies, R. W., Waring, R. B., Ray, J. A., Brown, T. A., and Scazzocchio, C. Making ends meet: a model for RNA splicing in fungal mitochondria, *Nature, 300*: 719, 1982.
35. Michel, F., Jacquier, A., and Dujon, B. Comparison of fungal mitochondrial introns reveals extensive homologies in RNA secondary structures, *Biochemie, 64*: 867, 1982.
36. Dounda, J. A., and Szostak, J. W. RNA-catalyzed synthesis of complementary-strand RNA, *Nature, 339*: 519, 1989.
37. Murphy, F. L., and Cech, T. R. Alteration of substrate specificity for the endoribonucleolytic cleavage of RNA by the Tetrahymena ribozyme, *Proc. Natl. Acad. Sci. USA, 86*: 9218, 1989.
38. Herschlag, D., and Cech, T. R. DNA cleavage catalyzed by the ribozyme from Tetrahymena, *Nature, 344*: 405, 1990.
39. Joyce, G. F. Amplification, mutation and selection of catalytic RNA, *Gene, 82*: 83, 1989.
40. Chamberlin, M., and Ryan, T. Bacteriophage DNA-dependent RNA polymerases, *The Enzymes* (P. Boyer, ed.), Academic Press, New York, 1982, pp. 87–108.
41. Joyce, G. F., and Inoue, T. Structure of the catalytic core of the *Tetrahymena* ribozyme as indicated by reactive abbreviated forms of the molecule, *Nucleic Acids Res., 15*: 9825, 1987.

Index